Membrane Shape and Biological Function

Editors

José Carlos Bozelli Jr
Department of Biochemistry and Biomedical Sciences
McMaster University, Canada

Richard M. Epand
Department of Biochemistry and Biomedical Sciences
McMaster University, Canada

CRC Press
Taylor & Francis Group
Boca Raton London New York

CRC Press is an imprint of the
Taylor & Francis Group, an **informa** business
A SCIENCE PUBLISHERS BOOK

Cover credit: Image reproduced by kind courtesy of Dr. Maikel C. Rheinstädter (co-author in Chapter 9).

First edition published 2025
by CRC Press
2385 NW Executive Center Drive, Suite 320, Boca Raton FL 33431

and by CRC Press
4 Park Square, Milton Park, Abingdon, Oxon, OX14 4RN

CRC Press is an imprint of Taylor & Francis Group, LLC

Library of Congress Cataloging-in-Publication Data (applied for)

ISBN: 978-1-032-26314-4 (hbk)
ISBN: 978-1-032-26315-1 (pbk)
ISBN: 978-1-003-28765-0 (ebk)

DOI: 10.1201/9781003287650

Typeset in Times New Roman
by Prime Publishing Services

Preface

Have you ever wondered why different cells and organelles maintain different shapes? Or why do they change their shape in response to different stimuli?

The key lies in the dynamic world of biological membranes—a realm as fascinating as it is crucial to our understanding of life. The field of membrane shape and its biological function has recently experienced a renaissance, propelled by advanced experimental and computational methods. A pivotal area of advancement lies in comprehending how the shape of the membrane modulates biological processes by influencing molecular properties and vice versa. These advancements, at the exciting crossroads of science, promise not only to deepen our fundamental understanding of life at the molecular level but also to unlock new possibilities in medicine and biotechnology.

Membrane Shape and Biological Function is your compass on this exhilarating journey of discovery. As we navigate through the dynamic world of biological membranes, we invite you to explore how the shape and physical properties of membranes profoundly influence a myriad of cellular processes.

Your adventure begins in Chapter 1, where the intriguing realm of lipid bilayers unfolds. Here, you'll discover how subtle deformations in the membrane affect the spatial distribution of various lipids, setting the stage for a deeper understanding of membrane dynamics and their critical role in biology. Chapter 2 delves into the intricate dance of interactions between lipids and proteins, revealing how the flexibility of lipid surfaces dictates the conformation and function of membrane proteins. Each chapter builds upon the last as the journey progresses, creating a comprehensive mosaic of knowledge. From the complexity of non-lamellar phases in Chapter 3 to the exploration of lipid shape and membrane interactions in Chapter 4, we delve deeper into the heart of lipid behavior. Chapter 5 forms a crucial link between fundamental biophysics and cellular biology, exploring the mechanics of lipid bilayers in shaping cellular processes. Subsequent chapters cover computational methods in membrane remodeling (Chapter 6), the intriguing world of bicontinuous inverted cubic phases (Chapter 7), and protein-membrane interactions (Chapter 8). In-depth explorations continue with the study of the mechanical properties of membranes (Chapter 9) and the pivotal role of the Golgi apparatus (Chapter 10). Chapter 11 bridges biophysics and cellular biology through the lens of inositol phospholipids, while Chapter 12 offers insights into the complexities of membrane fusion. Finally, Chapter 13 propels you into the exciting world of lipid bilayers as platforms for understanding biological memory and the development of neuromorphic computing, showcasing the far-reaching applications of lipid membrane science.

Whether you are a student embarking on your scientific journey, an experienced researcher, or simply a curious mind captivated by the wonders of biology, this book is a gateway to a world of discovery. It's not merely a collection of facts and theories; it is an invitation to marvel at the elegance and complexity of nature's design. Through these pages, we hope to inspire not only an enhanced understanding but also a sense of wonder and awe in the face of the intricate world of biological membranes.

Welcome to ***Membrane Shape and Biological Function***—a journey that promises to enlighten, inspire, and transform your understanding of life's most fundamental aspects. As you delve into this captivating world, may you find enlightenment and inspiration in the intricate dance of lipids and proteins, the essence of life.

Hamilton, ON, Canada
January 2024

José Carlos Bozelli Jr

In Memoriam and Acknowledgement

This book, *Membrane Shape and Biological Function* is dedicated to the extraordinary lives and contributions of two remarkable scientists in the field of lipid membrane research whom we lost during this project: Professor Richard M. Epand and Dr. Luis A. Bagatolli. We acknowledge with profound sadness the untimely passing of Dr. Luis A. Bagatolli. A visionary in lipid membrane research, Luis's expertise in microscopy and cell surface organization left a lasting mark on the scientific community. His battle with cancer prematurely ended a career full of promise and prevented his planned contributions to this work.

This book, however, is a heartfelt homage to my esteemed colleague and co-editor, Professor Richard M. Epand, whose passing has left a lasting mark on both the book and those who knew him. His relentless pursuit of understanding the intricacies of cellular membranes led to crucial advancements in lipid membrane biophysics and biochemistry. With over 700 papers and 33,000 citations, Richard's academic journey, starting from his Ph.D. at Columbia University to his transformative work at McMaster University, was marked by pioneering studies and innovative insights. His exploration of membrane curvature, antimicrobial peptides, and lipid polymorphism revolutionized our understanding of cellular processes and has profound implications in medicine and biotechnology. His pioneering work and visionary ideas have shaped much of the content herein.

Richard's contributions extended beyond the realm of academia; he was a mentor, a thought leader, and a friend to many. His significant contributions to editorial boards and societies like the Biophysical Society of Canada demonstrate his commitment to advancing the field of lipid membranes. His recognition with the Avanti Award and as a fellow of various scientific societies speaks volumes about his global impact.

Completing this book in his absence was a challenging journey, but it was guided every step by the high standards he set and the profound insights he provided. I hope this book not only serves as a valuable resource for those in the field but also as a fitting tribute to Richard's remarkable career and his lasting impact on the study of lipid membranes. His absence will be profoundly felt, but his legacy will continue to inspire and guide future explorations into the mysteries of biological membrane function and structure.

In honoring Richard M. Epand and remembering Luis A. Bagatolli, we celebrate their profound scientific contributions and dedication to unraveling the complexities of lipid membranes. Their legacies will continue to guide and inspire future generations in the quest to understand the fundamental processes of life. May their memory continue to be a beacon of inspiration and excellence in the field.

Contents

Abbreviations

AFM : Atomic Force Microscopy;
AH : Amphipathic Helices;
A_L : Area per lipid;
ALPS : Amphipathic Lipid Packing Sensor motif;
ALR : Autophagic Lysosome Reformation;
AMP : Antimicrobial Peptide;
AMPA : α-Amino-3-Hydroxy-5-Methyl-4-Isoxazolepropionic Acid;
Arf1 : ADP ribosylation factor 1;
BAR : Proteins from the Bin/Amphiphysin/RVS family;
CCP : Clathrin-Coated Pits;
Cer : Ceramide;
CERT : Ceramide Transfer Protein;
Cho : Cholesterol;
CMCS : Constant Mean Curvature Surfaces;
COPII : Coat Protein Complex II;
C_S : Bilayer Spontaneous Curvature;
DAG : Diacylglycerol;
DCV : Dense Core Vesicles;
DIB : Droplet Interface Bilayer;
dEIS : Dynamic Electrochemical Impedance Spectroscopy;
DLPC : 1,2-Dilauroyl-Phosphatidylcholine;
DMPC : 1,2-Dimyristoyl-Phosphatidylcholine;
DMPE : 1,2-Dimyristoyl-Phosphatidylethanolamine;
DOPC : 1,2-Dioleoyl-Phosphatidylcholine;
DOPE : 1,2-Dioleoyl-Phosphatidylethanolamine;
DOPhPC : 1,2-Di-O-Phytanoyl-sn-Glycero-3-Phospho-Choline;
DOPE-Me : N-Mono-methylated 1,2-Dioloeoyl-Phosphatidylethanolamine;
DPhPC : 1,2-Diphytanoyl-Phosphatidylcholine;
DPPC : 1,2-Dipalmitoyl-Phosphatidylcholine;
DPD : Dissipative Particle Dynamics;
DSPC : 1,2-Distearoyl-Phosphatidylcholine;
ER : Endoplasmic Reticulum;
ERES : Endoplasmic Reticulum Exit Sites;
ERGIC : ER-Golgi Intermediate Compartment;

ESR	:	Electron Spin Resonance;
FAPP1	:	Four-Phosphate Adaptor Protein 1;
FAST	:	Fusion-Associated Small Transmembrane Proteins;
FERM	:	Four-Point One-Ezrin-Radixin-Moesin Domains;
FMM	:	Fluid-Mosaic Model;
FRET	:	Förster Resonance Energy Transfer;
FSM	:	Flexible Surface Model;
GAP	:	GTPase-Activating Protein;
GEF	:	Guanine Nucleotide Exchange Factors;
GI-SAXS	:	Grazing-Incidence Small-Angle X-Ray Scattering;
GLTPH	:	Glycolipid Transfer Protein Homology;
GlcCer	:	Glucosylceramide;
GPCR	:	G-Protein-Coupled Receptor;
GRAB	:	GRIP Related Arf Binding;
GSL	:	Glycosphingolipid;
GUV	:	Giant Unilamellar Vesicles;
H	:	Mean Curvature;
H_I	:	Normal Hexagonal Phase;
H_{II}	:	Inverse Hexagonal Phase;
HD	:	Hemifusion diaphragm;
HMMM	:	Highly Mobile Membrane Mimetics;
HOPS	:	Homotypic Vacuole Fusion Protein Sorting;
INPP5	:	Phosphatidylinositol 4,5-Bisphosphate Phosphatases;
ITP	:	Inverse Translocation Peptides;
κ	:	Bending Modulus;
$\bar{\kappa}$	:	Modulus of Gaussian Curvature;
K	:	Gaussian Curvature;
K_A	:	Area Elastic Modulus;
LacCer	:	Lactosylceramide;
L_C	:	Liquid Crystalline;
L_D	:	Liquid Disordered;
LLPS	:	Liquid-Liquid Phase Separation;
LNP	:	Lipid Nanoparticles;
L_O	:	Liquid Ordered;
LPAAT	:	Lysophosphatidic Acid Acyltransferase;
LPC	:	Lysophosphatidylcholine;
MA	:	Micropipette Aspiration;
MCS	:	Membrane Contact Sites;
MI	:	Inactive Metarhodopsin-I;
MII	:	Active Metarhodopsin-II;
MM	:	Mattress Model;
MD	:	Molecular Dynamics;
MSP	:	Major Sperm Domain;
NGC	:	Negative Gaussian Curvature;
NMR	:	Nuclear Magnetic Resonance;
NSE	:	Neutron Spin Echo;

NT	:	Nanotubes;
ORP	:	OSBP-Related Protein;
OSBP	:	Oxysterol-Binding Protein;
PA	:	Phosphatidic Acid;
PAH	:	Poly(Allylamine Hydrochloride);
PC	:	Phosphatidylcholine;
PE	:	Phosphatidylethanolamine;
PEG	:	Polyethylene Glycol;
PG	:	Phosphatidylglycerol;
PH	:	Pleckstrin Homology Domain;
PI	:	Phosphatidylinositol;
PI(4)P	:	Phosphatidylinositol-4-Phosphate;
PI(4,5)P2	:	Phosphatidylinositol-(4,5)-Bisphosphate;
PI4K	:	Phosphatidylinositol 4-Kinases;
PIP5K	:	Phosphatidylinositol-4-Phosphate 5-Kinases;
PITP	:	Phosphatidylinositol Transfer Protein;
PI3KC3	:	Class III Phosphatidylinositol 3-Kinase;
PL	:	Glycerophospholipids;
PLA	:	Phospholipase A;
PLC	:	Phospholipase C;
PLD	:	Phospholipase D;
POPC	:	1-Palmitoyl-2-Oleoyl-Phosphatidylcholine;
POPE	:	1-Palmitoyl-2-Oleoyl-Phosphatidylethanolamine;
POPG	:	1-Palmitoyl-2-Oleoyl-Phosphatidylglycerol;
PM	:	Plasma Membrane;
PPE	:	Poly Phenylene Ethylene;
PS	:	Phosphatidylserine;
pSM	:	Palmitoyl Sphingomyelin;
$\mathbf{Q_{II}}$	:	bicontinuous inverted cubic Phases;
RBC	:	Red Blood Cell;
RDM	:	Rod Disk Membranes;
Sac1	:	Phosphatidylinositol 4-Phosphate Phosphatases;
SAXS	:	Small-Angle X-Ray Scattering;
SM	:	Sphingomyelin;
SMS	:	SM Synthase;
SOPC	:	1-Stearoyl-2-Oleoyl-Phosphatidylcholine;
SREBP	:	Sterol Regulatory Element-Binding Protein;
SUV	:	Small Unilamellar Vesicle;
TGN	:	Trans-Golgi Network;
TMD	:	Transmembrane Domain;
VAP-A/B	:	Vesicle-Associated Membrane Protein-Associated Proteins A/B;
VAP	:	Vesicle-Associated Membrane Protein-Associated Proteins;
XDS	:	X-Ray Diffuse Scattering;
XRD	:	X-Ray Diffraction.

1

The Effect of Membrane Deformation on the Spatial Distribution of Lipids

David W. Allender[1,2] and *Michael Schick*[2,*]

The Basic Idea

Many biological processes bring about deformations in bilayer lipid membranes. This book discusses many such processes. Our intent in this chapter is to present a simple picture of how these deformations affect the distribution of lipids around them. First we have to describe the deformation of the two leaflets of the membrane itself. The shape of each leaflet is conveniently determined by specifying the heights, $h^I(x, y)$ and $h^O(x, y)$, of the inner and outer leaflets above some point, (x,y), in an external reference plane. Second, we note that the lipids that comprise the membrane can be characterized by their shape, that is, by the cross-sectional area of their head group compared to the cross-sectional area of their hydrocarbon tails. Those lipids for which these areas are very similar are said to be "cylindrical"; those with relatively small head groups compared to their tails are "cone-like", and those with large heads compared to their tails are 'inverted cone-like". We make the reasonable assumption (Leibler 1986; Leibler and Andelman 1987; Seifert 1997) that the free energy of the system will be reduced if lipids of a given shape are located in regions of the membrane of a similar shape. Lastly, we note that cholesterol can affect the shape of lipids (Sodt et al. 2016). Presumably, it does this by intercalating between the lipid tails, making the shape of phospholipids more cone-like.

With these basic ideas, let us consider a simple example that exemplifies them and that will require almost no calculation: the creation of a pore, or hole, in a lipid bilayer. The creation of such a pore is interesting in that it should normally be very

[1] Department of Physics Kent State University Kent OH 44242.
Emails: dallende@kent.edu
[2] Department of Physics University of Washington Seattle WA 98195.
* Corresponding author: schick@uw.edu

difficult to do if the membrane is to be stable. On the other hand, if pores are ever to occur, as they do regularly, in fusion, for example, pore creation cannot be too difficult. How are these two requirements reconciled?

Let us consider the creation of a circular pore of radius R in a membrane with a surface tension, i.e. energy per unit area, of σ, at a temperature T. The energy of the pore consists of a reduction, $\pi R^2\sigma$, due to the removal of energetically expensive area and an increase in the cost to make the perimeter, $2\pi R\lambda$, where λ is the energy per unit length, line tension, of the perimeter. Thus

$$\begin{aligned} E(R) &= 2\pi R\lambda - \pi R^2\sigma \\ &= \pi\lambda^2/\sigma - \pi\sigma(\lambda/\sigma - R)^2. \end{aligned} \tag{1}$$

We see that to make a pore initially costs energy but that at a radius $R^* = \lambda/\sigma$, the energy of creation reaches a maximum, $E(R^*) = \pi\lambda^2/\sigma$. Thus, there is an energy barrier. Once this is crossed, the pore expands. The probability of obtaining this energy via a thermal fluctuation is proportional to $\exp[-\pi\lambda^2/\sigma k_B T]$. This probability depends quite sensitively on the line tension. Thus, a small change in the line tension will have a large effect on the probability of pore formation. If such small changes can be effected, then this helps explain how the membrane can both be very stable and yet undergo fusion when necessary.

Let us look more closely at the line tension. A schematic cross-section of a bilayer with a pore of radius large compared to the membrane thickness, is shown in Figure 1.

One sees that this edge favors lipids with large heads and small tails. That also means it disfavors cholesterol that would increase the tail area, particularly sphingomyelin. As a consequence, we expect that an increase in the amount of cholesterol in the membrane will increase the line tension and therefore greatly decrease the probability of pore formation. This is, indeed, the case (Zimmerberg, J. Personal communication).

Let us consider another simple example: the deformation of the plasma membrane by the insertion of a protein, Piezo1, a tension-sensitive ion channel. This protein deforms the membrane, pushing it toward the cytosol, as shown in Figure 2.

Molecular dynamics simulations have been carried out on this system (Buyan et al. 2020; Buyan et al. 2022). From them, smooth, cylindrically symmetric functions representing the cross-section of the membrane could be constructed. They are shown in Figure 3. This profile is cylindrically symmetric about the z-axis. The protein pushes the membrane towards the cytosol, causing the leaflets to wrap around it, with the outer leaflet closer to the protein. One sees that, in the outer leaf, as the protein is approached, cone-shaped lipids will be energetically favored. In addition, the amount of cholesterol, which makes the cone-shaped lipids even more cone-shaped, will also be increased. In the inner leaf, the reverse is true. As the protein approaches, inverted cone-shaped lipids are favored, and cholesterol will deplete. This illustrates a more general point: when the two leaflets of a deformed membrane are essentially parallel, as they are in Figure 3 then whatever happens to the cholesterol

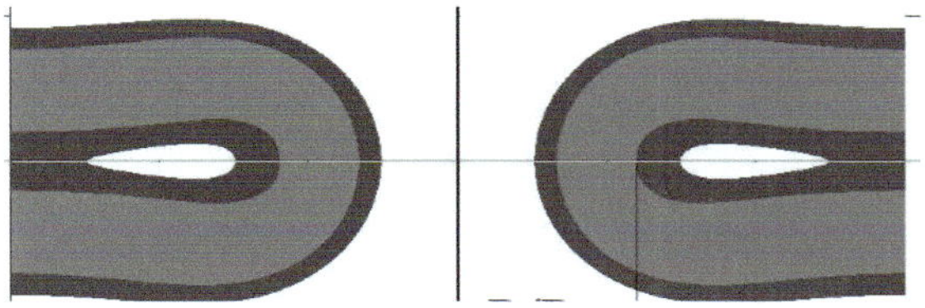

Figure 1. Schematic representation of the cross-section of a bilayer that ends at an edge.

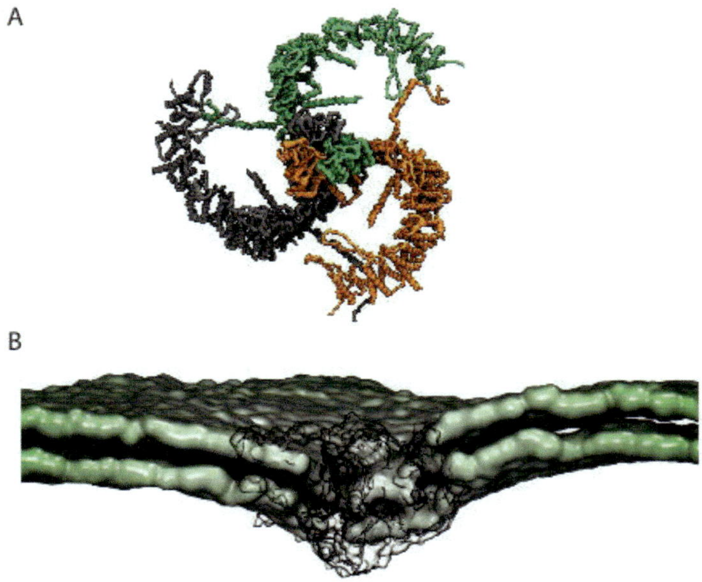

Figure 2. Model of Pizo1 embedded in a bilayer membrane. (A) top view, i.e. looking from outside the cell towards its interior, (B) side view. Reprinted from Buyan et al. (2022) with permission from Elsevier.

mole fraction in one leaflet will be the opposite of what happens to the cholesterol mole fraction in the other leaflet.

Now that the general ideas have been presented qualitatively, we want to make them quantitative. To do so, we turn to a model of the plasma membrane we have employed previously (Allender et al. 2019; Allender at al. 2020).

Model Plasma Membrane

Our model plasma membrane consists of two leaflets, each with an area A. The outer leaflet consists of palmitoyloleoylphosphatidylcholine (PC), palmitoylsphingomyelin (SM), and cholesterol. An important feature is that SM has two saturated chains while PC has one saturated and one unsaturated chain. Their mole fractions in the outer leaf are denoted y_{PCo}, y_{SM} and y_{Co}, respectively. They sum to unity. We take the ratio of PC to SM to be unity (Zachowski 1993). The inner leaflet consists of

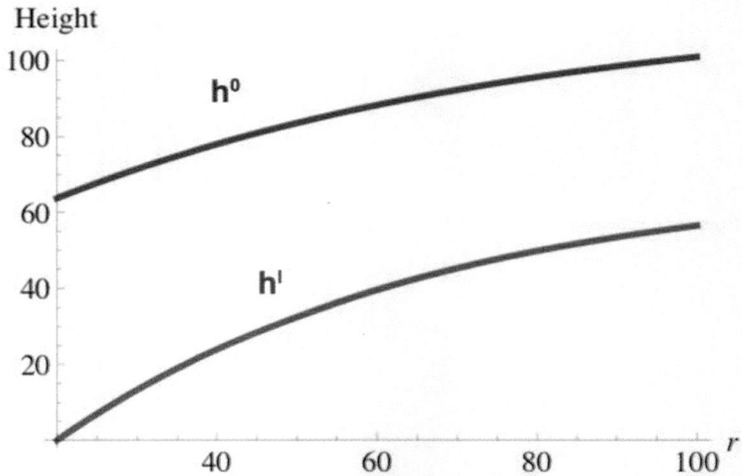

Figure 3. Smooth functions, obtained from the simulation data, that represent the profile of a bilayer membrane with a Piezo1 protein inserted along the z-axis of a cylindrical coordinate system. Both the height of the membrane and the distance from the z-axis are in Angstroms.

palmitoyloleoylphosphatidylethanolamine (PE), palmitoyloleoylphosphatidylserine (PS), PC, and cholesterol. In contrast to the outer leaflet, all phospholipids in the inner leaflet have one unsaturated chain. Their mole fractions are denoted y_{PE}, y_{PS}, y_{PCi}, and y_{Ci}. They sum to unity. We take the ratios of PE to PS to PC to be 5:3:1 (Zachowski 1993).

We take the area of all lipids, except for cholesterol, to be the same, $a = 0.7\ \text{nm}^2$. The area per lipid of cholesterol is taken to be smaller, $v \times a$, with $v = 0.6$ (Phillips 1971). Because cholesterol can flip flop between leaflets rapidly, we require that the chemical potential of cholesterol in the two leaflets be equal. We do not expect, in general, that the amount of cholesterol in each leaflet will be the same. As a result of the difference in area per molecule of cholesterol and the other lipids, and the constraint of equal areas of the two leaflets, the numbers of lipids in each leaflet will not be the same. To see this, we write the areas of the two leaflets:

$$A = N_{tot}^{O} a[1 - (1 - v)y_{Co}] \tag{2}$$

for the outer leaflet, with N_{tot}^{O} the total number of molecules in the outer leaflet, and similarly:

$$A = N_{tot}^{I} a[1 - (1 - v)y_{Ci}], \tag{3}$$

for the inner leaflet.

Equating these two expressions for the area, we find:

$$\frac{N_{tot}^{I}}{N_{tot}^{O}} = \frac{[1 - (1 - v)y_{Co}]}{[1 - (1 - v)y_{Ci}]} \tag{4}$$

We also note that the number densities of the lipids in the two leaflets are:

$$\rho^O(y_{Co}) = \frac{N^O_{tot}}{A} = \frac{1}{a[1-(1-v)y_{Co}]} \text{ and}$$

$$\rho^I(y_{Ci}) = \frac{N^I_{tot}}{A} = \frac{1}{a[1-(1-v)y_{Ci}]}, \tag{5}$$

The Free Energy

The free energy of the bilayer consists of several parts. First, there is the regular solution piece (Furman et al. 1977), a mean-field approximation to the free energy of the pairwise interactions between lipids and their entropy of mixing. We will assume that there are no direct, pairwise interactions between lipids in different leaflets. We can write the bilayer-free energy in the form:

$$F_{rs}(T,\{y_i\}) = \int d^2r\rho^O(y_{Co})f^O_{rs}(T, y_{SM}, y_{PCo}, y_{Co})$$

$$+ \int d^2r\rho^I(y_{Ci})f^I_{rs}(T, y_{PE}, y_{PS}, y_{PCi}, y_{Ci}), \tag{6}$$

where f^O and f^I are free energies per particle. They can be written:

$$f^O_{rs} = f^O_{int} + f^O_{ent}$$

$$f^O_{int} = 6\epsilon_{SM,PC}y_{SM}y_{PCo} + 6\epsilon_{SM,C}y_{SM}y_{Co} + 6\epsilon_{PC,C}y_{PCo}y_{Co}. \tag{7}$$

$$f^O_{ent} = k_BT\,[y_{SM}\ln(y_{SM}) + y_{P\,Co}\ln(y_{PCo}) + y_{Co}\ln(y_{Co})], \tag{8}$$

and

$$f^I_{rs} = f^I_{int} + f^I_{ent}$$

$$f^I_{int} = 6\epsilon_{PE,PC}y_{PE}y_{PCi} + 6\epsilon_{PE,C}y_{PE}y_{Ci} + 6\epsilon_{PE,PS}y_{PE}y_{PS}$$

$$+ 6\epsilon_{PS,PC}y_{PS}y_{PCi} + 6\epsilon_{PS,C}y_{PS}y_{Ci} + 6\epsilon_{PC,C}y_{PCi}y_{ci}, \tag{9}$$

$$f^I_{ent} = k_BT\,[y_{PE}\ln(y_{PE}) + y_{PCI}\ln(y_{PCI})$$

$$+ y_{PS}\ln(y_{PS}) + y_{Ci}\ln(y_{Ci})], \tag{10}$$

Terms in the free energy that are linear in the component mole fractions, such as the free energy of translation (Uline et al. 2010), have been ignored as they affect only the absolute values of the component chemical potentials. The epsilons in the above are interaction energies. The weakest interactions are those between unsaturated lipids, and we take them as zero: $\epsilon_{PE,PS} = \epsilon_{PE,PC} = \epsilon_{PS,PC} = 0$. The interaction between saturated SM and unsaturated PC is strong and repulsive, $\epsilon_{SM,PC}/k_BT = 0.32$. All the other interactions involve cholesterol: the repulsive $\epsilon_{P\,C,C}/k_BT = 0.20$ and $\epsilon_{P\,S,C}/k_BT = 0.16$ and the attractive $\epsilon_{P\,E,C}/k_BT = -0.15$. These values are the same as those used in reference (Allender 2020) and are similar to those in (Almeida 2009). All of these values come from fitting experimental data to mean-field calculations of Ising models and then adjusting the interactions in these models as mean-field theories overestimate their transition temperatures by a known amount (Allender 2020). Lastly, to approximate the SM, cholesterol interaction of the simulation, we take $\epsilon_{SM,C}/k_BT = -0.14$. (Buyan et al. 2022). Finally, for the bending modulus of the simulated bilayer, we choose $\kappa_B/k_BT = 22$, a reasonable value for a lipid bilayer with no cytoskeleton (Rawicz et al. 2000).

If the bilayer were flat, the state of the membrane would be determined by five independent mole fractions, which can be chosen to be y_{SM}, y_{Co}, y_{PE}, y_{PS}, and y_{Ci}. These could be determined from the three given ratios $y_{SM}/y_{PCo} = 1$, $y_{PE}:y_{PS}:y_{PCi} = 5:3:1$, from the equality of the cholesterol chemical potentials in the two leaflets, and from the total mole fraction of cholesterol in the bilayer. We take this to be 0.4 (vanMeer 2011). We denote the values of the lipid mole fractions by an overbar, $\bar{y}_i$. These values are independent of position in the monolayer. The free energy of the flat bilayer also contains the constant $A\sigma$, where σ is the bilayer surface tension. We shall ignore this constant.

When the bilayer is not flat, the mole fractions of the components are spatially dependent, $y_i(\mathbf{r}) = \bar{y}_i + \delta y_i(\mathbf{r})$, as are the heights of the two leaflets $h^I(\mathbf{r}) = \bar{h}^I + \delta h^I(\mathbf{r})$, and $h^O(\mathbf{r}) = \bar{h}^O + \delta h^O(\mathbf{r})$. This brings about several additional terms in free energy. There is a gradient squared term that penalizes variations in the spatial distribution of the component mole fractions:

$$F_{gs}(\{y_i\}) = \int d^2r \frac{b}{2} \sum_{\alpha} (\nabla \delta y_\alpha)^2 + \int d^2r \frac{b}{2} \sum_{\gamma} (\nabla \delta y_\gamma)^2, \tag{11}$$

with α taking the values of the outer leaf, SM, PCo, and Co, and γ the values of the inner leaf PE, PS, PCi, and Ci. We take (Dan and Safran 1998) the energy cost $b = 5k_BT$.

Due to the additional surface area of the non-flat membrane, there is a free energy contribution

$$F_{st}(h^I, h^O) = \int d^2r \left[\frac{\sigma^I}{2} [\nabla \delta h^I(\mathbf{r})]^2 + \frac{\sigma^O}{2} [\nabla \delta h^O(\mathbf{r})]^2 \right]. \tag{12}$$

Finally, we include the contribution from the bending of the membrane. It is:

$$\begin{aligned} F_{bend}(h^I, h^O, \{y_i\}) &= \int d^2r \left\{ \frac{\kappa_m^I}{2} [H^I(\mathbf{r}) - H_0^I[y_{PS}(\mathbf{r}), y_{PE}(\mathbf{r}), y_{PCi}(\mathbf{r}), y_{Ci}(\mathbf{r})]^2 \right. \\ &+ \left. \frac{\kappa_m^O}{2} [H^O(\mathbf{r}) - H_0^O(y_{SM}(\mathbf{r}), y_{Co}(\mathbf{r}), y_{PCo}(\mathbf{r})]^2 \right\}, \end{aligned} \tag{13}$$

where κ_m^I and κ_m^O are the monolayer bending moduli, $H^I(\mathbf{r})$ and $H^O(\mathbf{r})$ are the total curvatures (twice the mean curvatures) of the inner leaflet and of the outer leaflet, and the $H_0^I(\mathbf{r})$ and $H_0^O(\mathbf{r})$ are the so-called "spontaneous curvatures" of the leaflets that depend on the local membrane composition.

A short paragraph on curvature is appropriate here. The curvature of a surface is directly related to the way in which a normal to the surface varies as one moves from one place on the surface to another. As the direction of the surface normal is only specified within a sign, one must be clear as to what sign one is taking. In this chapter we shall choose the direction of the normal to the outer leaflet and the direction of the normal to the inner leaflet to be the same. In particular we choose them to point from the inside of the cell to the outside. We note that many authors prefer normals that point in opposite directions, with each normal extending from the water side of the leaflet to the oil side. Why do they do this? After all, there is a clear downside, namely that if the two leaflets are deformed in essentially the same way so that they remain almost parallel, with this choice of normals, the curvatures of the two nearly identical leaves will be the negative of one another. The benefit of the choice is in the spontaneous curvature of a molecule or, more properly, the spontaneous curvature of a monolayer of a given kind of molecule. Consider PE, which has a small head group compared to its tail area. If PE is in the inner leaf, it will bend that leaf in one way, while if it is in the outer leaf, it will bend it in the opposite way. If one chooses the normals to be in opposite directions, then the curvatures that PE will tend to impose are the same sign in either leaf. Thus, the authors can talk about how PE has a negative spontaneous curvature irrespective of which leaf it is in. Were this choice to be made, then one would have to replace, in Equation (13) $H^O(\mathbf{r})$ by $-H^O(\mathbf{r})$.

But to repeat, in this chapter, we choose the normals for both leaflets to point from the inside of the cell to the outside. As a consequence, two nearly parallel deformed leaflets have almost the same curvature. For a somewhat spherical cell, both their curvatures would be negative. However, PE will not have a definite spontaneous curvature but rather a negative one in the inner leaflet and a positive one in the outer leaflet.

We now return to the free energy. If the curvature is not too large, the contribution from the bending energy can be approximated as:

$$\begin{aligned} F_{bend}(h^I, h^O) &\approx \int d^2r \left\{ \frac{\kappa_m^I}{2} [\nabla^2 \delta h^I(\mathbf{r}) - H_0^I(y_{PS}(\mathbf{r}), y_{PE}(\mathbf{r}), y_{PCi}(\mathbf{r}), y_{Ci}(\mathbf{r}))]^2 \right. \\ &+ \left. \frac{\kappa_m^O}{2} [\nabla^2 \delta h^O(\mathbf{r}) - H_0^O(y_{SM}(\mathbf{r}), y_{Co}(\mathbf{r}), y_{PCo}(\mathbf{r}))]^2 \right\}. \end{aligned} \quad (14)$$

Thus, the total free energy is

$$F_{bi}(T, \{y_i\}, h^I, h^O) = F_{rs}(T, \{y_i\}) + F_{gs}(\{y_i\}) + F_{st}(h^I, h^O) + F_{bend}(h^I, h^O, \{y_i\}), \quad (15)$$

with the four pieces given by Equations (6, 11, 12, 14).

The product of the leaflets' spontaneous curvatures, H_0^O and H_0^I and their bending moduli, κ_m^O and κ_m^I was obtained from simulation in reference (Allender et al. 2019). They were found to depend on the mole fraction of cholesterol in the leaf. In the relevant range of cholesterol mole fraction, it was determined that in the exoplasmic leaf, the product of monolayer bending modulus and spontaneous curvature was well fit by the form:

$$\begin{aligned} \frac{\kappa_m^O H_0^O}{k_B T} &= \frac{y_{SM}}{y_{SM} + y_{PCo}}(b_{SM} y_{Co} - a_{SM}) + \frac{y_{PCo}}{y_{SM} + y_{PCo}}(b_{PC} y_{Co} - a_{PC}) \\ 0.25 &\leq y_{Co} \leq 0.6, \end{aligned} \tag{16}$$

where, in units of nm^{-1}, $a_{SM} = 6.438$, $b_{SM} = 18.58$, $a_{PC} = 1.341$, and $b_{PC} = 9.267$. The a_{SM} and a_{PC} are simply fitting parameters. The b_{SM} and b_{PC} are also fitting parameters which give information about the effect of cholesterol on the spontaneous curvatures of these molecules. Note that Equation (16) is simply a mole—fraction weighted average of the spontaneous curvatures of the phospholipids only and not cholesterol. However, these curvatures depend on the amount of cholesterol present. The contribution of cholesterol to the leaflet's spontaneous curvature is only through its effect on the phospholipids (Sodt et al. 2016); it tends to enlarge the area taken up by their acyl chains (Allender et al. 2019).

A similar fit to the product of the bending modulus and spontaneous curvature of the cytoplasmic leaf was found to be:

$$\begin{aligned} \frac{\kappa_m^I H_0^I}{k_B T} &= -\left[\frac{y_{PE}}{y_{PE} + y_{PS} + y_{PCi}}(b_{PE} y_{Ci} - a_{PE})\right. \\ &+ \frac{y_{PS}}{y_{PE} + y_{PS} + y_{PCi}}(b_{PS} y_{Ci} - a_{PS}) \\ &+ \left.\frac{y_{PCi}}{y_{PE} + y_{PS} + y_{PCi}}(b_{PC} y_{Ci} - a_{PC})\right] \\ 0.25 &\leq y_{Ci} \leq 0.6, \end{aligned} \tag{17}$$

with $a_{PE} = -3.156$, $b_{PE} = 2.71$, and $a_{PS} = -1.223$, $b_{PS} = 3.35$, again, all in nm^{-1}. Note the minus sign in Equation (17) that does not appear in Equation (16). This reflects the fact, noted earlier; that a molecule with a given architecture will tend to bend the inner leaflet in the opposite direction it would bend the outer leaflet. It should also be noted from the magnitude of the various fitting coefficients b_i that the increase of the spontaneous curvature of saturated SM due to the presence of cholesterol is much larger than its effect on any of the unsaturated lipids.

To proceed, we expand the free energy about its value for a flat membrane up to second order in the small quantities $\delta y_i(\mathbf{r})$, $\delta h^O(\mathbf{r})$, and $\delta h^I(\mathbf{r})$. This is straightforward but tedious. It is convenient to express these small quantities in terms of their Fourier transforms:

$$\begin{aligned} \delta y_i(\mathbf{k}) &= \frac{1}{A}\int d^2 r \delta y_i(\mathbf{r}) \exp[-i\mathbf{k}\cdot\mathbf{r}], \\ \delta y_i(\mathbf{r}) &= \frac{A}{(2\pi)^2}\int d^2 k\, \delta y_i(\mathbf{k}) \exp[i\mathbf{k}\cdot\mathbf{r}], \end{aligned} \tag{18}$$

and similarly for $\delta h^I(\mathbf{k})$ and $\delta h^O(\mathbf{k})$.

Some simplification arises from the fact that the mole fractions of the components in the outer leaf sum to unity so that δy_{PCo} can be eliminated from:

$$\delta y_{PCo} = -\delta y_{SM} - \delta y_{Co}. \tag{19}$$

Similarly:

$$\delta y_{PCi} = -\delta y_{PE} - \delta y_{PS} - \delta y_{Ci}. \tag{20}$$

Then, the free energy takes the form:

$$F_{bi}[T, \{y_i\}, \{h\}] = F_{bi}(T, \{\bar{y}_i\}, \bar{h}\,] + \delta\tilde{F}_{mem}$$

$$\begin{aligned} \delta\tilde{F}_{mem} &= \frac{A}{(2\pi)^2}\int d^2k \Bigg\{ \sum_{i,j} V_{i,j}\delta y_i(\mathbf{k})\delta y_j(-\mathbf{k}) + \sum_{m,n} V_{m,n}\delta y_m(\mathbf{k})\delta y_n(-\mathbf{k}) \\ &+ \kappa_m^O k^2 \sum_i C_i^O \delta y_i(\mathbf{k})\delta h^O(-\mathbf{k}) + \kappa_m^I k^2 \sum_m C_m^I \delta y_m(\mathbf{k})\delta h^I(-\mathbf{k}) \\ &+ \frac{k^4}{2}\kappa_m^O|\delta h^O(\mathbf{k})|^2 + \frac{k^4}{2}\kappa_m^I|\delta h^I(\mathbf{k})|^2 \Bigg\}, \end{aligned} \tag{21}$$

where the indices i and j take the values *SM*, C_o, and m and n take the values *PE*, *PS*, and C_i. Explicit expressions for all matrix elements in the above are given in Appendix B of reference (Buyan et al. 2022).

For further analysis, we distinguish two different systems for application: (i) the membrane deformation is known and (ii) the membrane deformation is unknown and due to random thermal fluctuations. If the form of the membrane deformation is known, then the term quadratic in the δh is a constant and can be ignored. We then minimize the free energy with respect to the composition fluctuations $\delta y_a(\mathbf{k})$ and $\delta y_\mu(\mathbf{k})$. This yields the simple set of two linear inhomogeneous equations:

$$\sum_j V_{i,j}\delta y_j(\mathbf{k}) = -\frac{\kappa_m^O}{2}k^2 h^O(\mathbf{k})C_i^O, \tag{22}$$

and the set of three linear, inhomogeneous, equations:

$$\sum_n V_{m,n}\delta y_n(\mathbf{k}) = -\frac{\kappa_m^I}{2}k^2 h^I(\mathbf{k})C_m^I. \tag{23}$$

Note that the composition fluctuations, $\delta y(\mathbf{k})$, are driven by the curvature, $-k^2h(\mathbf{k})$.

Application to Piezo1: The Deformation is Known

With the above theoretical formalism in hand, let us consider the system of the ion channel Piezo1 inserted into the plasma membrane. The system was considered by Molecular Dynamics simulation (Buyan et al. 2022). As noted earlier and shown in Figure 2, the protein pushes the membrane inward toward the cytoskeleton. The simulation results for the mole fractions of the lipids in the presence of the protein are shown in Figure 4A for the outer leaflet and 4B for the inner leaflet.

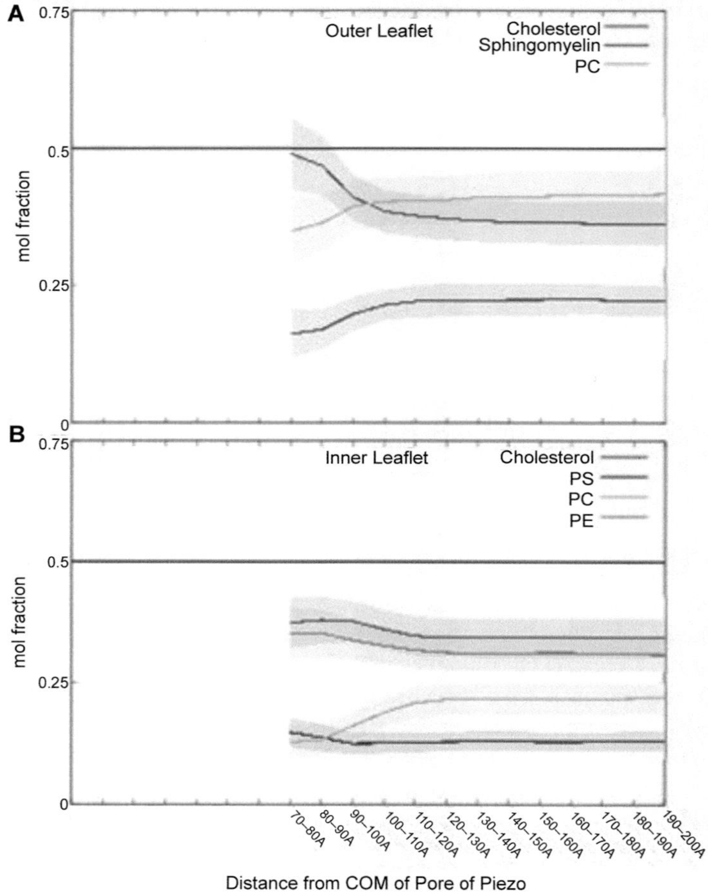

Figure 4. Molecular Dynamics results for the lipid distribution in the outer leaf of the plasma membrane (A). From top to bottom on the right, the mol fractions are for PC, cholesterol, and sphingomyelin. In the inner leaf, (B), from top to bottom on the right, the mol fractions are for cholesterol,PE, PC,and PS. The Piezo1 is at the origin. Reprinted from Buyan et al. (2022) with permission from Elsevier.

One notes several things. In the outer leaflet, the mole fraction of cholesterol increases markedly as the protein is approached, whereas the SM and PC decrease. The changes in the inner leaf are generally much smaller, with the decrease in PC being the exception. The mole fraction of cholesterol decreases in the inner leaf in the same spatial region in which it increases in the outer leaf.

The analytic theory illuminates these results. To apply the theory, the height of the outer leaflet, obtained from an angular average of results from the simulation, was fit to a smooth, cylindrically symmetric, and analytic function:

$$h^O(r) = -70.26\exp[-.41r/25] - 1.4\exp[-.0035(r/25)^2] + 23 \tag{24}$$

where $h^O(r)$ and r are in Angstroms.

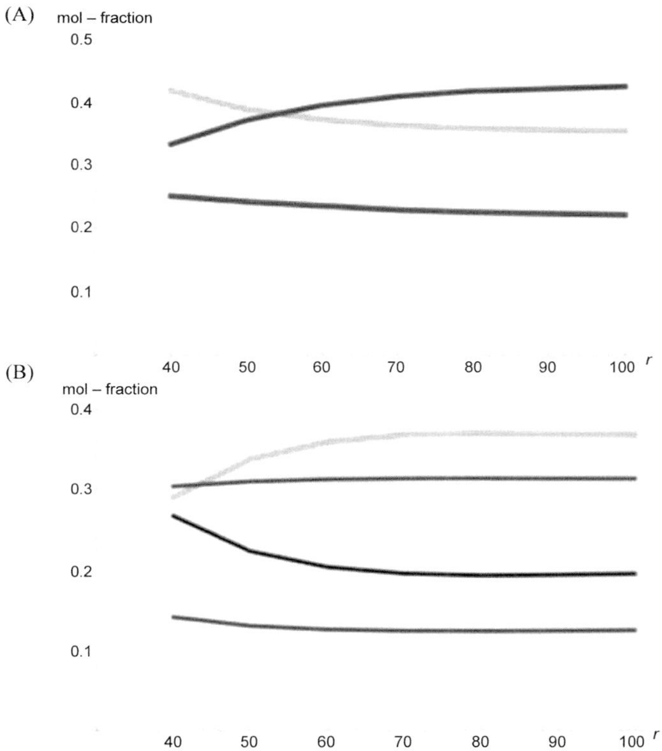

Figure 5. (A) Results from the theory for the outer leaflet. From top to bottom on the right, mole fractions are for PC, cholesterol, and SM. The distance, r, from the protein is in Angstroms. (B) Results from the theory for the inner leaflet. From top to bottom, mole fractions comprise cholesterol, PE, PC, and PS. The distance, r, from the protein is in Angstroms. Reprinted from Buyan et al. (2022) with permission from Elsevier.

From this fit to the outer leaf, we determine a fit to the inner leaflet such that the membrane had a constant thickness, a reasonable assumption given that the free energy cost of a variation in thickness is expected to be large. The thickness was set to 43.1Å. That fit was given by:

$$h^I(r) = -16.6836 \exp[-.214436r/25] - 93.1964 \exp[-.57983r/25] - 20.1. \quad (25)$$

These are the two functions plotted in Figure 3. From the functions $h^O(r)$ and $h^I(r)$, we obtain the curvatures, in cylindrical coordinates:

$$\nabla^2 h^O(r) = \frac{d^2}{dr^2} h^O(r) + \frac{1}{r}\frac{d}{dr} h^O(r), \quad (26)$$

and similarly for the inner leaflet.

As the protein is approached, the curvature is dominated by the contribution from the encircling of the protein by the membrane, a contribution that is positive.

With an analytic expression for the height of the outer leaf, one Fourier transforms it to obtain $h(\mathbf{k})$ and then solves the linear Equations (22) and (23) for the $y_i(\mathbf{k})$. After

Fourier transforming again, one obtains the desired mole fraction variations $\delta y_i(\mathbf{r})$. The results of the theory are shown in Figure 5A for the outer leaflet and Figure 5B for the inner leaflet.

As anticipated earlier, we see that the cholesterol in the outer leaf increases as the protein is approached. This is because the membrane wraps around the protein, which favors, in the outer leaf, lipids with small heads and large tails, and cholesterol to make the tails even larger. As the mole fraction of cholesterol increases, the sum of PC and SM must decrease. As the direct interaction between PC and cholesterol is repulsive, the mole fraction of PC decreases. The direct interaction between SM and cholesterol, however, is attractive. As a result of the competing effects, the mole fraction of SM does not change much.

In the inner leaflet, lipids with large heads and small tails are favored by wrapping the membrane about the protein. Hence, as anticipated, cholesterol is repelled to make the tails even smaller. The changes in the mole fraction of PS and PE are small, in agreement with the simulations. The reason for this is that while changes in mole fraction of cholesterol have a large effect on saturated lipids, like SM, such changes have a much smaller effect on disordered unsaturated lipids, like PS and PE. The reason for the large decrease in the mole fraction of PC in the inner leaflet observed in the simulation is not clear. However, the physical origins of the main features of the simulation results are illuminated by the analytic theory.

Membrane Deformations Due Simply to Thermal Fluctuations; The Deformation is Unknown

In the previous section, we have considered membrane deformations that are given, brought about by pore formation, or protein insertion, other external factors. We turn now to the equally interesting case in which the deviations from a flat membrane are simply due to thermal fluctuations. Now, we note that the thermal membrane deformations are small, i.e., 2π divided by the wavevector, which is much greater than the membrane thickness. Then, the condition that the membrane is of constant thickness implies:

$$\delta h^O(\mathbf{k}) = \delta h^I(\mathbf{k}) \equiv \delta h(\mathbf{k}). \tag{27}$$

In this case, the height deviations, $h(\mathbf{k})$, are not given externally but must be integrated over in the partition function. The free energy, δF_{mem}, is obtained from $\delta\tilde{F}_{mem}$ of Equation (21) according to:

$$\exp[-\delta F_{mem}(T, A, \{y_i\}]/k_BT = \mathcal{K}\Pi_{\mathbf{k}} \int dh(\mathbf{k}) \exp[-\delta\tilde{F}_{mem}(T, A, \{y_i\}, \{h(\mathbf{k})\})/k_BT], \tag{28}$$

where $\mathcal{K}$ is an uninteresting constant. As the dependence of $\tilde{F}_{mem}$ on the $h(\mathbf{k})$ is only quadratic, the integrals can be carried out explicitly. We note that the largest contribution to the integral comes from those $h(\mathbf{k})$ that make the exponent stationary:

$$h(\mathbf{k}) = -\frac{1}{2Ck^4}[\sum_\alpha D^O_\alpha \delta y_\alpha(\mathbf{k}) + \sum_\mu D^I_\mu \delta y_\mu(\mathbf{k})]. \tag{29}$$

Note that as this expression depends on lipid mole fractions in both leaves and as the free energy depends quadratically on the $h(\mathbf{k})$, the coupling of the mole fractions to the height fluctuations provides a coupling between the lipid mole fractions in the two leaflets.
The free energy δF_{mem} can be written in the form:

$$\frac{\delta F_{mem}}{k_B T} = \frac{A^2}{(2\pi)^2} \int d^2k \, \phi_i^T(\mathbf{k}) \cdot M_{i,j}(\mathbf{k}) \cdot \phi_j(-\mathbf{k}), \tag{30}$$

where $\phi_I^T(\mathbf{k})$ is the row matrix:

$$\phi^T(\mathbf{k}) = (\delta y_{SM}(\mathbf{k}), \delta y_{Co}(\mathbf{k}), \delta y_{PE}(\mathbf{k}), \delta y_{PS}(\mathbf{k}), \delta y_{Ci}(\mathbf{k})), \tag{31}$$

ϕ_j is the corresponding column matrix, and $M_{i,j}(\mathbf{k})$ is a five-by-five symmetric matrix. Its fifteen independent elements are given in the Supplementary Material to Ref. (Allender et al. 2020).
From the matrix $M_{i,j}(\mathbf{k})$, one obtains directly the structure factors

$$S_{\alpha,\beta}(\mathbf{k}) \equiv \frac{A}{a} < \delta y_\alpha(\mathbf{k}) \delta y_\beta(-\mathbf{k}) >, \tag{32}$$

$$= M^{-1}_{\alpha,\beta}(\mathbf{k}) \tag{33}$$

where the angular brackets denote an ensemble average. While the first moments of the mole fraction deviations tell us nothing about a fluid phase, i.e.,

$$< \delta y_i(\mathbf{k}) > = 0, \text{ all i}, \tag{34}$$

the second moments, the structure factors, tell us a great deal.

For this application of the theory, we choose the interaction strengths to be the same as in Ref. (Allender et al. 2020). These are the same as we used earlier in this chapter, with the exception of $\epsilon_{SM,C}$. Earlier, this interaction was chosen to better fit those of a Molecular Dynamics simulation. Here, we take $\epsilon_{SM,C} = -0.58$ as determined from a fit to experimental data (Allender et al. 2020). We also take the bilayer bending modulus as $\kappa_B/k_BT = 44$, a value obtained from mammalian red blood cells (Evans 1983). It is reasonable to expect a larger κ_B in these cells than was used in the previous section due to the presence of the cytoskeleton, which stiffens the membrane.

First, consider a normal fluid. We see from Equation (32) that the structure factor tells us about the probability of thermal excitation in the composition of some components. In a normal fluid, any variation of the composition of a component costs energy $\epsilon = c_1 + c_2k^2$, where k is the wavevector of the variation, and c_1 and c_2 are constants. As the amount of excitation is inversely proportional to the energy of that excitation (i.e., those with the smallest energies are most easily excited), one finds the fluid is characterized by the usual Ornstein-Zernike structure factor, one independent of the direction of the wavevector:

$$S_{OZ}(k) \propto \frac{1}{c_1 + c_2 k^2}. \tag{35}$$

The plasma membrane is not, however, an ordinary fluid. The energy of an excitation not only contains constant terms c_1 coming from the interactions and entropy and a penalty for gradients in the composition, c_2k^2, but also an energy arising from the bending of the membrane and the coupling of the membrane height to the composition. This coupling favors variations in the composition so that molecules of a given shape go to the places in the membrane of that shape. The form of this energy is: (Allender 2020)

$$\epsilon(k) \propto -\frac{k^2}{((\sigma/\kappa_B) + k^2)} \tag{36}$$

Adding this term to the others causes the excitation energy of a composition variation to exhibit a minimum for wave vectors on the order of $(\sigma/\kappa_B)^{1/2}$. As a consequence, the structure factor has a peak at a wavenumber of this order. In particular, consider $S_{1,1}(k) = S_{SM,SM}(k)$. This structure factor tells us about the relative importance of the various Fourier components of the SM density distribution. As stated above, in a normal fluid phase, $S_{SM,SM}(k)$ would be expected to have a peak at $k = 0$, which would imply that the most important spatial variation in its density distribution would have an infinite wavelength; that is, the density distribution of SM would be essentially uniform. But what does the structure factor, $S_{SM,SM}$ look like in our model of the plasma membrane? It is shown in Figure 6 plotted as a function of the dimensionless wavevector $q \equiv k(\kappa_B/\sigma_B)^{1/2}$.

As we see, the structure factor does have a peak, not at a zero wavevector, but at a non-zero wavevector. This means that the SM distribution is not uniform. Rather, it is found in regions with a characteristic length of a half wavelength,

$$\frac{\lambda}{2} = \frac{\pi}{k} = \left(\frac{\kappa_m}{\sigma_m}\right)^{1/2}\frac{\pi}{q} = \left(\frac{\kappa_B}{\sigma_B}\right)^{1/2}\frac{\pi}{q}, \tag{37}$$

where we have taken the monolayer bending modulus and surface tension, κ_m and σ_m, to be equal to one half of the values of the bilayer membrane κ_B and σ_B.

We note that the natural length that enters in the above comes from the properties of the bilayer: $L_0 \equiv (\kappa_B/\sigma_B)^{1/2}$. For the plasma membrane, we take $\sigma_B = 4.7 \times 10^{-3} k_BT/nm^2$ from Dai and Sheetz (1995), and a bending modulus $\kappa_B = 44k_BT$ (Evans 1983). Then, one obtains L_0, which is approximately 97 nm. A peak in $S_{SM,SM}$ at $q = 5$ corresponds, then, to $\lambda/2$ of about 61 nm.

We pause to note the significance of this length. For variations over distances less than L_0, the energy of a membrane deformation is dominated by the bending energy. Hence, the spontaneous curvature, and the lipid composition upon which it depends, are of importance. For variations over distances greater than L_0, the energy of a membrane deformation is dominated by the surface tension, σ_B. Bending energies and lipid shapes are not as important.

All structure factors of the form $S_{a,a}$ show the same characteristic peak at a small, non-zero, dimensionless wavevector q. Thus, according to our model, the plasma membrane is structured with a characteristic size L_0. It is a microemulsion. To gain more insight into the structure, we look at $S_{SM,Co}$. It is shown in Figure 7.

This structure factor also peaks at a value of q of about five. This indicates that the concentrations of SM and cholesterol are correlated; they are found together. The regions are denoted liquid-ordered-like or lo-like after the well-ordered SM tails. The SM-PC structure factor is negative, which indicates that SM and PC in the outer

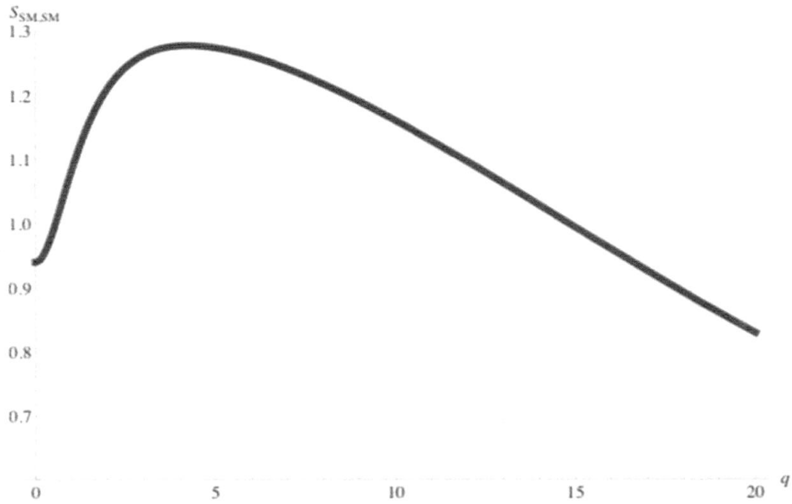

Figure 6. The structure factor $S_{SM,SM}(q)$ is shown as a function of the dimensionless wavevector q.

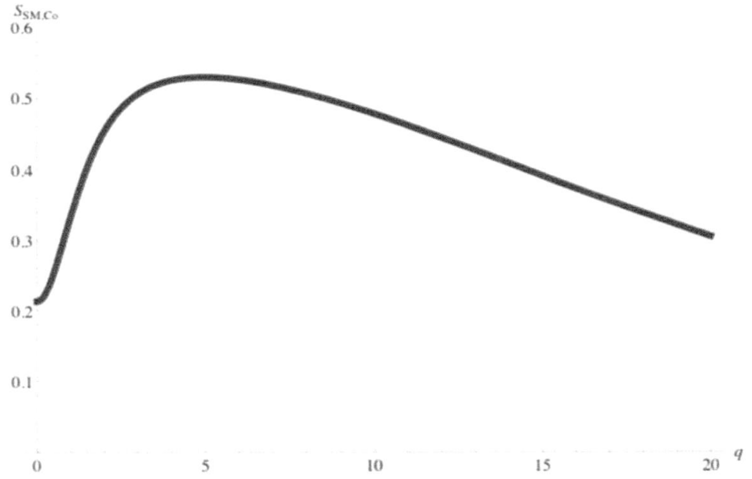

Figure 7. Structure factor $S_{SM,Co}(q)$ plotted as a function of the dimensionless wavevector q.

leaf are anti-correlated. The PC regions are denoted liquid-disordered-like, or ld-like, after the disordered tails of PC. Because there are equal mole fractions of SM and PC in the outer leaf in our model, and because the cholesterol favors the SM, there will be somewhat more lo-like regions than ld-like. Both regions exhibit a characteristic size of about 70 nm. Monte Carlo simulations of a model that, like ours, couples lipid composition to membrane curvature have been carried out (Sadeghi et al. 2014). As in our model, the relative amounts of the lo and ld regions are comparable in the simulation. Snapshots vividly show the disordered liquid with domains of two kinds, each of which exhibits the same characteristic width.

And what of the inner leaflet? How are the mole fractions of the components there correlated with the lo- and ld-like regions in the outer leaflet? One thing can be said without calculation. If the membrane distortion in the outer leaf favors lipids with small heads and large tails, and therefore cholesterol which enlarges the tail area, then that same region on the inner leaf must disfavor such lipids and cholesterol. We can predict, therefore, that whereas PC is found in the outer leaflet in an ld-like region, in the inner leaflet, it will be found below the lo-like region. Similarly, as cholesterol in the outer leaflet is found predominantly in the lo-like region, it is expected to be found in the inner leaf predominantly below the ld-like region in the outer leaf. The only lipids we cannot place this way are the PS and the PE in the inner leaf. The structure factor $S_{1,3}(q) = S_{Sm,P\ S}(q)$ is shown in Figure 8 as a function of the dimensionless wavevector q. This shows that the locations of the PS will be predominantly below that of the SM in the outer leaf. This correlation had been observed earlier in the literature (Raghupathy et al. 2015; Skotland and Sandvig 2019).

Finally, the structure factor $S_{SM,PE}$ shows that the PE in the inner leaf is found predominantly below the PC in the outer leaf. Thus, we find that the plasma membrane consists of two different kinds of regions: one with SM and cholesterol in the outer leaf and PS and PC in the inner leaf, the other with PC in the outer leaf and PE and cholesterol in the inner leaf. All of the structural factors between lipids in the outer and inner leaflets are consistent with this assignment.

We identify the domains with SM and cholesterol in the outer leaf and PS and PC in the inner leaf with the "rafts" that were posited long ago (Simons and vanMeer 1988) and which have attracted much attention ever since (Lingwood and Simons 2010). We stress that our result shows the plasma membrane to be a one-phase system. It is a disordered liquid phase, but one in which the liquid has structure. The membrane is not a two-phase system in which lo-like and ld-like domains are in two-phase coexistence. Although this picture often appears in the literature, we

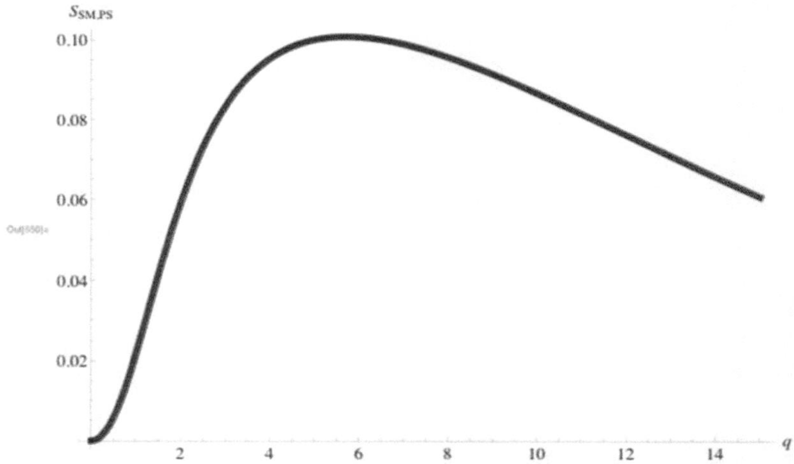

Figure 8. Structure factor $S_{SM,P\ S}(q)$ is shown as a function of the dimensionless wavevector q.

believe that it is not a tenable one and have discussed this previously (Allender et al. 2022). Two strong arguments against this view are that phase separation in the mammalian plasma membrane has not been observed under conditions of biological temperatures and surface tensions; and that, were the system to be phase separated, there would not be many raft domains, but only two, each of them one-half the size of the cell. In contrast, in our picture, there are many domains of one kind embedded in the background of another domain. The characteristic size, L_0, of the embedded domains, is given by the membrane's elastic properties, $L_0 = (\kappa_B/\sigma_B)^{1/2}$ (Schick 2012).

Summary

We have considered one effect of membrane deformation on the distribution of lipids, namely the coupling of membrane curvature to lipid concentration: the idea that lipids of a given shape tend to be drawn to regions of the membrane that are particularly compatible with that shape. Although there are other mechanisms that certainly affect lipid distribution near membrane deformations, such as direct lipid-protein interactions, the mechanism we have emphasized is quite general. We have shown that when the membrane remains of rather constant thickness, what occurs in the two leaflets is much the opposite of one another; that is, if at some location one leaflet favors lipids with small heads and large tails, and thus favors cholesterol, which increases the tail area then at that same location in the other leaflet, lipids with large heads and small tails, and thus less cholesterol, will be favored. We have also seen that this effect is stronger in the outer leaflet with the large concentration of saturated SM, on whose tails the effect of cholesterol is particularly strong, than in the inner leaflet, which is dominated by unsaturated lipids, on whose tails the effect of cholesterol is weaker. We believe that this simple mechanism provides much insight into how the deformation of membranes affects lipid distribution. It also provides a firm foundation for the concept of membrane rafts.

References

Allender, D.W., A.J. Soft and M. Schick 2019 Cholesterol-dependent bending energy is important in cholesterol distribution of the plasma membrane. Biophys. J. 116: 2356–2366.

Allender, D.W., H. Giang and M. Schick. 2020. Model plasma membrane exhibits a microemulsion in both leaves providing a foundation for "rafts". Biophys. J. 118: 1019–1031.

Almeida, P.F.F. 2009. Thermodynamics of lipid interactions in complex bilayers Biopchimica et Biophysica Acta 1788: 71–85.

Buyan, A., C.D. Cox, J. Barnoud et al. 2020. Piezo1 forms specific, functionally important interactions with phosphoinositides and cholesterol. Biophys. J. 119: 1683–1697.

Buyan, A.J., D.W. Allender, B. Corry et al. 2022. Lipid redistribution in the highly curved footprint of Piezo1 Biophys. J. 122: 1900–1913.

Dai, J. and M. Sheetz 1995 Mechanical properties of neuronal growth membranes studied by tether formation with laser optical tweezers Biophys. J. 68: 988–996,

Dan, N. and S.A. Safran. 1998. Effect of lipid characteristics on the structure of transmembrane proteins, Biophys. J. 75: 1410–1414.

Evans, E.A. 1983 Bending elastic modulus of red blood cell membrane derived from buckling instability in micropipette aspiration tests Biophys. J. 43: 27–30.

Furman, D., S. Dattagupta and R.B. Griffiths. 1977. Global phase diagram for a three-component model. Phys. Rev. B 15: 441–464.

Leibler, S. 1986. Curvature instability in membranes. J. Phys. France 47: 507–516.
Leibler, S. and D.A. Andelman. 1987. Ordered and curved mesostructures in membranes and amphiphilic films. J. Physique 48: 2-13-2018.
Lingwood, D. and K. Simons. 2010. Lipid rafts as a membrane-organizing principle, Science 327: 46–50.
Phillips, M. 1971. The physical state of phospholipids and cholesterol in monolayers, bilayers, and membranes. Prog. Surf. Membr. Sci. 5: 139–222.
Phillips, M.C. 1972 The physical state of phospholipids and cholesterol in monolayers, bilayers, and membranes. Prog. Surf. Membrane Sci. 5: 139–222.
Raghupathy, R., A.A. Anilkumar, A. Polley et al. 2015. Transbilayer lipid interactions mediate nanoclustering of lipid-anchored proteins. Cell 161: 581–594.
Rawicz, W., K.C. Olbrich, T. McIntosh et al. 2000. Effect of chain length and unsaturation on elasticity of lipid bilayers, Biophys. J. 79: 328–339.
Sadeghi, S., M. Mueller and R.L.C. Vink. 2014. Raft formation in lipid bilayers coupled to curvature. Biophys. J. 107: 1591–1600.
Schick, M. 2012. Membrane heterogeneity: Manifestation of a curvature-induced microemulsion. Phys. Rev. B 85: 031902-1-031902-4.
Seifert, U. 1997. Configuration of fluid membranes and vesicles. Advances in Physics 46: 13–137.
Simons, K. and G. van Meer. 1998. Lipid sorting in epithelial cells Biochemistry 27: 6197–6202.
Skotland, T. and K. Sandvig. 2019. The role of PS 18:0/18:1 in membrane function, Nature Comm. 101: 1–10.
Sodt, A.J., R.M. Venables, E. Lyman et al. 2016. Nonadditive compositional curvature energetics of lipid bilayers. Phys. Rev. Lett. 117: 138104-1-138104-6.
Uline, M.J., G.S. Longo, M. Schick et al. 2010. Calculating partition coefficients of chain anchors in liquid-ordered and liquid-disordered phases. Biophys. J. 98: 1883–1892.
vanMeer, G. 2011. Dynamic transbilayer lipid asymmetry. Cold Spring Harb. Perspect. Biol. 3: 1–11.
Zachowski, A. 1993. Phospholipids in animal eukaryotic membranes: transverse asymmetry and movement. Biochemical. J. 294: 1–14.

2

Flexible Surface Model for Lipid-Protein Interactions

Steven D. E. Fried[1] and *Michael F. Brown*[2,3,*]

Introduction

Recently, the idea of nonspecific mechanical interactions driving physiological processes at the membrane level has gained increasing traction (Phillips et al. 2009; Brown 2012; Soubias and Gawrisch 2012; Brown 2017; Bassereau et al. 2018; Corradi et al. 2019; Lorent et al. 2020). A defining moment in the development of these theories was the discovery that membrane lipid composition is precisely regulated *in vivo* and appears to play key functional roles in a variety of biophysical processes (Cronan and Gelmann 1975; Farías et al. 1975; Jensen and Schutzbach 1984; McElhaney 1984; Spector and Yorek 1985; Lindblom, Brentel et al. 1986; Morein, Andersson et al. 1996; Dowhan 1997; Epand 1998; Lewis and McElhaney 2013). Given the recent interest in lipid modulation of signal transduction (Schmick and Bastiaens 2014), including that of G-protein-coupled receptors (Casiraghi et al. 2016), viral budding (Paulino et al. 2019; Mehregan et al. 2022), and functional significance of cholesterol (Allender et al. 2019; Chakraborty et al. 2020; Doole et al. 2022; Gupta et al. 2022; Doole et al. 2023) and lipid rafts (Sodt et al. 2014; Palacios-Rápalo et al. 2021; Vona et al. 2021), questions arise about the mechanistic roles of the lipids in each of these scenarios. Do the lipids primarily act as an inert scaffold for proteins within the bilayer, or do they play an active role in membrane biochemistry?

This question was originally addressed by the fluid-mosaic model (FMM) of Singer and Nicholson (1972), which has since become the accepted standard model. It was proposed that biomembranes are composed of a fluid lipid bilayer solvent containing a two-dimensional solution of amphipathic proteins. Integral membrane

[1] Department of Chemistry, Stanford University, Stanford, California 94305, USA.
Email: sdefried@stanford.edu
[2] Department of Chemistry and Biochemistry, University of Arizona, Tucson, Arizona 85721, USA.
[3] Department of Physics, University of Arizona, Tucson, Arizona 85721, USA.
* Corresponding author: mfbrown@arizona.edu

proteins are embedded in the lipid bilayer and solvated by the hydrocarbon chains of the lipid molecules. This fluid mosaic of oriented membrane lipids and proteins allows the mobility required to achieve biological functions that rely on molecular motions. Notably, the FMM assumes either weak coupling by nonspecific associations of lipids with integral membrane proteins or strong coupling due to chemically specific interactions with either the head groups or the acyl chains. Based on the FMM, the coupling interactions between the phospholipids and proteins of membranes are assumed to be similar to the interactions among the lipids themselves: they "do not interact strongly" and "appear to be largely independent" (Singer and Nicolson 1972).

Although the FMM has long been a mainstay of biochemistry textbooks, it faces significant problems due to its assumption of protein-lipid independence. One such example that challenges the simple framework of the FMM is the visual receptor rhodopsin, a G-protein-coupled receptor. Visual rhodopsin holds a preeminent position in the development of the FMM as the first membrane protein for which rotational (Cone 1972; Liebman et al. 1974) and translational diffusion in the membrane was observed (Liebman and Entine 1974; Poo and Cone 1974). Yet, it also provides a counterexample to the view of weak proteolipid coupling, as its activation depends strikingly on the membrane lipids (Baldwin and Hubbell 1985; Gibson and Brown 1993; Soubias et al. 2010; Brown 2017). The properties of the bilayer that cause these effects were first characterized using spin-label electron spin resonance (ESR) (Hubbell and McConnell 1971; Devaux and McConnell 1972; Yang et al. 2012) and fluorescence depolarization measurements (Mitchell et al. 1990) and subsequently by order parameters determined using solid-state ^{2}H NMR spectroscopy (Thurmond et al. 1993; Brown 1996; Soubias et al. 2014). Various lipid-protein coupling mechanisms have been proposed to describe these membrane effects, including hydrophobic mismatch (Killian 1998), bilayer deformation (Soubias and Gawrisch 2012), lipid phase domains (Engelman 2005), and curvature (Brown 1994, 2012). Regardless of the precise mechanism, the significant dependence of rhodopsin on its membrane environment indicates that the simplistic assumptions underlying the FMM are severely outdated.

In what follows, a new biomembrane model is detailed that comes as a significant update and replacement to the FMM and which has received experimental support from our work (Botelho et al. 2002; Huber et al. 2004; Botelho et al. 2006; Brown 2012; Fried et al. 2021) and others (Soubias et al. 2010; Soubias and Gawrisch 2012; Horn et al. 2014; Soubias et al. 2014) in recent years. The basic idea is that the membrane curvature—the so-called language of shape (Hyde et al. 1997)—plays a role in governing the energetics of lipid-protein couplings (Huang and Mason 1986; Gibson and Brown 1993; Mouritsen and Bloom 1993; Brown 1994; Killian 1998; Matsuzaki et al. 1998; Lewis and Cafiso 1999; Nielsen and Andersen 2000; Davies et al. 2001; Hong and Tamm 2004; Jensen and Mouritsen 2004). We have proposed that the free energy of elastic membrane curvature deformation affects lipid-protein interactions that modulate key biological functions, such as signaling and ion transport. Protein conformational changes and stability (Brown 1994; Perozo et al. 2002; Andersen and Koeppe 2007; Phillips et al. 2009; Phillips et al. 2012), folding (Moon and Fleming 2011), and membrane fusion (Siegel et al. 2006) can all involve elastic restructuring of the lipid bilayer. For example, it is known that many plant and animal lipids tend to form nonlamellar phases. In addition to liquid–liquid immiscibility due to ordered or disordered lipids, the curvature elasticity of the liquid-crystalline lipids can be an important factor. A related hypothesis of homeostasis of spontaneous curvature was put forth in the 1980s for integral membrane proteins (Wiedmann et al. 1988; Brown 1997, 2012) and for microbial growth (Gruner 1985; Lindblom et al. 1986). It is presently an alternative to the FMM, and in the case of membrane proteins (Brown 1994; Botelho et al. 2002), it is called the flexible surface model (FSM).

We hypothesize that both strong out-of-plane and in-plane couplings are implicated in functional lipid-protein interactions (Brown 1994, 2017) and formulated by a curvature force field (Figure 1). This hypothesis is experimentally tested by observing lipid influences on visual rhodopsin, a canonical member of the highly pharmacologically targeted G-protein-coupled receptors (GPCRs) (Latorraca et al. 2017; Santos et al. 2017). A strong yet reversible coupling between the membrane-bound receptor and the lipids is observed using molecular spectroscopy (UV-visible and infrared). These discoveries reveal that both acyl chain unsaturation and polar headgroup size are interchangeable in controlling proteolipid coupling. Moreover, strong lipid-protein interactions are consistently implied, which demonstrates the insufficiency of the standard FMM and instead points to a flexible surface model (FSM) for connecting the deformation of a continuous membrane surface (Brown 2012, 2017) and protein function (Figure 1). The FSM is based on the Helfrich formulation of membrane bending energy (Helfrich 1973). However, unlike the original treatment, which was primarily focused on the mechanical properties of the membrane itself, the FSM extends these ideas to lipid-protein interactions. Here, we consider the mesoscopic interactions of the lipids with proteins within the membrane bilayer in terms of the continuum elastic picture. Through its introduction of membrane curvature strain, the FSM differs from previously considered hydrophobic matching (Mouritsen and Bloom 1984) between Hookean lipid molecules and the protein hydrophobic domains (Figure 2). The strength of the FSM is that it is a minimal model based in its original inception around experimental data linking lipid composition to rhodopsin conformational energetics (Beach et al. 1984; Wiedmann

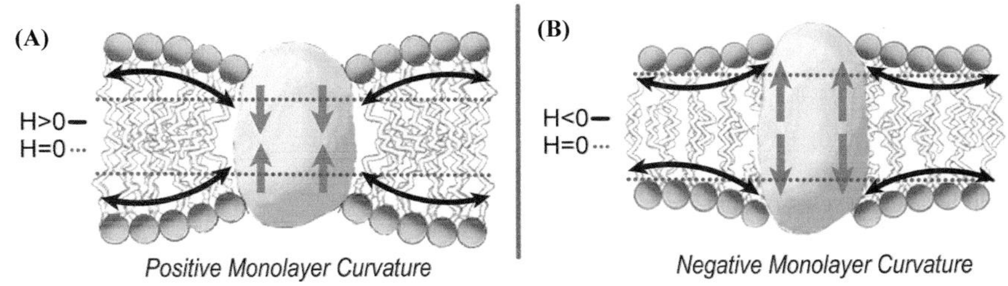

Figure 1. The flexible surface model (FSM) description of lipid-protein interactions considers out-of-plane and in-plane deformations of the bilayer lipids. Protein expansion or compression is coupled to the deformation of the bilayer curvature, formulated as bending a neutral plane where the area per lipid is constant. (A) Mean curvature ($H = (1/2)(1/r_1 + 1/r_2)$ for curvature radii r_1 and r_2) of each monolayer may be positive or (B) negative depending on whether the surface is convex or concave with respect to the aqueous interface. Figure adapted with permission from Fried et al. (2021).

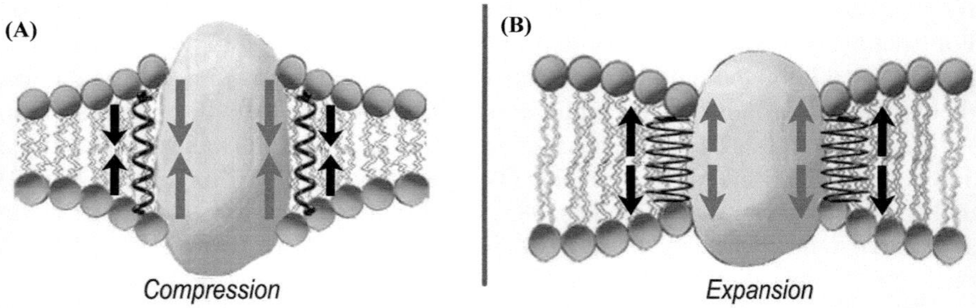

Figure 2. Alternative mattress model (MM) framework (Mouritsen and Bloom 1984), where the membrane deformation free energy increases with a hydrophobic mismatch between the protein and lipids. (A) A restorative force in the lipid bilayer exerts compression or (B) expansion forces at the proteolipid boundary, which can couple to protein conformational changes. Figure adapted with permission from Fried et al. (2021).

et al. 1988; Gibson and Brown 1991; Brown 1994). More recent research from a theory perspective has filled in many of the physical details of these mechanisms with advantages toward predicting and modeling mesoscopic proteolipid coupling (Grabe et al. 2003; Argudo et al. 2017). The FSM framework, in view of the future challenges on the horizon in biochemistry, should enable innovations in pharmaceutical sciences and bioengineering with regard to membrane-bound receptors, channels, virus budding, and fusion.

Shifting the Rhodopsin Activation Equilibrium Involves Membrane Curvature Forces

The G-protein-coupled receptor (GPCR) rhodopsin is an ideal model system for studying the influences of nonlamellar-forming lipids on the conformations of integral membrane proteins. Rhodopsin, as a prototypical Class-A GPCR, undergoes large conformational changes associated with the equilibrium between inactive metarhodopsin-I (MI) and active metarhodopsin-II (MII) (Altenbach et al. 2008; Choe et al. 2011; Hofmann and Lamb 2023). Moreover, this conformational equilibrium is readily quantified by absorption in the UV-visible region following the conversion of its cofactor retinal from an inverse agonist to an agonist by light (Lewis and Kliger 2000). The fact that rhodopsin is light-activated means that conformational states can be quantified shortly after an activating visible pulse using a transient-absorption approach (Thorgeirsson et al. 1993; Jäger et al. 1998; Szundi et al. 1998; Szundi et al. 2016; Szundi et al. 2017), thus affording extraordinary control that captures this metastable equilibrium prior to conversion to other photoproduct states (e.g., metarhodopsin-III or opsin) (Bartl and Vogel 2007).

In an experiment quantifying the effects of nonlamellar-forming lipids on the conformational energetics of bovine rhodopsin (Fried et al. 2021), spectral changes of rhodopsin in recombinant lipid membranes were recorded following its photoexcitation with a λ = 477 nm light source. Absorption changes were measured using a multichannel detector in time increments spanning 10 μs to 50 ms (Figure 3). Time-resolved spectroscopic measurements were made for rhodopsin in a variety of lipid systems, including the native rod disk membranes (RDM), 1-palmitoyl-

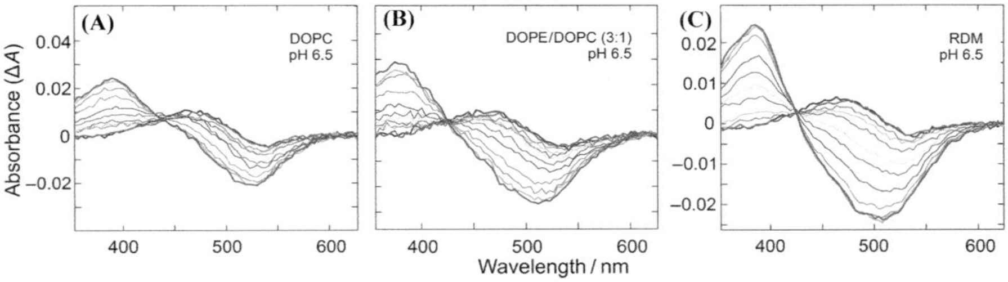

Figure 3. Time-resolved electronic (UV-visible) spectroscopy reveals how neutral lipid substitutions influence the metarhodopsin conformational equilibrium. Difference spectra (light minus dark) monitor rhodopsin activation in (A) DOPC, (B) DOPE/DOPC (3:1), and (C) native RDM lipid environments (30ºC). The recombinant lipid/protein ratio was 100:1. The various pump-probe delays were 10, 20, 50, 100, 200, 500 μs, 1, 2, 5, 10, 20, and 50 ms in ascending order. Note that a shift to the MII state (λ_{max} = 380 nm) occurs with increasing phosphoethanolamine headgroups. Figure adapted with permission from Fried et al. (2021).

2-oleoyl-phosphatidylcholine (POPC), 1,2-dioleoyl-phosphatidylcholine (DOPC), and a 3:1 mixture of 1,2-dioleoyl-phosphatidylethanolamine (DOPE) with DOPC. Lipids were chosen to represent a diversity of acyl chain types (i.e., saturated palmitoyl groups versus unsaturated oleoyl groups) to distinguish between the effects of large (i.e., phosphocholine) and small (i.e., phosphoethanolamine) headgroups and to represent lipid systems with nonlamellar phase transitions in the low-temperature regime (Gruner et al. 1988; Gruner 1989). A sample of 100% DOPE could not be used due to the formation of an inverted cubic phase under standard conditions (Shyamsunder et al. 1988).

It was observed that the equilibrium fraction of MII was highly regulated by the lipid environment, an effect reproducible with pH and temperature (Fried et al. 2021). The metarhodopsin equilibrium was most forward-shifted in the native RDMs, indicating lipid conditions that reversibly favor the MII state. This was followed by the 3:1 DOPE/DOPC lipid environment, pure DOPC (Figure 3), and pure POPC (not shown). In relation to the lipid structures, the bulky PC headgroups backshift the equilibrium to the inactive MI state relative to the smaller PE headgroups. Another trend is found for lipid unsaturation: highly unsaturated (1,2-dioleoyl) lipids stabilize the active MII state more than saturated palmitoyl acyl chains. This trend continues to the native RDMs with abundant polyunsaturated ω-3 fatty acids, such as docosahexaenoic acid (Gibson et al. 1999). Various models have been introduced to describe lipid-protein interactions analyzed here for their connection to the new experimental findings.

Curvature-Based Elasticity Governs Lipid-Protein Interactions in Biomembranes

A number of mechanisms for proteolipid coupling may exist in a biomembrane, making the adoption of a unified, self-consistent model challenging. Still, this chapter will show that putting aside the molecular view of membranes in favor of a mesoscopic surface depiction is often useful toward describing the major types of lipid-protein interactions. In the FSM, the membrane is considered in terms of differential geometry as an elastic sheet capable of bending and stretching according to the Helfrich free energy model (Helfrich 1973). Two primary energetic contributions of lipid-protein coupling are considered: the membrane elastic deformation energy, ΔG^{o}_{el}, and solvation energy due to the proteolipid interfacial tension, ΔG^{o}_{solv} (Brown 1994; Botelho et al. 2002). Adding both contributions, we obtain the total lipid-protein coupling energy: $\Delta G^{o}_{LP} = \Delta G^{o}_{el} + \Delta G^{o}_{solv}$. It will shortly be demonstrated how both terms logically appear from the FSM framework and may be quantified by a continuum mechanics approach.

At the heart of the FSM is the essential idea of spontaneous curvature as an inherent property of any bendable surface, including lipid membranes. The monolayer spontaneous curvature of a membrane lipid bilayer describes the balance of attractive and repulsive forces yielding the lateral pressure profile (Figure 4). If one assumes the optimal separation of the polar head groups exceeds the chains—as for single-chain detergents, lysolipids, or double-chain gangliosides with large head groups—there exists a positive, spontaneous curvature (concave toward hydrocarbon). Small micelles or normal hexagonal phases (H_I) are thus formed. At high surfactant concentrations, micelles are forced to pack into a cubic lattice with long-range translational order (micellar cubic phase). Alternatively, if the optimal headgroup separation is the same as the chains, the spontaneous curvature is zero, as it is for saturated phosphatidylcholines. The planar bilayer is then stabilized as in standard textbook representations. Finally, assuming the optimal headgroup separation is less than the chains, we find that the spontaneous curvature is negative (concave toward water), as in saturated or polyunsaturated phosphatidylethanolamines. Reverse hexagonal phases (H_{II}) or bicontinuous cubic phases are thereby favored. In the latter instance, there are two non-interpenetrating aqueous regions divided by a lipid film

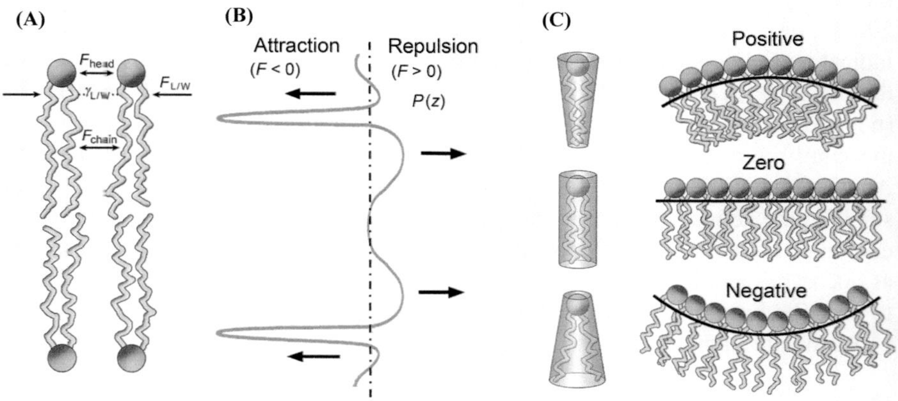

Figure 4. The chemical properties of membrane lipids in a bilayer give rise to a spontaneous monolayer curvature. (A) Attractive force (pressure) acting at the lipid/water interface ($F_{L/W}$) due to the hydrophobic effect balances the repulsive force (pressure) of the headgroups (F_{head}) and acyl chains (F_{chain}). (B) Matching profile of lateral pressure versus depth along the bilayer normal. (C) Depiction of molecular packing showing lipids with different head groups and acyl chains inscribed within their geometrical shapes. The monolayer spontaneous (intrinsic) curvature at the lipids arises from an imbalance of lateral forces: positive (toward hydrocarbon), zero (planar), or negative (toward water). Figure adapted with permission from Brown (2012).

draped onto a lattice with cubic symmetry (so-called plumber's nightmare) (Brown 2017).

Deforming a lipid monolayer away from its spontaneous curvature results in curvature elastic strain, increasing the free energy of the system. Accordingly, coupling the membrane curvature to protein conformational changes that induce membrane deformations can be considered. The FSM thus extends the fluid-mosaic model of membranes by explaining how the membrane lipids influence protein functional states through nonspecific biophysical or material properties. Of course, changes in protein shape can affect the hydrophobic matching of the bilayer thickness to the intramembranous protein surface, accompanied by changes in membrane curvature. An energy penalty in the membrane bilayer thus results from altering the curvature of each opposed monolayer (leaflet) away from its respective spontaneous (intrinsic) curvature. Release of the curvature frustration coupled to a protein shape change can drive the appearance of new functional states triggered by environmental cues or changes in allosteric coupling to ligands (agonists or antagonists). Clearly, disregarding the molecular details is both the strength and weakness of the continuum picture. The FSM explains previously mysterious findings for various membrane proteins, such as GPCRs and ion channels, in a universal way (Brown 2017).

Language of Shape Expresses Protein Coupling to Membrane Curvature Elastic Strain

Next, we consider the physics underlying the FSM as it relates to lipid-protein coupling. In the FSM, we consider a balance of two main energies of the lipid membrane: the elastic energy of curvature and the solvation energy arising from surface tension. We first draw our attention to the elastic energy of curvature associated with monolayer bending, a significant feature of the FSM. Any point on a two-dimensional surface is described by two principal curvatures along either axis, expressed as c_1 and c_2, where $c_1 = 1/r_1$ and $c_2 = 1/r_2$ are the radii of curvature r_1 and r_2 (Figure 5). In a language of shape (Hyde et al. 1997), both principal curvatures may be combined into a mean curvature H, where $H = (1/2)\,(c_1 + c_2)$, and a Gaussian curvature $K = c_1c_2$ that is invariant to bending on a connected surface free of stretching, as discussed by Ninham and coworkers (Hyde et al. 1997). As a result, the elastic energy describing the curvature deformation of a surface is expressed in terms of both the Gaussian curvature and the displacement of the mean curvature H from the spontaneous curvature H_0 at the energy minimum (Helfrich 1973). Through modeling the elastic bending energy with a harmonic (Hookean) potential and making these variable substitutions, the free energy of bending of a neutral surface S with $H_0 = 0$ into a curved shape described by (r_1, r_2) is given by the integral (Helfrich 1973):

$$\oint\!\!\oint_S g_c\, dA = \oint\!\!\oint_S ((\kappa/2)\,(1/r_1 + 1/r_2\,)^2 + \bar{\kappa}/(r_1\, r_2))dA. \qquad (1)$$

In the above formula, the free energy density g_c is related to the mean curvature H through the bending modulus κ and to the Gaussian curvature K by the modulus of Gaussian curvature $\bar{\kappa}$.

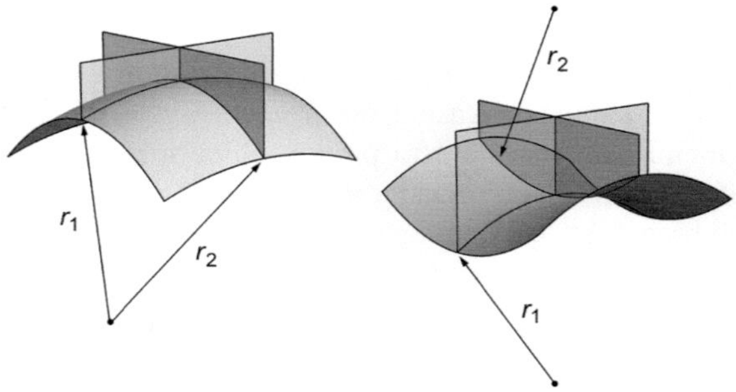

Figure 5. Differential geometry expresses curvature elastic strain as an important component of the membrane free energy. The deformation of a monolayer surface is described by two principal curvatures, $c_1 = 1/r_1$ and $c_2 = 1/r_2$, shown here with the same or opposite signs (Brown 2017).

One can then separate the Gaussian curvature term $1/r_1r_2$ to arrive at a simple expression for the free energy density at any point. In the more generalizable case where H_0 is allowed to be nonzero, this expression reads (Helfrich 1973):

$$g_c = \kappa\,(H - H_0)^2 + \bar{\kappa}\,K. \tag{2}$$

The reader should note that the natural consequence of this equation is a driving force on the surface to deform to a mean spontaneous curvature H_0, minimizing the free energy. So far, we have considered curvature strain associated with the deformation of a single surface, as in the case of a lipid monolayer. In a lipid bilayer comprised of two monolayers, one needs to consider the deformations of each monolayer leaflet separately, where the curvature energy is not minimized because the chain packing energy opposes it. Hence, a bilayer exhibiting frustration, or non-minimized curvature free energy, can work on protein systems at the membrane through coupling between membrane deformation and protein conformational changes (Brown 1994).

Now, the FSM predicts that the free energy of curvature plays a principal role in governing the protein chemical equilibria that involve changes in membrane curvature. If one considers a transition from a state with mean curvature H_1 to one with mean curvature H_2, the free energy density of curvature associated with this deformation is calculated as $\Delta g_c = \kappa(H_2 - H_0)^2 - \kappa(H_1 - H_0)^2$. Recasting this expression, we then obtain $\Delta g_c = -2\kappa(H_2 - H_1)H_0 + \kappa(H_2^2 - H_1^2)$. Next, we take the curvature of the initial state H_1 as zero (a feature of the inactive MI state) so that to first order with respect to the final curvature H_2 (Brown 1994, 2017):

$$\Delta G^{\circ}_{el}/N_A = \Delta G^{\circ}_{c}/N_A = -2\kappa(A_L H_2 N_L)H_0. \tag{3}$$

In this equation, ΔG°_{el} is the standard-state free energy change due to curvature elastic deformation per mole of protein for a given protein conformational change (e.g., the MI to MII change of rhodopsin). Additionally, A_L describes the cross-sectional area per lipid, N_A is the Avogadro constant, and N_L is the number of lipids per protein. Here, we make the simplifying approximation that the value of H_0 is provided by the radius of curvature (R_W) of the inverse hexagonal phase (H_{II}) cylinders of the lipids under conditions of dual solvent stress (Rand et al. 1990).

For a conformational change of a membrane protein like rhodopsin, the equilibrium constant is $K = \exp(-\Delta G^{o}/RT)$. As a consequence, a linear relationship exists between ln K and the spontaneous curvature R_0 of a membrane system, with a positive slope given by the expression $2\kappa(A_L H_2 N_L)N_A/RT$. The experimental observation of such a linear relationship in the case of rhodopsin, using UV-visible spectroscopy to quantify the protein conformational equilibrium, is strong evidence in favor of coupling to the monolayer curvature strain (Botelho et al. 2002; Brown 2017; Fried et al. 2021). According to the linear model of Equation 3, varying the membrane spontaneous curvature (Kollmitzer et al. 2013) by altering the recombinant lipid bilayer composition of POPC, DOPC, and DOPE/DOPC (3:1) systematically adjusts the metarhodopsin equilibrium. Using the above formula, one may estimate the final monolayer curvature H_2 corresponding to the MII state. As an example, it may be assumed that $A_L = 70$ Å^2 and $N_L = 50$ lipids per rhodopsin per monolayer with a monolayer bending modulus $\kappa \approx 10\ k_B T$ (Botelho et al. 2002), which compares well to the bilayer value determined from NMR experiments and molecular dynamics simulations (Doktorova et al. 2023). In the vicinity of the membrane isoelectric point (pI = 4.3 for RDMs and 7.3 for membrane recombinants) (Tsui et al. 1990), we obtain a final radius of curvature of ~ 100–200 Å. By contrast, under more alkaline conditions (pH 8), the monolayer takes on a greater negative charge, causing a larger radius of curvature (~ 400 Å) in the active MII state (Fried et al. 2021).

Membrane Elastic Curvature is Balanced by Proteolipid Interfacial Tension

Additionally, the energy of hydrophobic packing interactions at the lipid-protein interface (Brown 1994; Marsh 2008) is governed by both the strength of the protein-lipid interactions at the interface (i.e., surface tension γ_{LP}) and the surface area A_P of the protein interacting with the bilayer. Provided a change in state occurs that alters the protein surface area by $\Delta A_P \equiv A_{P2} - A_{P1}$, the molar free energy change due to solvation is:

$$\Delta G^{o}_{solv}/N_A = \gamma_{LP}\,\Delta A_P. \tag{4}$$

Still another way of viewing this type of protein-lipid interfacial tension, responsible for sealing the protein into the membrane, is through the concept of a Laplace pressure. A curved interface exists between the protein and membrane lipids, causing a surface tension-induced pressure differential analogous to capillary action and related phenomena (Brown 1994). The pressure differential between the protein and lipids will then be given by $\Delta P \equiv P_P - P_L = \gamma_{LP}\,(1/r_1 + 1/r_2)$, where γ_{LP} describes the proteolipid surface tension and r_1 and r_2 describe the two principal radii of curvature. For simplicity, one can assume a cylindrical transmembrane protein shape, wherein one of the radii of curvature becomes infinite, so the Laplace pressure becomes $\Delta P = \gamma_{LP}/r$. Integrating this pressure over the surface area displacement undergone during a protein conformational change gives a surface work term equivalent to a free energy change (Brown 1994).

Hence, for a change of a protein with intramembranous surface area A_{P1} to a state with area A_{P2}, the free energy due to chain packing is given by Equation 4.

Equivalently, this same term can be expressed through a dependence on hydrophobic mismatch: $\Delta G_{solv} = \sigma_L g_{mis} \Delta|L_L - L_P|$ (Marsh 2008). Here, σ_L describes the lipid width (related to the cross-sectional area A_L by $\sigma_L = \sqrt{A_L}$), g_{mis} describes the free energy density of hydrophobic mismatch, L_L describes the lipid length, and L_P is one-half the protein hydrophobic length. In consequence, $\Delta|L_L - L_P|$ describes the difference in hydrophobic mismatch between the two states of the proteolipid system. Note the analogy to Equation 4, where g_{mis} is a descriptor of the surface tension γ_{LP} and in this term $\sigma_L \Delta|L_L - L_P| = \Delta A_P$, assuming longitudinal expansion in the protein and negligible deformation of the lipids themselves.

Combining both membrane curvature and solvation energy (i.e., interfacial tension) contributions, we thus obtain the following expression for lipid-protein coupling energies (Brown 2017):

$$\Delta G^{o}_{LP}/N_A = -2\kappa(A_L H_2 N_L)H_0 + \gamma_{LP}\Delta A_P. \tag{5}$$

By considering membrane surface properties in the flexible surface model, we determine two major energetic contributions to protein-lipid interactions: the curvature elastic energy ΔG^{o}_{el} described by the bending modulus κ, and chain packing (solvation) energy ΔG^{o}_{sol} described by the surface tension γ_{LP} at the proteolipid interface. A balance exists between these two terms for changes such as the conversion of MI to MII, which entails a structural expansion of rhodopsin. The first (elastic) term is negative and corresponds to a favorable driving force for the conformational change, as the membrane lipids adopt a structure closer to their inherent, spontaneous curvature. The second (solvation) term, on the other hand, is positive, assuming hydrophobic mismatch at the proteolipid interface increases upon MII formation. It follows that a balance of both elastic and solvation energies modulates rhodopsin light activation in the FSM.

Membrane Thickness or Area Elastic Deformation Gives Another Perspective

An alternative is that elastic energy may exist in the form of membrane thickness or area deformation (Gibson and Brown 1993; Brown 1994). Maintaining the continuum mechanics view outlined in the FSM makes it possible to take a further step by allowing stretching of the continuous membrane surface, treated with a harmonic approximation described by an area elastic modulus, K_A (Kinnun et al. 2015). For incompressible materials, a correspondence exists between area deformation (Gibson and Brown 1993) and thickness deformation (Mouritsen and Bloom 1984), which has been described in the literature as a hydrophobic mismatch (Killian 1998; Marsh 2008). According to Gibson and Brown (Gibson and Brown 1993), the molar free energy per lipid molecule in a monolayer due to elastic area deformation is formulated in analogy to Hooke's law:

$$G_t/N_A = (1/2)K_A(A_L - A_{L0})^2/A_{L0}, \tag{6}$$

Here, the free energy density comes in terms of the lipid-water interfacial tension, which is related to the area elastic modulus. In addition, A_L describes the perturbed surface area per lipid, and A_{L0} is the non-perturbed surface area per lipid. It is assumed that perturbations to the deformed lipid area are sufficiently small that a harmonic approximation is applicable, where K_A equals $2\gamma_{LW}$ for a monolayer (half that of a bilayer, $4\gamma_{LW}$) (Gibson and Brown 1993; Botelho et al. 2002). By relating this type of membrane deformation to the area per lipid and area elastic modulus, we obtain an expression that is closely tied to experimental observables, namely the bilayer surface tension and lipid order parameters that are measurable by solid-state ^{2}H NMR spectroscopy (Kinnun et al. 2015).

An equivalent formulation is the so-called mattress model (MM) of Bloom and Mouritsen based on a thickness elastic strain due to stretching or compressions of individual lipids in response to hydrophobic mismatch (Mouritsen and Bloom 1984; Jensen and Mouritsen 2004; Marsh 2008) (Figure 2). Once again, we recognize that lipid bilayer area deformation must be accompanied by a corresponding deformation in thickness. The distortion of a material is generally described by Young's modulus and Poisson's ratio. Provided the lipids are incompressible, then the area and thickness deformation must be inversely related. The connection between transverse and longitudinal strain is given by the Poisson ratio $v = 1/2$ (Kinnun et al. 2015). With a harmonic approximation, one may then consider individual lipids as Hookean objects whose longitudinal stretching or compression gives a thickness-mediated elastic energy G_t proportional to the square of the lipid displacement (Marsh 2008):

$$G_t/N_A = (1/2)\, K_t A_{L0} (L - L_0)^2/L_0. \tag{7}$$

In the above equation, K_t is the thickness elastic modulus (or Young's modulus) describing the longitudinal compression of individual lipids in a monolayer, where L denotes the perturbed lipid length and L_0 is the unperturbed lipid length (55). The thickness modulus is related to the area elastic modulus K_A by $K_t = 4v^2\, K_A/L_0$ for a monolayer, where v is the Poisson ratio describing the connection between transverse and longitudinal strain of a lipid. For any incompressible yet deformable material such as a lipid film (monolayer or bilayer), then $v = 1/2$, and the moduli are related by $K_t = K_A/L_0$ for a monolayer or equivalently $K_t = K_A/D_B$ for a bilayer.

In the latter case, we substitute $D_B = 2L$ and $D_{B0} = 2L_0$ for the thickness of two opposed lipids in place of L and L_0, obtaining an isomorphous form of Equation 7 whose free energy is twice that of the monolayer (Marsh 2008; Kinnun et al. 2015). Accordingly, for small deformations, a direct relation is found between Equations 6 and 7 in terms of the lipid lengths (Marsh 2008) or cross-sectional areas (Gibson and Brown 1993). In the hydrophobic mismatch or mattress model (MM), the free energy inherent to the area or thickness deformation of a membrane (Equations 6–7) is the only elastic energy term in coupling the conformational equilibria of the protein. If we consider a transition from a membrane of lipid length L_1 to L_2, the molar free energy change per protein then reads:

$$\Delta G^{o}_{el}/N_A = \Delta G^{o}_{t}/N_A = K_t\, A_{L0}\, N_L((\Delta L^2/2L_0) - \Delta L). \tag{8}$$

Here, we define $\Delta L^2 \equiv L_2^2 - L_1^2$ and $\Delta L \equiv L_2 - L_1$. To first order, this free energy change due to membrane thickness deformation is given by $\Delta G_t^{o} = -\, K_t N_A N_L A_{L0} \Delta L$. Consequently, the hydrophobic matching framework produces an alternate term describing membrane elastic energy based on thickness or area deformation and depends on membrane lipid length. Combining this elastic term and the solvation contribution ΔG°_{solv} yields the total free energy of lipid-protein coupling (Mouritsen and Bloom 1984; Marsh 2008).

The reader should note that thus far in the hydrophobic matching or MM treatment of lipid stretching, only the free energy of the individual lipids has been considered. In a simplified case, the concept of a persistence length may be introduced to describe the effective range over which lipids deform due to hydrophobic mismatch (Jensen and Mouritsen 2004). Introducing a persistence length brings us back full circle to the concept of membrane curvature. As individual membrane lipids extend or compress according to their proximity to a transmembrane protein, the membrane surface has a continuous deformation described by differential geometry. Instead of introducing this curvature through a persistence length, we propose the FSM as directly formulated based on curvature elastic strain rather than purely thickness elastic strain.

To recapitulate at this point, we conclude that there exist at least three possible coupling interactions between the bilayer and integral membrane proteins: curvature, lipid-protein interfacial tension, and thickness/area deformation, which is also related to the lipid-water interfacial tension (Gibson and Brown 1993; Brown 1994). When comparing the various proposed models, we note that the FMM assumes independent protein and lipid behavior—it includes none of these terms (Singer and Nicolson 1972). The MM contains ΔG°_{t} and ΔG°_{solv} terms in accordance with a hydrophobic mismatch and longitudinal tension on the lipids (Equations 4, 6–7) (Mouritsen and Bloom 1984). By considering membrane curvature as a source of elastic energy, the FSM contains the ΔG°_{c} and ΔG°_{solv} terms (Equations 4–5). Notably, by virtue of its consideration of bending energy (Brown 2012, 2017), only the FSM describes both in- and out-of-plane coupling interactions in the membrane (Fried et al. 2021).

Membrane Surface Electrostatics Also Affect the Protein Energy Landscape

Another aspect is that as the composition of the surrounding lipid bilayer changes, a protein such as rhodopsin encounters influences from the monolayer curvature and the membrane surface electrostatics. We first note that rhodopsin activation depends on proton uptake by Glu[134] of the E(D)RY motif, breaking the second of two ionic locks in forming a G-protein binding cavity (Fried et al. 2022). This process is described by the reaction mechanism $\mathrm{Rh} + h\nu \rightarrow \mathrm{MI} \rightleftharpoons \mathrm{MII_a} \rightleftharpoons \mathrm{MII_b} + \mathrm{H_3O^+} \rightleftharpoons \mathrm{MII_bH^+}$, where the MII substates are denoted by $\mathrm{MII_a}$ and $\mathrm{MII_b}$ (Jäger et al. 1998; Mahalingam et al. 2008), and thus pK_A shifts of Glu[134] will couple to the energetics of MII formation.

More generally, membrane electrostatics influence local pH-buffering at the bilayer surface due to titratable groups of the protein and the lipid headgroups, which for rhodopsin may recruit hydronium ions to protonate Glu[134] of the E(D)RY motif (Figure 6A). Specifically, charged phosphatidylserine (PS) and zwitterionic phosphatidylethanolamine (PE) are predominantly located in the cytoplasmic leaflet of the retinal disk membranes (RDM), while charge-neutral phosphatidylcholine (PC) and sphingomyelin (SM) lipids mainly reside in the extracellular leaflet

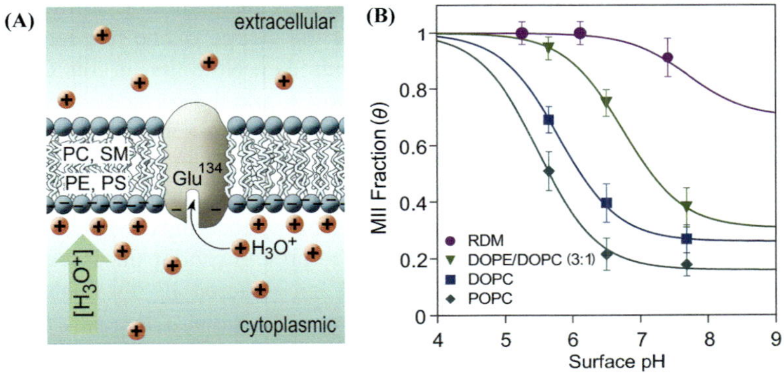

Figure 6. Membrane curvature and surface electrostatics both govern metarhodopsin-II (MII) formation and may be distinguished from one another. (A) Illustration of how negatively charged functional groups of lipids or proteins can lead to an electrical double-layer of hydronium ions at the membrane surface. Greater local hydronium concentration drives the protonation of Glu^{134}, yielding the formation of the active MII state. (B) The fraction of MII state versus local pH at the membrane surface was calculated using the Gouy-Chapman model for various lipid environments (Fried et al. 2021). Figure adapted with permission from Fried et al. (2021).

(Tsui et al. 1986; Tsui et al. 1990). The charge of the lipid headgroups themselves can be a significant inducer of monolayer curvature, as like charges repel one another (Epand 1998). Moreover, for a surface containing titratable groups, the pH will be locally buffered according to Poisson-Boltzmann statistics, changing the effective hydronium concentration at the membrane surface versus the bulk pH. It follows that the true pK_A of rhodopsin's Glu^{134} in a particular lipid membrane will differ from the *apparent* pK_A value based on the bulk solution pH. As a result, these hydronium recruitment influences must be separated from those of curvature to isolate the effects of elastic strain alone.

Notably, the local pH near the membrane can be calculated from the surface potential to yield a Henderson-Hasselbalch titration curve in terms of the *intrinsic* pK_A value. In order to account for the electrostatic influences of the lipids, protein, and aqueous buffer on the MI-MII equilibrium (Parkes and Liebman 1984), a Poisson-Boltzmann equation for a planar surface is used to obtain the surface (local) pH. This approach can be further extended in terms of the electrochemical potential μ_i by including the electrostatic potential ψ, leading to $\bar{\mu}_i \equiv \mu_i + \mu_i^{el}$, where $\mu_i^{el} = z_i F\psi$ is the electrical energy, z_i is the ion charge, and F is Faraday's constant. Because the electrochemical potential for hydronium atoms must be equal at the membrane surface ($\psi = \psi_0$) as in the bulk solution ($\psi = 0$), the potentials are equated to yield the Nernst equation, which reads:

$$\mathrm{pH}_{\mathrm{surface}} = \mathrm{pH}_{\mathrm{bulk}} + zF\psi_0/2.303RT. \tag{9}$$

Although the electrostatic surface potential of the membrane is not directly measurable in these experiments, it can be related to surface charge density by the Poisson-Boltzmann equation, which gives the Gouy-Chapman model:

$$\sigma = (C^{1/2}/A)\ \sinh(zF\psi_0/2RT). \tag{10}$$

Here, σ is the membrane surface charge density, ψ_0 is the membrane surface potential, C is the molarity of cations in the buffer, z is the charge of the respective cations, and $A = 134.8\ M^{1/2}$ is the Gouy-Chapman parameter. Note that in this formula, the electrical potential ψ_0 is not an additive function, while the surface charge density σ can be separated into $\sigma = \sigma_L + \sigma_P$ for the lipid and protein components. The surface charge density for the lipids is negligible in the case of neutral POPC, POPE, and DOPE lipids. On the other hand, for native retinal disk membranes, the surface charge density is appreciable due to the charged phosphoserine (PS) lipid headgroups (Hubbell 1990; Tsui et al. 1990) as well as the protein rhodopsin. Because surface charge density σ is dependent on local pH encountered by titratable groups, a nonlinear system of equations results between the Gouy-Chapman model with directly calculated charge density σ and the Nernst equation equating the hydronium electrochemical potential at the membrane and in the bulk solution (Fried et al. 2021). The two independent variables are the local pH and surface potential ψ_0. Such a system can be iteratively solved (Wang et al. 2002) for a given bulk pH to yield the local pH at the membrane surface.

This analysis was used to plot the equilibrium MII conformations (determined from UV-visible spectra of rhodopsin, Figure 3) versus locally buffered surface pH (Figure 6B). This method already compensates for lipid surface electrostatic effects on proton uptake, yet a significant influence of the membrane lipid composition on rhodopsin activation is still observed. Increasing either lipid saturation or the proportion of phosphatidylcholine lipids backshifts the equilibrium to the inactive MI state compared to the native rod disk membranes. The largest pK_A shift corresponding to modulation of the MI-MII equilibrium is observed in the case of POPC membranes, whose phospholipids include one saturated palmitoyl (16:0) and one unsaturated oleoyl (18:1) fatty acid combined with a phosphocholine headgroup. The introduction of an additional unsaturated fatty acid group in the case of DOPC lipids gives less of a shift, while including the smaller unmethylated DOPE lipids is most similar to the activity of the native rod disk membranes.

Clearly, some of these variations may be explained by local differences in hydronium concentration, yet the same significant trend occurs when surface pH effects are accounted for with the Gouy-Chapman model (Botelho et al. 2002; Wang et al. 2002). The negatively charged PS lipids increase the apparent pK_A of rhodopsin in the RDMs due to the greater local concentration of hydronium ions near the membrane surface that drives the rhodopsin equilibrium forward to protonated MII (Tsui et al. 1990; DeLange et al. 1997; Gibson et al. 1999; Wang et al. 2002). Even so, this effect can only partially account for the RDM results, and it has no bearing on membranes with zwitterionic headgroups, such as POPC, DOPC, and DOPE. When the localized variations in hydronium concentration are included, i.e., the MII fraction is plotted versus calculated surface pH rather than bulk pH, variations in the pK_A of more than two pH units persist between rhodopsin in the POPC and RDM lipid systems. Electrostatic properties of the membrane are thus insufficient to account for the variations in rhodopsin activation seen between POPC, DOPC, DOPE, and RDM lipids. Instead, we turn to a mesoscopic, mechanical explanation of these findings based on the lipid geometries of the individual membrane system (e.g., headgroup size and unsaturation).

New Biomembrane Model Includes Strong Nonspecific Proteolipid Coupling

The results described herein lead us to a picture in which cellular membrane function is governed by mechanical properties of the lipid bilayer (Lindblom and Rilfors 1989; Lee 2004; Andersen and Koeppe 2007; van Meer et al. 2008; Phillips et al. 2009; Honerkamp-Smith et al. 2009; Soubias and Gawrisch 2012; Brown 2017; Bassereau et al. 2018; Corradi et al. 2019; Lorent et al. 2020). They point to strong in- and out-of-plane coupling interactions controlling the activation of rhodopsin, which supersede the earlier fluid-mosaic model (FMM). Notably, the FMM does not consider this strong proteolipid coupling, but rather the in-plane fluidity that yields rotational and translational diffusion of membrane proteins such as rhodopsin (Cone 1972; Liebman and Entine 1974; Liebman et al. 1974; Poo and Cone 1974). Because membranes are complex fluids, they have a polymorphism that entails additional components of the stress tensor for describing shape transitions in terms of principal curvatures. These aspects illuminate the insufficiency of the classical FMM, which includes only weak in-plane proteolipid coupling that governs the gel-to-liquid-crystalline transition (Singer and Nicolson 1972). On the other hand, the new FSM emphasizes strong, out-of-plane interactions, giving rise to monolayer curvature associated with lamellar to nonlamellar phase transitions of the lipids.

Coming back to rhodopsin, the following is the current evidence supporting the view of the out-of-plane coupling of proteins to the lipid monolayer curvature. First, the shifting of the pH-titration curve describing MII formation through protonation of Glu^{134} shows a reversible modulation of the metarhodopsin equilibrium by membrane lipids. The membrane lipid composition alters the pK_A of Glu^{134} by over two pH units, so a lower pH is needed to reversibly shift the metarhodopsin equilibrium to MII when native nonlamellar-forming lipids are absent. Second, the modulation of the metarhodopsin equilibrium due to a variety of lipid substitutions reveals a lack of chemical specificity to lipid type. Small lipid headgroups can be replaced by unsaturated acyl chains and vice versa. This observation rules out the possibility of specific proteolipid interactions, such as those in lipid rafts (Engelman 2005; van Meer et al. 2008), as the dominant governing factor. Both alterations to lipid headgroup (i.e., PE versus PC) and degree of unsaturation (i.e., (1-palmitoyl-2-oleoyl) versus (1,2-dioleoyl) lipids) have analogous effects on the metarhodopsin equilibrium. Furthermore, the combined influences of both have the greatest impact on rhodopsin energetics, strongly indicating the role of spontaneous curvature.

Likewise, these findings also have implications for the hydrophobic matching framework of proteolipid interactions, wherein stretching or compression of the lipid chains increases the membrane potential energy to drive protein conformational changes (Mouritsen and Bloom 1984). Here, the proteolipid coupling is strong and is only directed out of the plane. Because membrane expansion or compression requires an input of energy that can be modeled by Hooke's law, varying the equilibrium length of the lipid chains will affect the stability of expanded or contracted protein conformations. While hydrophobic matching can account for the effects of lipid acyl chain length (e.g., number of carbons) on rhodopsin activation (Botelho et al. 2006), it predicts no effects of monolayer spontaneous curvature on rhodopsin activation, a property emerging from the combined effects of both lipid unsaturation and headgroup bulkiness on lipid-protein interactions (Gibson and Brown 1993). As an alternate explanation, one may consider the lamellar versus nonlamellar polymorphisms for lipid bilayers (Figure 7A) (Lindblom and Rilfors 1989; Seddon 1990). Among the canonical nonlamellar forms of lipids is the inverted hexagonal (H_{II}) phase, favored by unsaturated and small headgroup lipids, in which pure DOPE can be found under standard biochemical conditions (Seddon 1990). The tendency of the membrane to adopt a nonlamellar configuration may couple to conformational changes of membrane proteins and is consistent with the results herein (Brown 2012, 2017). Both increasing lipid unsaturation and decreasing lipid headgroup size, which lowers the free energy inherent to the H_{II} phase, independently drive the conformational change of rhodopsin to the MII active state for the same reason—a tendency of the membrane to adopt a negative spontaneous monolayer curvature (Figure 7B).

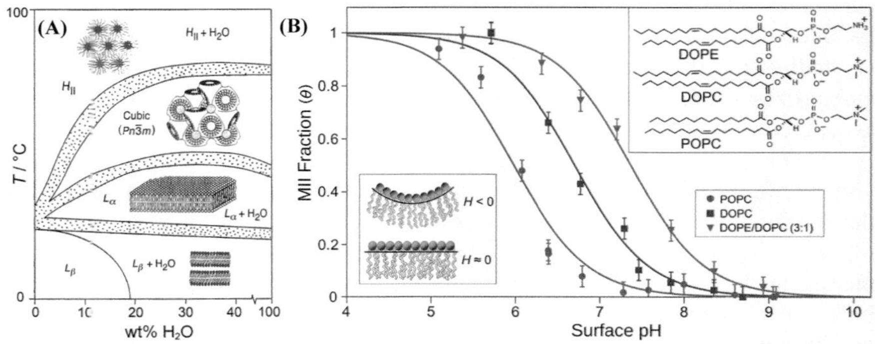

Figure 7. Spontaneous curvature of lipid soft matter explains the coupling of the membrane curvature elastic strain to rhodopsin conformational energetics. (A) The phase diagram shows several possible representative lipid microstructures with a greater tendency to curve toward water (negative spontaneous curvature) as the temperature rises and hydration decreases. (B) Active MII fraction (θ) of rhodopsin is plotted versus surface pH at 10°C. Note that the rhodopsin activation equilibrium undergoes a pK_A shift depending on both the lipid headgroups and acyl chains. Decreasing headgroup size (PE versus PC) or increasing acyl chain unsaturation (oleoyl versus palmitoyl groups) increases the active MII population in visual signaling. Figure adapted with permission from Fried et al. (2021) and Brown (2017).

Flexible Surface Model Explains Effects of Nonlamellar-Forming Lipids on Rhodopsin

In light of these observations, it is important to consider how the current data align with our proposed flexible surface model (Gibson and Brown 1993; Brown 1994, 2017; Fried et al. 2021) of lipid-protein interactions (Figure 1). The FSM predicts that conformational changes during rhodopsin activation should be governed by the elastic energy of curvature of the lipid membrane (Brown 2012), as treated using differential geometry (Brown 2017). Lipid composition gives rise to a spontaneous (intrinsic) curvature of the membrane, describing the free energy minimum. In turn, the bending of the membrane to adopt the spontaneous curvature couples to protein conformational changes (Brown 1994) and biochemical catalytic events (Fried et al. 2021) at the membrane level. Formulated in terms of bending a neutral plane with a constant area per lipid, the FSM considers the lateral forces between lipids in addition to longitudinal (vertical) stresses. The FSM thus considers a Hookean, elastic behavior in two directions (in-plane and out-of-plane), as opposed to the MM assumption of purely out-of-plane Hookean behavior (Figure 1). The FSM accounts for the same phenomena, such as lipid length effects, detailed in the hydrophobic matching framework through the inclusion of a proteolipid solvation term that balances the elastic energy term contained in the monolayer curvature (Brown 2012, 2017).

More extensive data sets at varying pH for membrane-bound rhodopsin consistently demonstrate that increasing lipid saturation and the size of the phospholipid headgroups back shifts the rhodopsin equilibrium to the inactive MI state (Figure 7B). According to the FSM, both lipid saturation and headgroup size modulate the free energy of MII formation by altering the intrinsic, or spontaneous, curvature (H_0) of the membrane due to imbalances in the lateral pressure profiles of the lipids (Cantor 1999; Marsh 2008; Brown 2012). Highly unsaturated lipids experience stronger lateral repulsive forces between the acyl chains, leading to an intrinsic negative curvature of a monolayer leaflet toward water. Analogously, decreasing the size of the phospholipid headgroup also gives a lateral pressure imbalance, where the headgroup repulsions are smaller in magnitude than the acyl chain repulsions. Hence, both unsaturated acyl chains and small phospholipid headgroups contribute to a tendency of the membrane to adopt a negative monolayer curvature deformation ($H < 0$) toward the water, which explains the chemically nonspecific influence of the lipid soft matter on rhodopsin activation.

In the case of visual rhodopsin, it is found that DOPE lipids, which are both highly unsaturated and possess smaller phosphoethanolamine headgroups, provide a more stabilizing environment to MII formation than the larger headgroup DOPC and POPC lipids (Botelho et al. 2002; Soubias et al. 2010). Based on the FSM, it therefore follows that the formation of active-state MII is promoted by a negative membrane curvature of deformation. Altering the lipid composition to increase the lateral pressure between the lipid acyl chains and decrease lateral pressure between headgroups will diminish the free energy cost of membrane deformation, driving the metarhodopsin activation equilibrium forward. On the other hand, greater lipid saturation and headgroup size decrease absolute membrane intrinsic curvature, such that the free energy cost of MII formation is high. These trends constitute design principles due to natural selection that can be used to predict and assess the mechanisms of diverse lipid types on rhodopsin activation (Gibson and Brown 1993), with future work aimed at a deeper understanding of how electrostatic contributions to curvature balance other known surface charge effects (Tsui et al. 1990; Gibson and Brown 1991). It follows that a combination of mesoscopic-scale bilayer deformation properties in addition to specific lipid interactions modulates rhodopsin activation (Lindblom and Rilfors 1989; Bloom et al. 1991; Lee 2004; Andersen and Koeppe 2007; van Meer et al. 2008; Honerkamp-Smith et al. 2009; Salas-Estrada et al. 2018). In contrast to the standard biochemical view of GPCR activation governed completely by agonists and antagonists, mechanical properties of the membrane (Khelashvili et al. 2012; Soubias and Gawrisch 2012) play equally important roles in rhodopsin modulation. It is expected that the results acquired herein for the FSM have applicability in describing the membrane protein interactions of other GPCRs and potentially other membrane proteins and peptides as well (Perozo et al. 2002; Andersen and Koeppe 2007).

Greatest Challenge

How protein-lipid interactions affect biophysical processes at the membrane level is a major challenge that involves bilayer fusion, protein oligomerization, cellular permeability, membrane trafficking, and conformational changes of receptors and ion channels. The essential question is whether the fluid-mosaic model in its original form—which only considers weak proteolipid couplings—gives a fully consistent description of membrane structure, dynamics, and function. Our proposal of a flexible surface model (FSM) is based on the Helfrich curvature free energy and explains functional lipid-protein interactions through a balance of membrane elastic deformation with the proteolipid hydrophobic matching. In the FSM, curvature frustration from a mismatch of the monolayer spontaneous curvature to the proteolipid interface is coupled to membrane protein conformational changes, giving a unifying biophysical paradigm. Achieving this understanding requires further experimental and computational testing of the effects of lipid substitutions on key functions of additional biomembrane systems.

Conclusions and Outlook

The flexible surface model (FSM) improves upon the standard fluid-mosaic model by incorporating chemically nonspecific proteolipid couplings out of the plane of the lipid bilayer (Brown 1994, 2017). Treated mechanically from a mesoscopic perspective, the FSM uses the language of shape to formulate free energy according to monolayer curvature strain. This framework has the potential to explain a variety of phenomena in membrane biochemistry, as demonstrated here in the case of GPCR activation. Further core problems of interest in biophysics that the FSM can shed light on include the role of membrane asymmetry in proteolipid coupling, as well as the origin and function of spatial organization in the membrane (Botelho et al. 2006; Eicher et al. 2018). Membrane fusion is another relevant problem for which monolayer bending energy should play a significant role and which is germane to a number of high-profile research areas, including endo/exocytosis and the fusing or budding of enveloped viruses such as influenza, HIV, and coronaviruses (Liang et al. 2016). Understanding these processes in terms of mechanical stress fields can potentially open new avenues in terms of therapeutic design and engineering of lipid nanoparticles or membrane scaffolds for diverse research or medicinal purposes, demanding further exploration into these significant areas of application.

Acknowledgments

This research was supported by grants to M. F. B. from the NIH (R01EY026041) and NSF (MCB 1817862 and CHE 1904125). S. D. E. F. is supported by an NSF Graduate Research Fellowship under Grant DGE-1656518. We dedicate this chapter of the book to the life and work of Professor Richard M. Epand. M. F. B. is especially thankful to Richard Epand for his many innovative scientific contributions, warmth and gentle collegiality, and being an eternal treasure to biophysics.

References

Allender, D., A.J. Sodt and M. Schick. 2019. Cholesterol-dependent bending energy is important in cholesterol distribution of the plasma membrane. Biophys. J. 116: 2356–2366.

Altenbach, C., A.K. Kusnetzow, O.P. Ernst et al. 2008. High-resolution distance mapping in rhodopsin reveals the pattern of helix movement due to activation. Proc. Natl. Acad. Sci. U. S. A. 105: 7439–7444.

Andersen, O.S. and R.E. Koeppe. 2007. Bilayer thickness and membrane protein function: an energetic perspective. Annu. Rev. Biophys. Biomol. Struct. 36: 107–130.

Argudo, D., N.P. Bethel, F.V. Marcoline et al. 2017. New continuum approaches for determining protein-induced membrane deformations. Biophys. J. 112: 2159–2172.

Baldwin, P.A. and W.L. Hubbell. 1985. Effects of lipid environment on the light-induced conformational changes of rhodopsin. 2. Roles of lipid chain length, unsaturation, and phase state. Biochemistry 24: 2633–2639.

Bartl, F.J. and R. Vogel. 2007. Structural and functional properties of metarhodopsin III: Recent spectroscopic studies on deactivation pathways of rhodopsin. Phys. Chem. Chem. Phys. 9: 1648–1658.

Bassereau, P., R. Jin, T. Baumgart et al. 2018. The 2018 biomembrane curvature and remodeling roadmap. J. Phys. D 51: 343001.

Beach, J.M., R.D. Pates, J.F. Ellena et al. 1984. Flash photolysis studies of rhodopsin function in recombinant membranes. Biophys. J. 45: 292a.

Bloom, M., E. Evans and O.G. Mouritsen. 1991. Physical properties of the fluid lipid-bilayer component of cell membranes: a perspective. Q. Rev. Biophys. 24: 293–397.

Botelho, A.V., N.J. Gibson, R.L. Thurmond et al. 2002. Conformational energetics of rhodopsin modulated by nonlamellar-forming lipids. Biochemistry 41: 6354–6368.

Botelho, A.V., T. Huber, T.P. Sakmar et al. 2006. Curvature and hydrophobic forces drive oligomerization and modulate activity of rhodopsin in membranes. Biophys. J. 91: 4464–4477.

Brown, M.F. 1994. Modulation of rhodopsin function by properties of the membrane bilayer. Chem. Phys. Lipids 73: 159–180.

Brown, M.F. 1996. Membrane structure and dynamics studied with NMR spectroscopy. pp. 175–252. *In*: Merz, K. Jr. and B. Roux (eds.). Biological Membranes. A Molecular Perspective from Computation and Experiment. Basel: Birkhäuser.

Brown, M.F. 1997. Influence of non-lamellar forming lipids on rhodopsin. Curr. Top. Membr. 44: 285–356.

Brown, M.F. 2012. Curvature forces in membrane lipid-protein interactions. Biochemistry 51: 9782–9795.

Brown, M.F. 2017. Soft matter in lipid-protein interactions. Annu. Rev. Biophys. 46: 379–410.

Cantor, R.S. 1999. The influence of membrane lateral pressures on simple geometric models of protein conformational equilibria. Chem. Phys. Lipids 101: 45–56.

Casiraghi, M., M. Damian, E. Lescop et al. 2016. Functional modulation of a G protein-coupled receptor conformational landscape in a lipid bilayer. J. Am. Chem. Soc. 138: 11170–11175.

Chakraborty, S., M. Doktorova, T.R. Molugu et al. 2020. How cholesterol stiffens unsaturated lipid membranes. Proc. Natl. Acad. Sci U.S.A. 117: 21896–21905.

Choe, H.W., Y.J. Kim, J.H. Park et al. 2011. Crystal structure of metarhodopsin II. Nature 471: 651–655.

Cone, R.A. 1972. Rotational diffusion of rhodopsin in the visual receptor membrane. Nature New Biol. 236: 39–43.

Corradi, V., B.I. Sejdiu, H. Mesa-Galloso et al. 2019. Emerging diversity in lipid–protein interactions. Chem. Rev. 119: 5775–5848.

Cronan, Jr., J.E. and E.P. Gelmann. 1975. Physical properties of membrane lipids: biological relevance and regulation. Bateriol. Rev. 39: 232–256.

Davies, S.M.A., R.M. Epand, R. Kraayenhof et al. 2001. Regulation of CTP: Phosphocholine cytidylyltransferase activity by the physical properties of lipid membranes: an important role for stored curvature strain energy. Biochemistry 40: 10522–10531.

DeLange, F., M. Merkx, P.H.M. Bovee-Geurts et al. 1997. Modulation of the metarhodopsin I/metarhodopsin II equilibrium of bovine rhodopsin by ionic strength. Evidence for a surface-charge effect. Eur. J. Biochem. 243: 174–180.

Devaux, P. and H.M. McConnell. 1972. Lateral diffusion in spin-labelled phosphatidylcholine multilayers. J. Am. Chem. Soc. 94: 4475–4481.

Doktorova, M., G. Khelashvili, R. Ashkar et al. 2023. Molecular simulations and NMR reveal how lipid fluctuations affect membrane mechanics. Biophys. J. 122: 984–1002.

Doole, F.T., S. Gupta, T. Kumarage et al. 2023. Biophysics of membrane stiffening by cholesterol and phosphatidylinositol 4,5-bisphosphate (PIP2). pp. 61–85. *In*: Dantsker, A.R. (ed.). Cholesterol and PI(4,5)P2 in Vital Biological Functions: From Coexistence to Crosstalk. Cham: Springer International Publishing.

Doole, F.T., T. Kumarage, R. Ashkar et al. 2022. Cholesterol stiffening of lipid membranes. J. Membr. Biol. 255: 385–405.

Dowhan, W. 1997. Molecular basis for membrane phospholipid diversity: why are there so many lipids? Annu. Rev. Biochem. 66: 199–232.

Eicher, B., D. Marquardt, F.A. Heberle et al. 2018. Intrinsic curvature-mediated transbilayer coupling in asymmetric lipid vesicles. Biophys. J. 114: 146–157.

Engelman, D.M. 2005. Membranes are more mosaic than fluid. Nature 438: 578–80.

Epand, R.M. 1998. Lipid polymorphism and protein-lipid interactions. Biochim. Biophys. Acta 1376: 353–368.

Farías, R.N., B. Bloj, R.D. Morero et al. 1975. Regulation of allosteric membrane-bound enzymes through changes in membrane lipid composition. Biochim. Biophys. Acta 415: 231–251.

Fried, S.D.E., K.S.K. Hewage, A.R. Eitel et al. 2022. Hydration-mediated G-protein–coupled receptor activation. Proc. Natl. Acad. Sci U.S.A. 119: e2117349119.

Fried, S.D.E., J.W. Lewis, I. Szundi et al. 2021. Membrane curvature revisited—the archetype of rhodopsin studied by time-resolved electronic spectroscopy. Biophys. J. 120: 440–452.

Gibson, N.J. and M.F. Brown. 1991. Membrane lipid influences on the energetics of the metarhodopsin I and metarhodopsin II conformational states of rhodopsin probed by flash photolysis. Photochem. Photobiol. 54: 985–992.

Gibson, N.J. and M.F. Brown. 1993. Lipid headgroup and acyl chain composition modulate the MI-MII equilibrium of rhodopsin in recombinant membranes. Biochemistry 32: 2438–2454.

Gibson, S.K., J.H. Parkes and P.A. Liebman. 1999. Phosphorylation alters the pH-dependent active state equilibrium of rhodopsin by modulating the membrane surface potential. Biochemistry 38: 11103–11114.

Grabe, M., J. Neu, G. Oster et al. 2003. Protein interactions and membrane geometry. Biophys. J. 84: 854–868.

Gruner, S.M. 1985. Intrinsic curvature hypothesis for biomembrane lipid composition: a role for nonbilayer lipids. Proc. Natl. Acad. Sci. U.S.A. 82: 3665–3669.

Gruner, S.M. 1989. Stability of lyotropic phases with curved interfaces. J. Phys. Chem. 93: 7562–7570.

Gruner, S.M., M.W. Tate, G.L. Kirk et al. 1988. X-ray diffraction study of the polymorphic behavior of N-methylated dioleoylphosphatidylethanolamine. Biochemistry 27: 2853–2866.

Gupta, S., F.T. Doole, T. Kumarage et al. 2022. Cholesterol stiffening of lipid membranes and drug interactions: insights from neutron spin echo and deuterium NMR spectroscopy. pp. 771–796. *In*: Bukiya, A.N. and A.M. Dopico (eds.). Cholesterol. Academic Press.

Helfrich, W 1973. Elastic properties of lipid bilayers: theory and possible experiments. Z. Naturforsch. 28c: 693–703.

Hofmann, K.P. and T.D. Lamb. 2023. Rhodopsin, light-sensor of vision. Prog. Retin. Eye Res. 93: 101116.

Honerkamp-Smith, A.R., S.L. Veatch and S.L. Keller. 2009. An introduction to critical points for biophysicists; observations of compositional heterogeneity in lipid membranes. Biochim. Biophys. Acta 1788: 53–63.

Hong, H. and L.K. Tamm. 2004. Elastic coupling of integral membrane protein stability to lipid bilayer forces Proc. Natl. Acad. Sci. U.S.A. 101: 4065–4070.

Horn, J.N., T.-C. Kao and A. Grossfield. 2014. Coarse-grained molecular dynamics provides insight into the interactions of lipids and cholesterol with rhodopsin. pp. 75–94. *In*: Filizola, M. (ed.). G Protein-Coupled Receptors—Modeling and Simulation. Dordrecht: Springer.

Huang, C. and J.T. Mason. 1986. Structure and properties of mixed-chain phospholipid assemblies. Biochim. Biophys. Acta 864: 423–470.

Hubbell, W.L. 1990. Transbilayer coupling mechanism for the formation of lipid asymmetry in biological membranes. Biophys. J. 57: 99–108.

Hubbell, W.L. and H.M. McConnell. 1971. Molecular motion in spin-labeled phospholipids and membranes. J. Am. Chem. Soc. 93: 314–326.

Huber, T., A.V. Botelho, K. Beyer et al. 2004. Membrane model for the GPCR prototype rhodopsin: hydrophobic interface and dynamical structure. Biophys. J. 86: 2078–2100.

Hyde, S.T., S. Andersson, K. Larsson et al. 1997. The Language of Shape. The Role of Curvature in Condensed Matter: Physics, Chemistry and Biology. Amsterdam: Elsevier.

Jäger, S., I. Szundi, J.W. Lewis et al. 1998. Effects of pH on rhodopsin photointermediates from lumirhodopsin to metarhodopsin II. Biochemistry 37: 6998–7005.

Jensen, M.Ø. and O.G. Mouritsen. 2004. Lipids do influence protein function—the hydrophobic matching hypothesis revisited. Biochim. Biophys. Acta 1666: 205–226.

Jensen, J.W. and J.S. Schutzbach. 1984. Activation of mannosyltransferase II by nonbilayer phospholipids. Biochemistry 23: 1115–1119.

Khelashvili, G., P.B.C. Albornoz, N. Johner et al. 2012. Why GPCRs behave differently in cubic and lamellar lipidic mesophases. J. Am. Chem. Soc. 134: 15858–15868.

Killian, J.A. 1998. Hydrophobic mismatch between proteins and lipids in membranes. Biochim. Biophys. Acta 1376: 401–416.

Kinnun, J.J., K.J. Mallikarjunaiah, H.I. Petrache et al. 2015. Elastic deformation and area per lipid of membranes: atomistic view from solid-state deuterium NMR spectroscopy. Biochim. Biophys. Acta 1848: 246–259.

Kollmitzer, B., P. Heftberger, M. Rappolt et al. 2013. Monolayer spontaneous curvature of raft-forming membrane lipids. Soft Matter 9: 10877–10884.

Latorraca, N.R., A.J. Venkatakrishnan and R.O. Dror. 2017. GPCR dynamics: structures in motion. Chem. Rev. 117: 139–155.

Lee, A.G. 2004. How lipids affect the activities of integral membrane proteins. Biochim. Biophys. Acta 1666: 62–87.

Lewis, J.R. and D.S. Cafiso. 1999. Correlation between the free energy of a channel-forming voltage-gated peptide and the spontaneous curvature of bilayer lipids. Biochemistry 38: 5932–5938.

Lewis, J.W. and D.S. Kliger. 2000. Absorption spectroscopy in studies of visual pigments: spectral and kinetic characterization of intermediates. Meth. Enzymol. 315: 164–178.

Lewis, R.N.A.H. and R.N. McElhaney. 2013. Membrane lipid phase transitions and phase organization studied by Fourier transform infrared spectroscopy. Biochim. Biophys. Acta 1828: 2347–2358.

Liang, R., J.M.J. Swanson, J.J. Madsen et al. 2016. Acid activation mechanism of the influenza A M2 proton channel. Proc. Natl. Acad. Sci U.S.A. 113: E6955–E6964.

Liebman, P.A. and G. Entine. 1974. Lateral diffusion of visual pigment in photoreceptor disk membranes. Science 185: 457–459.

Liebman, P.A., W.S. Jagger, M.W. Kaplan et al. 1974. Membrane structure changes in rod outer segments associated with rhodopsin bleaching. Nature 251: 31–36.

Lindblom, G., I. Brentel, M. Sjölund et al. 1986. Phase equilibria of membrane lipids from *Acholeplasma laidlawii*. The importance of a single lipid forming nonlamellar phases. Biochemistry 25: 7502–7510.

Lindblom, G. and L. Rilfors. 1989. Cubic phases and isotropic structures formed by membrane lipids—possible biological relevance. Biochim. Biophys. Acta 988: 221–256.

Lorent, J.H., K.R. Levental, L. Ganesan et al. 2020. Plasma membranes are asymmetric in lipid unsaturation, packing and protein shape. Nat. Chem. Biol. 16: 644–652.

Matsuzaki, K., K.-I. Sugishita, N. Ishibe et al. 1998. Relationship of membrane curvature to the formation of pores by magainin 2. Biochemistry 37: 11856–11863.

Mahalingam, M., K. Martinez-Mayorga, M.F. Brown et al. 2008. Two protonation switches control rhodopsin activation in membranes. Proc. Natl. Acad. Sci. U.S.A. 105: 17795–17800.

Marsh, D. 2008. Energetics of hydrophobic matching in lipid-protein interactions. Biophys. J. 94: 3996–4013.

McElhaney, R.N. 1984. The structure and function of the *Acholeplasma laidlawii* plasma membrane. Biochim. Biophys. Acta. 779: 1–42.

Mehregan, A., S. Pérez-Conesa, Y. Zhuang et al. 2022. Probing effects of the SARS-CoV-2 E protein on membrane curvature and intracellular calcium. Biochim. Biophys. Acta 1864: 183994.

Mitchell, D.C., M. Straume, J.L. Miller et al. 1990. Modulation of metarhodopsin formation by cholesterol-induced ordering of bilayer lipids. Biochemistry 29: 9143–9149.

Moon, C.P. and K.G. Fleming. 2011. Side-chain hydrophobicity scale derived from transmembrane protein folding into lipid bilayers. Proc. Natl. Acad. Sci U.S.A. 108: 10174–10177.

Morein, S., A.-S. Andersson, L. Rilfors et al. 1996. Wild-type *Escherichia coli* cells regulate the membrane lipid composition in a "window" between gel and non-lamellar structures. J. Biol. Chem. 271: 6801–6809.

Mouritsen, O.G. and M. Bloom. 1984. Mattress model of lipid-protein interactions in membranes. Biophys. J. 46: 141–153.

Mouritsen, O.G. and M. Bloom. 1993. Models of lipid-protein interactions in membranes. Annu. Rev. Biophys. Biomol. Struct. 22: 145–171.

Nielsen, C. and O.S. Andersen. 2000. Inclusion-induced bilayer deformations: effects of monolayer equililbrium curvature. Biophys. J. 79: 2583–2604.

Palacios-Rápalo, S.N., L.A. De Jesús-González, C.D. Cordero-Rivera et al. 2021. Cholesterol-rich lipid rafts as platforms for SARS-CoV-2 entry. Front. Immunol. 12: 796855.

Parkes, J.H and P.A. Liebman. 1984. Temperature and pH dependence of the metarhodopsin I-metarhodopsin II kinetics and equilibria in bovine rod disk membrane suspensions. Biochemistry 23: 5054–5061.

Paulino, J., X. Pang, I. Hung et al. 2019. Influenza A M2 channel clustering at high protein/lipid ratios: viral budding implications. Biophys. J. 116: 1075–1084.

Perozo, E., A. Kloda, D.M. Cortes et al. 2002. Physical principles underlying the transduction of bilayer deformation forces during mechanosensitive channel gating. Nat. Struct. Biol. 9: 696–703.

Phillips, R., T. Ursell, P. Wiggins et al. 2009. Emerging roles for lipids in shaping membrane-protein function. Nature 459: 379–385.

Phillips, R., J. Kondev, J. Theriot et al. 2012. Physical Biology of the Cell, Garland Science.

Poo, M. and R.A. Cone. 1974. Lateral diffusion of rhodopsin in the photoreceptor membrane. Nature 247: 438–441.

Rand, R.P., N.L. Fuller, S.M. Gruner et al. 1990. Membrane curvature, lipid segregation, and structural transitions for phospholipids under dual-solvent stress. Biochemistry 29: 76–87.

Salas-Estrada, L.A., N. Leioatts, T.D. Romo et al. 2018. Lipids alter rhodopsin function via ligand-like and solvent-like interactions. Biophys. J. 114: 355–367.

Santos, R., O. Ursu, A. Gaulton et al. 2017. A comprehensive map of molecular drug targets. Nat. Rev. Drug Disc. 16: 19–34.

Schmick, M. and P.I.H. Bastiaens. 2014. The interdependence of membrane shape and cellular signal processing. Cell 156: 1132–1138.

Seddon, J.M. 1990. Structure of the inverted hexagonal (H_{II}) phase, and non-lamellar phase transitions of lipids. Biochim. Biophys. Acta 1031: 1–69.

Shyamsunder, E., S.M. Gruner, M.W. Tate et al. 1988. Observation of inverted cubic phase in hydrated dioleoylphosphatidylethanolamine membranes. Biochemistry 27: 2332–2336.

Siegel, D.P., V. Cherezov, D.V. Greathouse et al. 2006. Transmembrane peptides stabilize inverted cubic phases in a biphasic length-dependent manner: implications for protein-induced membrane fusion. Biophys. J. 90: 200–211.

Singer, S.J. and G.L. Nicolson. 1972. The fluid mosaic model of the structure of cell membranes. Science 175: 720–731.

Sodt, A.J., M.L. Sandar, K. Gawrisch et al. 2014. The molecular structure of the liquid-ordered phase of lipid bilayers. J. Am. Chem. Soc. 136: 725–732.

Soubias, O. and K. Gawrisch. 2012. The role of the lipid matrix for structure and function of the GPCR rhodopsin. Biochim. Biophys. Acta 1818: 234–240.

Soubias, O., W.E. Teague, K.G. Hines et al. 2014. The role of membrane curvature elastic stress for function of rhodopsin-like G protein-coupled receptors. Biochimie 107: 28–32.

Soubias, O., W.E. Teague, K.G. Hines et al. 2010. Contribution of membrane elastic energy to rhodopsin function. Biophys. J. 99: 817–824.

Spector, A.A. and M.A. Yorek. 1985. Membrane lipid composition and cellular function. J. Lipid Res. 26: 1015–1035.

Szundi, I., C. Funatogawa, Y. Guo et al. 2017. Protein sequence and membrane lipid roles in the activation kinetics of bovine and human rhodopsins. Biophys. J. 113: 1934–1944.

Szundi, I., C. Funatogawa and D.S. Kliger. 2016. Complexity of bovine rhodopsin activation revealed at low temperature and alkaline pH. Biochemistry 55: 5095–5105.

Szundi, I., T.L. Mah, J.W. Lewis et al. 1998. Proton transfer reactions linked to rhodopsin activation. Biochemistry 37: 14237–14244.

Thorgeirsson, T.E., J.W. Lewis, S.E. Wallace-Williams et al. 1993. Effects of temperature on rhodopsin photointermediates from lumirhodopsin to metarhodopsin II. Biochemistry 32: 13861–13872.

Thurmond, R.L., G. Lindblom and M.F. Brown. 1993. Curvature, order, and dynamics of lipid hexagonal phases studied by deuterium NMR spectroscopy. Biochemistry 32: 5394–5410.

Tsui, F.C., D.M. Ojcius and W.L. Hubbell. 1986. The intrinsic pK_a values for phosphatidylserine and phosphatidylethanolamine in phosphatidylcholine host bilayers. Biophys. J. 49: 459–468.

Tsui, F.C., S.A. Sundberg and W.L. Hubbell. 1990. Distribution of charge on photoreceptor disc membranes and implications for charged lipid asymmetry. Biophys. J. 57: 85–97.

van Meer, G., D.R. Voelker and G.W. Feigenson. 2008. Membrane lipids: where they are and how they behave. Nat. Rev. Mol. Cell Biol. 9: 112–124.

Vona, R., E. Iessi and P. Matarrese. 2021. Role of cholesterol and lipid rafts in cancer signaling: a promising therapeutic opportunity? Front. Cell Dev. Biol. 9: 622908.

Wang, Y., A.V. Botelho, G.V. Martinez et al. 2002. Electrostatic properties of membrane lipids coupled to metarhodopsin II formation in visual transduction. J. Am. Chem. Soc. 124: 7690–7701.

Wiedmann, T.S., R.D. Pates, J.M. Beach et al. 1988. Lipid-protein interactions mediate the photochemical function of rhodopsin. Biochemistry 27: 6469–6474.

Yang, Z., Y. Liu, P. Borbat et al. 2012. Pulsed ESR dipolar spectroscopy for distance measurements in immobilized spin labeled proteins in liquid solution. J. Am. Chem. Soc. 134: 9950–9952.

3

Non-Lamellar Phases of Membrane Lipids

John M. Seddon[1,*] and *Arwen I. I. Tyler*[2]

Introduction

Membranes in living cells exist predominantly in the form of fluid bilayers. In terms of shape, these self-assembled lipid/protein membranes can be locally flat or exhibit static and/or dynamic regions of curvature due to asymmetry of lipid composition or physicochemical conditions across the membrane or due to thermally-activated fluctuations. A fundamental source of membrane curvature for a lipid monolayer can be a non-zero spontaneous curvature arising from a mismatch in the lateral packing between the lipid polar headgroups and the hydrocarbon chains. For a number of common lipid molecules, this packing mismatch can lead to the stabilization of non-lamellar phases based upon curved fluid monolayers or bilayers (for recent reviews, see Mezzenga et al. 2019; Conn and Seddon 2021).

Such lyotropic phases can exhibit 2-D or 3-D periodicity, and a variety of crystallographic plane group or space group symmetries have been observed in model systems. The formation of such non-lamellar structures in living cells is less clear-cut, although there is increasing evidence that suggests that they do play important roles in a range of biological processes (Mezzenga et al. 2019). Non-lamellar lipid phases are believed to be intimately involved in the mechanism of fat (triglyceride) digestion in the gut (Clulow et al. 2018; Salentinig et al. 2015). In any event, the localized dynamic shape changes during biological processes such as membrane fusion and fission are geometrically and topologically closely related to the local structures of certain non-lamellar phases. Thus, studies of the latter in model systems may yield valuable insights into the molecular rearrangements during these complex biological processes.

[1] Chemistry Department, Imperial College London, London W12 0BZ, UK.
[2] School of Food Science and Nutrition, University of Leeds, Leeds LS2 9JT, UK.
* Corresponding author: j.seddon@imperial.ac.uk

A further important aspect of self-assembled non-lamellar phases is that they have a large number of existing or potential biomedical applications, such as in membrane protein crystallization (Caffrey 2015; Landau and Rosenbusch 1996; Zabara et al. 2018), protein encapsulation (Conn and Drummond 2013), or drug/gene delivery to living cells (Angelova et al. 2011; Zhai et al. 2019). A recent review of cubosomes gives an overview of their potential biomedical applications (Barriga et al. 2019). It should be noted that dispersing lipids as lipid nanoparticles with incorporated mRNA has led to the recent development of the highly successful COVID-19 vaccine (Hou et al. 2021). This application is critically dependent on the self-assembly of the lipids in the nanoparticles, the interaction of the ionizable cationic lipid component with the encapsulated mRNA, and the changes induced by progressive lowering of the pH upon uptake by endosomes within cells.

In this chapter, we will briefly describe the self-assembly of lipids, the range of non-lamellar phase structures that may be expected to form, the relationship between lateral pressure profiles and membrane curvature elasticity, and the role of curvature and packing frustration in stabilizing curved membrane phases. We do not provide a comprehensive review of this rapidly evolving field, and we also do not cover here the important roles of proteins, peptides, nucleic acids, or nanoparticles in promoting and modulating membrane curvature.

Lipid Self-Assembly

The formation of lyotropic liquid-crystalline phases by amphiphilic molecules such as lipids in an aqueous solution is a thermodynamically spontaneous process driven by a combination of the hydrophobic effect acting on the hydrocarbon chains and the hydrophilicity of the polar headgroups, together with a variety of packing and curvature constraints. When the preferred cross-sectional area of the hydrated polar headgroup of a lipid is similar to that of the hydrocarbon chains, a flat monolayer or bilayer can form. These can be fluid, soft solid (gel phases), or crystalline.

However, when there is an imbalance between these preferred cross-sectional areas, there is a tendency for curved interfaces to form, leading to the formation of non-lamellar phases; in these cases, the hydrocarbon chains are generally in a disordered conformation, and the interfaces are fluid. For large, highly-hydrated headgroups, the monolayer tends to curve away from the water and toward the hydrocarbon chains (Type I), as shown on the left side of Figure 1.

This is taken to correspond to *positive* mean curvature (see below). An example would be a surfactant micelle, which may be spherical or cylindrical in shape. However, for lipids with relatively small, weakly hydrated headgroups and relatively bulky hydrophobic regions, the curvature can lie in the opposite direction, with *negative* mean curvature (Type II), as shown on the right side of Figure 1. This latter case leads to the formation of *inverse* lyotropic phases. Examples of lipids that tend to form such inverse phases are shown in Figure 2.

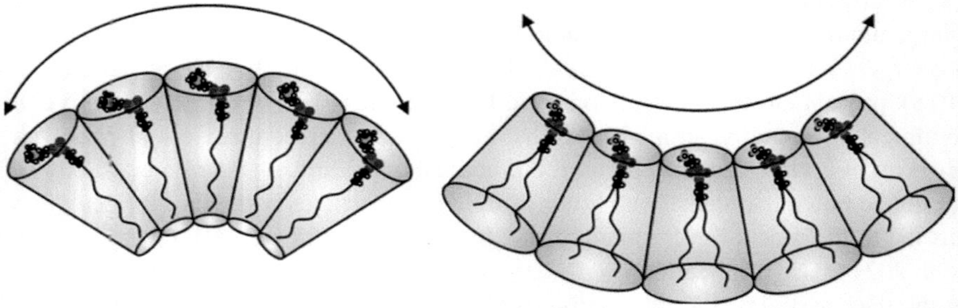

Figure 1. Examples of lipids displaying type I curvature (left) and Type II (inverse) curvature (right). Reproduced with permission from (Shearman et al. 2006).

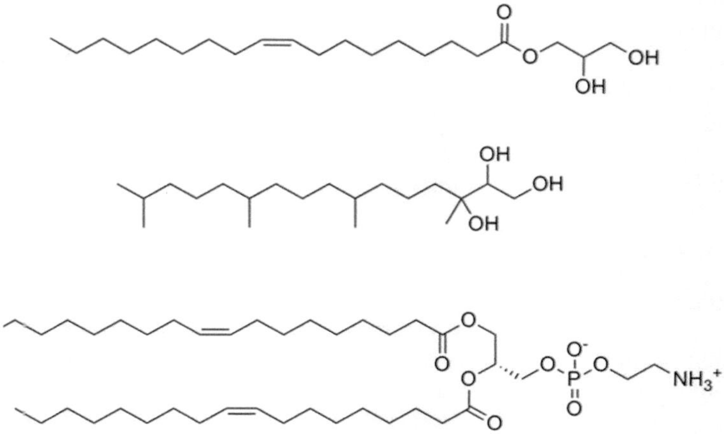

Figure 2. Examples of typical Type II-forming non-lamellar lipids: monoolein (top), phytantriol (middle), and dioleoyl phosphatidylethanolamine (bottom).

Membrane Curvature

In considering membrane curvature, we need to consider two types of interfacial curvature, the mean curvature *H* and the Gaussian curvature *K*. These are defined in terms of the principal curvatures c_1 and c_2 at any point on the surface by:

$$H = {}^1/2\ (c_1 + c_2)$$

$$K = c_1\ c_2$$

As described in Figure 1, the sign of *H* can be either positive or negative for a lipid monolayer. However, for a lipid bilayer, the sign of *H* is arbitrary and must be defined with respect to some external frame of reference, for example, a *z*-axis. A lipid monolayer may have a spontaneous curvature H_0, its ideal fully-relaxed mean curvature. On the other hand, a symmetric-composition lipid bilayer, with equal conditions on both sides, will normally have zero spontaneous curvature ($H_0 = 0$)

due to symmetry. If this symmetry is broken, for example, by having an asymmetric lipid composition between the two monolayer leaflets, then the bilayer will curve spontaneously toward one side or the other.

The sign of the Gaussian curvature K is of fundamental importance in defining the types of self-assembled structures and liquid-crystalline phases formed by lipids. When $K > 0$ (elliptic surfaces), the monolayer or bilayer tends to close on itself (for example, a spherical micelle or a spherical vesicle), and this leads to discontinuous lyotropic phases based on packings of discrete lipid aggregates. When $K = 0$ (parabolic surfaces), the lipid monolayers form planes (lamellar phases of bilayers) or cylinders (2-D hexagonal phases). The third possibility is that $K < 0$ (hyperbolic surfaces). In this case, the interfaces tend to form saddle surfaces, which can be extended to span space continuously (similar to a sponge). This leads to bicontinuous phases, most often of cubic symmetry. These structures are geometrically closely related to membrane channels and, hence, to membrane fusion and fission.

These different types of curvature and the resulting types of lyotropic phases may be combined into a schematic phase diagram (see Figure 3), showing the expected relative locations of the different fluid lyotropic phases to be expected as a function of the average interfacial mean curvature, measured at the polar – nonpolar interface of the lipid monolayer.

A range of thermodynamic parameters can modify the preferred mean curvature, such as the degree of hydration, the temperature, pH, salt concentration, or the hydrostatic pressure.

The fluid lamellar phase, consisting of a 1-D stack of fluid bilayers with zero mean curvature (and zero Gaussian curvature), occupies the center region of this "phase diagram". Moving rightwards leads to structures with increasingly positive mean curvature: these are the so-called "Type I" phases, adopted by many surfactants and other single-chain amphiphiles, and will not be further considered here.

In this chapter, we focus on the structures found moving leftwards toward increasingly negative average interfacial mean curvature, leading to inverse or "Type II" phases. The natural location for the inverse bicontinuous phases is immediately to the left of the flat lamellar phase in the region labeled 'Q_{II}' in Figure 3. This is because these phases can be described as draping a continuous fluid lipid bilayer onto

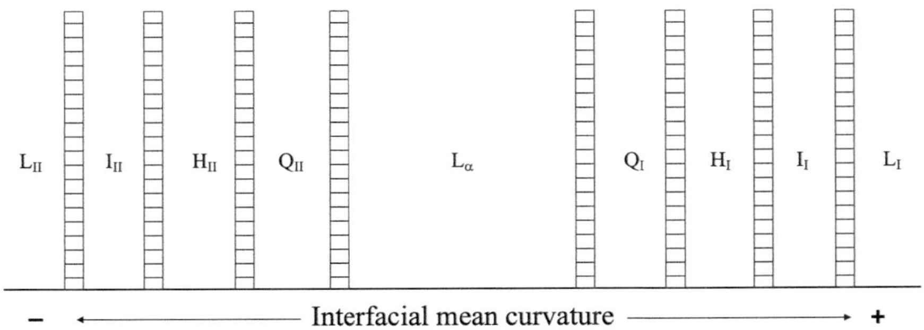

Figure 3. A schematic lyotropic "phase diagram" showing the expected sequence of phases as a function of the average interfacial mean curvature. Adapted from (Seddon and Templer 1993) and used with permission.

a periodic minimal surface of zero mean curvature and (mainly) negative Gaussian curvature. It has been shown that the polar-nonpolar interfaces are parallel to the minimal surface located at the center of the bilayer (Oka et al. 2018). It can be shown that this geometry leads to relatively modest values of negative mean curvature at the polar-nonpolar interfaces of the two monolayers, making up the curved bilayer within such inverse bicontinuous phases. The mean curvature at these interfaces varies continuously between being most negative above the saddle points and zero above the flat points of the underlying minimal surface. Note that bicontinuous phases of lower symmetry than cubic (for example, rhombohedral) can also be encountered in this region of the phase diagram but will not be considered here.

If the need for negative curvature becomes stronger, for example, due to increasing temperature, causing more chain conformational disorder and greater splay in the chain region. In that case, a transition to the more highly-curved 2-D inverse H_{II} phase may be induced. Lipid systems having a very large chain splay relative to their polar headgroups may require even more strongly negative mean curvature, found in the region labeled 'I_{II}' in Figure 3; this can be achieved by transforming to inverse *discontinuous* phases, made up of 3-D packings, usually of cubic or 3-D hexagonal symmetry, of quasi-spherical aggregates (inverse micelles). The L_{II} phase (sometimes denoted L_2) is an inverse micellar phase, with no long-range order, which is sometimes observed at low water content (see for example, Figure 6 below).

Lateral Pressure Profile, Curvature Elastic Energy, and Chain Packing Frustration

The curvature elastic energy per unit area g_c of a fluid lipid membrane is given by the Helfrich equation (Helfrich 1973):

$$g_c = 2\,\kappa(H - H_0)^2 + \kappa_G K$$

This expression contains two curvature moduli, the bending modulus κ, and the Gaussian modulus κ_G. It is straightforward to measure κ for a lipid bilayer (κ^b) using, for example, optical fluctuation analysis on giant unilamellar vesicles (Faizi et al. 2020) or molecular dynamics simulations (Khelashvili et al. 2013). It is a positive quantity, and typical values for phospholipid fluid bilayers are found to be in the region of 20 k_BT, implying a monolayer value in the region of 10 k_BT. Membrane charge has been found to increase the bending modulus of bilayers (Faizi et al. 2019). It has also been found experimentally that the bending moduli of *asymmetric* composition POPC/DOPC bilayers (Elani et al. 2015; Karamdad et al. 2016) and DMPC / DOPC bilayers (Lu et al. 2016) can be roughly twice as large as those of the corresponding symmetric bilayers. These findings have been confirmed and extended by a recent spin-echo spectroscopy study of asymmetric phospholipid bilayers enriched in amino phospholipids (phosphatidylethanolamine and phosphatidylserine) in the inner monolayer, which is found to have significantly larger bending moduli than the corresponding symmetric-composition bilayers (Frewein et al. 2023).

A theoretical study argues that the stiffening caused by asymmetry is due to differential stress within the two monolayers of the asymmetric bilayer (Foley et al. 2022; Hossein and Deserno 2020).

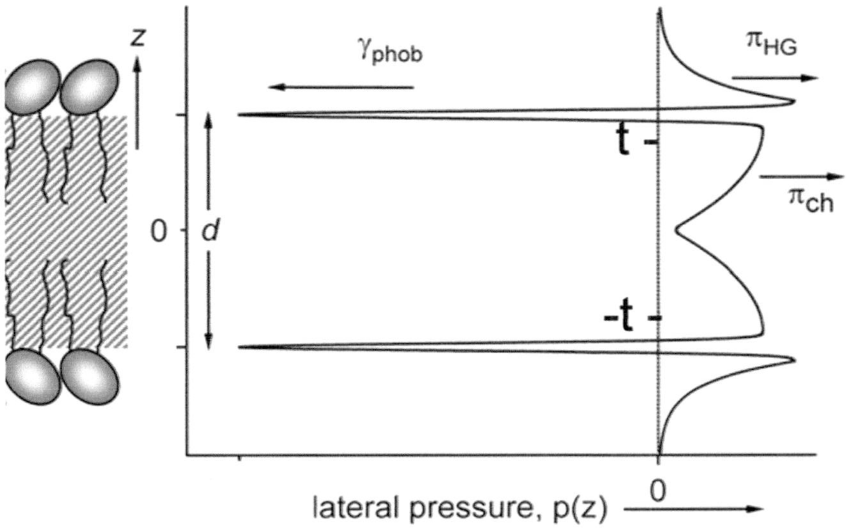

Figure 4. Expected form of the lateral pressure profile across a fluid lipid bilayer. Adapted from (Marsh 2007) and reproduced with permission.

The lateral pressure profile $p(z)$ across a lipid membrane plays a key role in determining the curvature elastic properties of the membrane and may affect the conformation and insertion of proteins into membranes (Marsh 2007). The expected form of such a profile across a symmetric-composition fluid lipid bilayer is shown in Figure 4.

For a lipid bilayer, the product of the bending modulus and the spontaneous mean curvature H_0 is equal to the first moment of the lateral pressure profile across the bilayer:

$$\kappa^b H_0 = \int p(z)\, z dz$$

For a symmetrical lipid composition bilayer, the integral must be zero, and therefore H_0 must equal zero, since κ^b is non-zero. Since the bending modulus is positive, the bilayer being flat ($H = 0$) will minimize this contribution to the curvature energy. On the other hand, for an asymmetric bilayer (as is typically the case for cell membranes), H_0 will be non-zero, and given by the integral on the right-hand side divided by the bilayer bending modulus, and the mean curvature H will tend to adopt a value of H_0.

The second moment of the lateral pressure profile across a lipid monolayer of thickness d_m gives the monolayer Gaussian curvature modulus:

$$\kappa_G^m = -\int_{-t}^{d_m - t} p(z) z^2 dz$$

Here, the origin of the integral is taken at the neutral surface, a distance t from the chain ends, which is the surface where there is no change in the area upon bending the monolayer (see the chapters by D. P. Siegel and G. Pabst in this volume).

The lateral pressure profile can be extracted from molecular dynamics simulations, which should, in principle, allow estimates of κ_G for lipid bilayers to be determined (Campelo et al. 2014). A recent monolayer value for DPPC at 325 K from coarse-grained simulations is estimated to be –3.4 k_BT (Kalutskii et al. 2023). Estimates for the value of κ_G for typical lipid bilayers from simulations fall in the range of –10 k_BT to –30 k_BT; however, this approach has a number of pitfalls, and has not yet been properly validated (Hu et al. 2012; Hu et al. 2013; Zelisko et al. 2017). Another potential method for determining κ_G is from studying phase-separated vesicles, whose shape depends on the difference between the Gaussian moduli of the two phase-separated domains (Julicher and Lipowsky 1993, 1996).

It should be noted the bilayer value of the modulus, κ_G^b, is not simply twice the monolayer value but is given by the more complex expression:

$$\kappa_G^b = 2\kappa_G^m - 4t\kappa^m H_o^m$$

This result follows by shifting the origin of integration from the neutral surface to the bilayer center by substituting $z^2 \rightarrow (z + t)^2$, expanding into three terms, and then simplifying.

Thus, even if κ_G^m (monolayer) is negative, κ_G^b (bilayer) can be positive if the monolayer spontaneous curvature H_0 is sufficiently negative, meaning that a saddle-surface deformation of the bilayer becomes energetically favored, potentially leading to the formation of inverse bicontinuous cubic phases. Continuum elastic theory shows that high curvature promotes the fusion of lipid bilayers by lowering the energy barriers both to stalk and to fusion pore formation (Golani and Schwarz 2023). Interestingly, a molecular lipid model finds that certain amphiphilic peptides can shift the Gaussian modulus of a lipid bilayer to a positive value, thereby triggering instability with respect to the formation of saddle curvature (Downing et al. 2020). Thus, amphiphilic peptides may favor the formation of inverse bicontinuous cubic phases or locally induce membrane fusion.

Non-Lamellar Phases, Phase Diagrams and Phase Stability

Inverse Bicontinuous Phases

Three inverse bicontinuous phases denoted the Q_{II}^G, Q_{II}^D, and Q_{II}^P phases, have been frequently observed in lipid-water systems (Figure 5).

These are based on fluid lipid bilayers draped onto the gyroid (G), the diamond (D), and primitive (P) minimal surfaces and have crystallographic space groups Ia3d, Pn3m, and Im3m, respectively. In all of these phases, the lipid bilayer subdivides space into two interpenetrating water networks. These inverse bicontinuous cubic phases normally show the phase sequence Q_{II}^G – Q_{II}^D – Q_{II}^P with increasing water content (Templer et al. 1998), which is consistent with a decreasing order of compactness of the associated minimal surfaces (assuming constant curvature). In other words, for the same distribution of Gaussian curvature, the amount of water required to fill the cubic phases increases in the order G < D < P. This order appears to be universally adopted in most phase diagrams, although it is rare for all three cubic phases to be adopted for a single lipid-water system. One of the few examples is

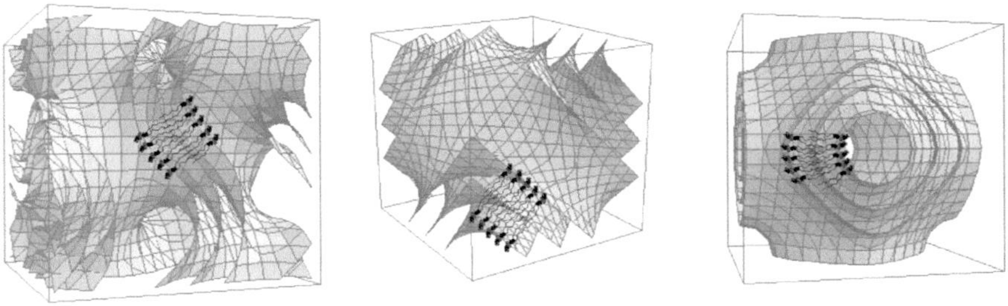

Figure 5. Sketches of the three inverse bicontinuous cubic phases Q_{II}^{G} (Ia3d), Q_{II}^{D} (Pn3m), and Q_{II}^{P} (Im3m), from left to right, based respectively on the G, D, and P underlying minimal surfaces, which sit at the midplane of the fluid lipid bilayer. Adapted from (Shearman et al. 2006) and used with permission.

the system of dilauroyl phosphatidylcholine/lauric acid/water (Templer et al. 1998), whose experimental phase diagram is shown in Figure 6. A theoretical study based on a membrane curvature model could reproduce this phase diagram remarkably well (Schwarz and Gompper 2000).

A counterintuitive finding is that for a number of amphiphiles, for example, the monoacylglycerols, isothermal transitions from the flat fluid lamellar L_α phase to inverse bicontinuous cubic phases can be induced by increasing the water content. A theoretical attempt to explain this intriguing behavior has attributed it to a competition between water-headgroup and headgroup-headgroup hydrogen bonding (Lee et al. 2008).

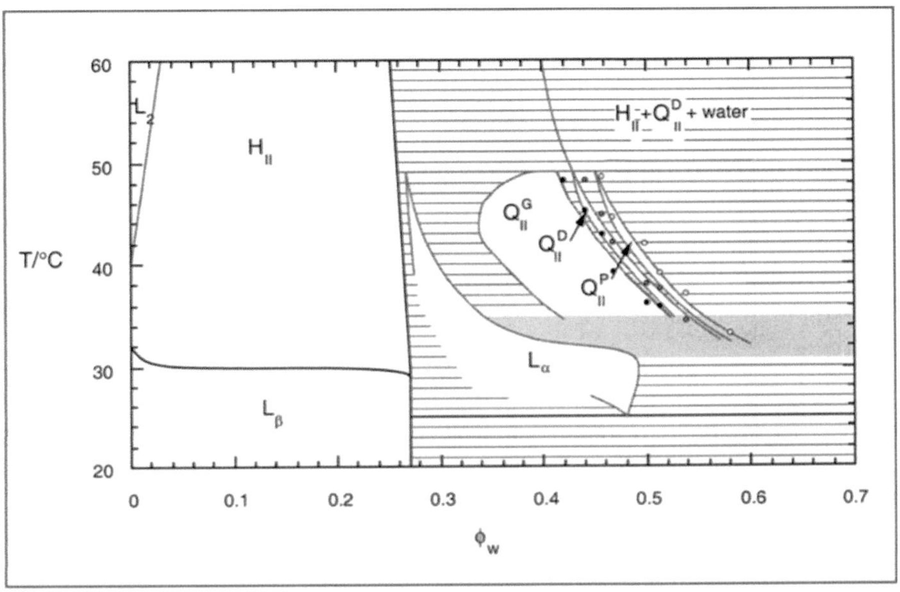

Figure 6. Pseudo-binary phase diagram of dilauroyl phosphatidylcholine/lauric acid (1:2) in water. Note that L_2 is used here to denote the inverse micellar L_{II} phase. Reproduced with permission from (Templer et al. 1998).

An interesting issue that has so far been little explored is whether *asymmetric* bicontinuous cubic phases, having different lipid compositions on the two sides of the bilayer, can be produced, and if so, what novel properties such structures would have. For example, the gyroid inverse cubic phase of space group Ia3d consists of two interwoven water networks of opposite chirality, separated by a continuous fluid bilayer. If the bilayer could be engineered to be asymmetric in composition, then the two water channels would no longer be equivalent in diameter, and such material might form the basis of a "chiral mesoporous sieve", which could discriminate between enantiomeric solutes. An example of such a chiral structure is formed when cytochrome *c*, a water-soluble protein that can be found in the mitochondria of cells, is incorporated into the inverse bicontinuous cubic phase of monoolein, which has previously been found to induce the formation of a non-centrosymmetric cubic phase of spacegroup $P4_332$ (Q^{212}) (Mazzoni et al. 2016). This structure is thought to be similar to that of the gyroid bicontinuous cubic phase, with one water channel replaced by inverse micelles at the junction points and with one molecule of cytochrome c sitting in the center of each inverse micelle, suggesting that cytochrome *c* partitions to regions of high inverse lipid mean curvature.

Swelling of Bicontinuous Phases

Inverse bicontinuous phases should be able to swell in water without any large chain packing costs. However, thermal fluctuations should destroy the long-range order beyond a certain limit, estimated to be approximately 300 Å (Bruinsma 1992). It is now clear that this limit can be exceeded, and much larger lattice parameters can be achieved. One approach is to add "lamellar-preferring" lipids to tune the interfacial curvature. Another approach is to add charged lipids to induce electrostatic swelling (Engblom 2000), which can induce swelling to nearly 500 Å with anionic lipids (Barriga et al. 2015; Tyler et al. 2015) or to as much as 650 Å with cationic lipids (Leung and Leal 2019). Swollen cubic phases of spacegroup Pn3m have been developed for protein crystallization (Zabara et al. 2018). Another approach to swelling cubic phases has been to add hydrophilic sugar amphiphiles to a monoglyceride. Adding 10 mol% octyl beta-glucoside to monoolein allowed it to swell the lattice parameter of the Pn3m cubic phase at 20 C from 107 Å to 153 Å (Angelov et al. 2003). Similarly, by adding sucrose stearate to a monolinolein Im3m cubic phase, it was possible to swell the water channels from 70 Å to 120 Å (Negrini and Mezzenga 2012).

Very highly-swollen bicontinuous cubic phases, with lattice parameters into the range of optical wavelengths, within the wing scale cells of certain butterflies act as templates for the formation of photonic crystals from chitin, leading to structural color and giving rise to the beautiful iridescent colors of the butterfly wings (Saranathan et al. 2010; Wilts et al. 2017; Wilts et al. 2012). The range of lyotropic phase structures related to chitin-based nanostructures in arthropods (such as insects) have been characterized (Saranathan et al. 2015).

Alignment of Cubic Phases on Solid Substrates

For certain applications of lipid cubic phases, for example, for use as sensors, it would be important to coat them as thin films on solid substrates. It is therefore important to understand how such lipid films will align, in other words, the alignment of the lattice vector directions of the cubic phase with respect to the perpendicular and parallel directions of the substrate. Lipid films designed to deliver siRNA to cells have been characterized by grazing-incidence small-angle x-ray scattering (GI-SAXS) (Kang and Leal 2016).

Thin films of monoolein and phytantriol in the Pn3m cubic phase on highly-ordered pyrolytic graphite have been successfully imaged by AFM (Rittman et al. 2013). The images show that in both cases, the cubic phase films align with the [111] direction of the cubic lattice directed along the perpendicular to the film. A further important finding was that only one network of water channels is imaged, implying that the other network is capped off by the bilayer.

A grazing-incidence small-angle x-ray scattering (GI-SAXS) study of spin-coated films of phytantriol, 500–1,500 nm thick, on hydrophobic silicon wafers, found that the Pn3m cubic phase aligned with its [111] direction perpendicular to the plane of the substrate (Richardson et al. 2014), in agreement with the AFM study discussed in the previous paragraph. On the other hand, the Ia3d cubic phase aligned with its [110] direction perpendicular to the substrate. These findings were in agreement with a theoretical model from these authors based on the minimization of the interfacial energy. A more recent GI-SAXS study has found that the Im3m cubic phase, based on the P minimal surface, aligns with its [001] direction perpendicular to the solid substrate, and the square nanoarray was also directly visualized by AFM (Parsons et al. 2022).

Inverse Hexagonal H_{II} Phase

One of the first inverse non-lamellar phases of lipids to be extensively studied was the inverse hexagonal H_{II} phase (Seddon 1990b). This liquid-crystalline structure consists of inverse cylindrical micelles packed onto a 2-D hexagonal lattice and has been extensively studied by X-ray diffraction (Frewein et al. 2019; Rappolt et al. 2003). The H_{II} phase only has a limited ability to swell in water without added hydrophobic solvents, and the lattice parameter is typically in the region of 60–70 Å. If the cylinders were circular in cross-section, they would only fill 91% of the available volume. In order to uniformly fill the hydrophobic region with fluid hydrocarbon chains, either some of the chains must deviate from their preferred length or the polar-nonpolar interfaces must deviate from circular in cross-section, deforming to being faceted into hexagons (see Figure 7).

Thus, there is a state of chain packing and/or curvature frustration within this phase. This frustration can be relieved by adding oils such as liquid alkanes, which partition into the chain region of the phase, allowing the interface to adopt its preferred mean curvature (Gruner 1985). Weak amphiphilic molecules such as alpha-tocopherol (vitamin E) also promote the formation of the H_{II} phase, but by a mechanism different from that of alkanes (Harper et al. 2020). It is found that alpha-tocopherol relieves chain *compressive* packing stress rather than the relief of chain *extension* stress induced by alkanes. The H_{II} phase of DOPE has been used to estimate the monolayer spontaneous curvature of a range of bilayer-forming lipids (Kollmitzer et al. 2013), and this approach has been refined and extended in a more recent paper (Kaltenegger et al. 2021).

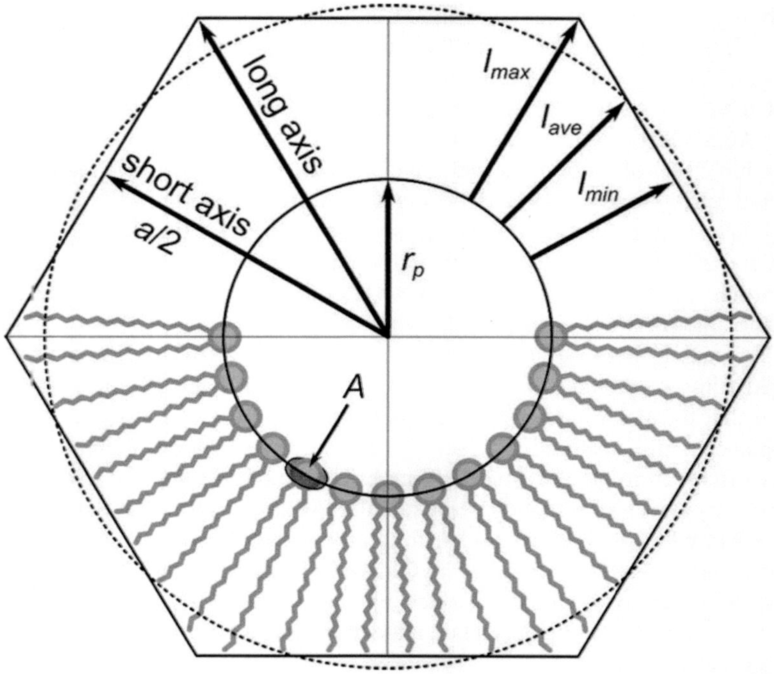

Figure 7. The unit cell of the inverse hexagonal H_{II} phase illustrates the packing/curvature frustration inherent in this phase. The key structural parameters are the lattice parameter *a*, the radius of the water core r_p, and the lipid cross-sectional headgroup area *A*. The lipid length varies between l_{min} and l_{max}. Reproduced with permission from (Harper et al. 2020).

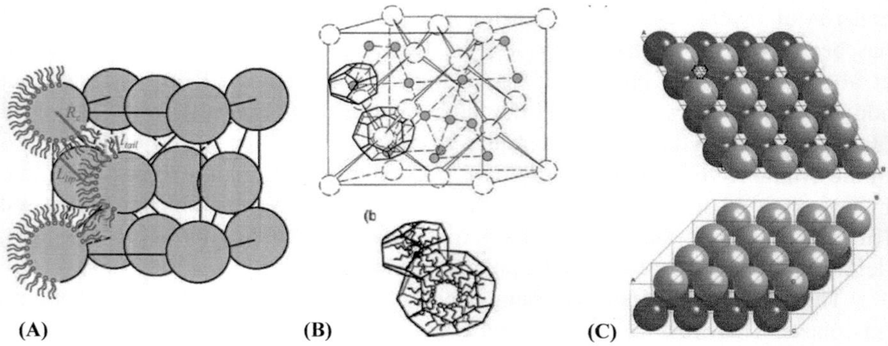

Figure 8. Ordered inverse micellar (I_{II}) phases: (A) cubic Fm3m, from (Martiel et al. 2013); (B) cubic Fd3m, from (Seddon et al. 2000); (C) 3-D hexagonal P6$_3$/mmc, from (Shearman et al. 2009). All reproduced with permission.

Inverse Discontinuous Phases

For lipid systems having a strong tendency for inverse mean curvature, the long inverse cylinders of the H_{II} phase may break up into short or even spherical inverse micellar particles, which can pack into 3-D lattices. To date, three such inverse, discontinuous ordered phases have been reported (see Figure 8): a cubic phase of spacegroup Fd3m (Luzzati et al. 1992; Seddon 1990a), a cubic phase of spacegroup Fm3m, whose structure is simply an *fcc* packing of identical spherical inverse micelles (Martiel et al. 2013), and a 3-D hexagonal phase of spacegroup $P6_3/mmc$ (Shearman et al. 2009).

The lattice parameters tend to be in the range of 100–150 Å, but the Fm3m phase can rise as high as 300 Å. The degree of chain packing frustration in these structures is high, which probably explains why they are less commonly observed than the H_{II} and bicontinuous cubic phases.

The Fd3m cubic phase has a peculiar structure: it consists of two different sizes of quasi-spherical micelles, eight larger ones and 16 smaller ones per unit cell (Luzzati et al. 1992). It turns out that this packing motif packing minimizes the degree of chain stretching required to fill space uniformly (Duesing et al. 1997; Rappolt et al. 2013). This formation of this phase is facilitated by the presence of a weakly amphiphilic molecule such as a long-chain fatty acid, fatty alcohol, or diacylglycerol, which tunes the preferred curvature to strongly negative values. However, the Fd3m phase can also be formed in purely binary lipid/water systems (Minamikawa and Hato 1998; Seddon et al. 1996), although both examples to date have been glycolipids rather than phospholipids.

One lipid system forming the Fd3m phase has been found to transform into the $P6_3/mmc$ phase by adding cholesterol (Shearman et al. 2009). The structure of this phase consists of an hcp packing of identical spherical inverse micelles. It is a more frustrated structure than the Fd3m cubic phase as it has a much larger variation in distance from the polar-nonpolar interface to the polyhedral surface, defining the volume available to each inverse micelle. It appears that the cholesterol can relieve the chain packing frustration because it is only weakly bound to the interface and can spend a certain amount of time partitioned into the hydrophobic region between the inverse micelles. A similar role is played by organic solvent molecules, such as isooctane, cyclohexane, or limonene, in stabilizing the Fm3m cubic phase (Martiel et al. 2013). In these cases, there is only a relatively narrow compositional window of stability: if too much organic solvent is added, the inverse micelles disorder into an inverse micellar (L_2) solution with no long-range order.

Lipid Nanoparticles: Cubosomes, Hexosomes and Micellosomes

It is quite straightforward to produce lipid nanoparticles (LNPs) in the region of 100–200 nm in size, whose internal structure is based upon various non-lamellar lipid phases (Gustafsson et al. 1996; Larsson 2004). One way to prepare LNPs is by sonication in the presence of a small amount of an amphiphilic diblock copolymer, whose function is to stabilize the LNPs against flocculation. A gentler and more reproducible method is to use microfluidics (see Figure 9), as discussed in a range of articles (Belliveau et al. 2012; Kim et al. 2018; Pilkington et al. 2021; Yaghmur and Hamad 2022).

Generally, the term "cubosomes" refers to LNPs based upon inverse bicontinuous cubic phases, "hexosomes" to LNPs based upon the inverse hexagonal H_{II} phase, and "micellosomes" to LNPs based upon inverse, discontinuous phases, typically cubic

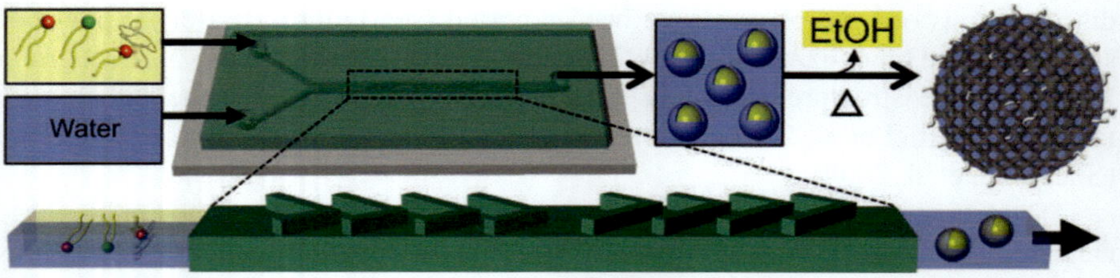

Figure 9. Schematic illustration of the production of monodisperse lipid cubosomes by microfluidics, mixing one solution of lipids in ethanol with an aqueous solution. The cubosomes have an inverse bicontinuous cubic internal structure, with a diameter of typically 200 nm and a low polydispersity index of 0.04. Reproduced with permission from (Kim et al. 2018).

(e.g., of spacegroup Fd3m), or 3-D hexagonal (e.g., of spacegroup $P6_3/mmc$). For bicontinuous cubosomes, it is believed that one of the aqueous networks is closed off, whereas the other is open to the surrounding aqueous solution (Larsson 2004). This suggestion is supported by cryo-electron microscopy studies (Demurtas et al. 2015) and a recent FRET study of cubosomes (Barriga et al. 2019).

Such cubosomes, hexosomes, or micellosomes are typically characterized by a combination of dynamic light scattering to gain information on the particle size and polydispersity, small-angle X-ray scattering (SAXS) to determine the internal ordering and symmetry, and cryo-electron microscopy to visualize the nanoparticles directly (Angelova et al. 2012; Yaghmur and Glatter 2009). The non-lamellar phase structures can be surprisingly well maintained, as assessed by the relative sharpness of the X-ray peaks from the dispersions.

The geometry and topology of different symmetry cubosomes as a function of their size have been explored within the framework of a Landau-Brazovskii-type model functional (Gozdz 2015). It is found that the internal structure may be extremely rich and depends upon the size of the cubosomes.

Lipid nanoparticles can be useful for drug delivery or for the delivery of nucleic acids to cells. For example, siRNA was encapsulated in hexosomes and was shown to induce gene silencing in cells (Zhen et al. 2012). This approach was extended to inverse bicontinuous cubic phase structures (Kang and Leal 2016). Applications of cubosomes and related lipid nanostructures have been reviewed a number of times (Mulet et al. 2013). It was shown that swollen lipidic cubosomes with water channel diameters of up to 171 Å could be formed by the incorporation of a combination of cholesterol and anionic phospholipids and that the channels were freely accessible to a protein, the lectin PHA-L, whose hydrodynamic radius is some 42 Å (Barriga et al. 2019). Functionalization of cubosomes (such as biotinylation or copper-free click chemistry) for active targeting and delivery of drugs has been described (Aleandri et al. 2015; Pramanik et al. 2022). Micellosomes with a cubic Fd3m internal structure have been fabricated (Wadsater et al. 2014), and their degradation by lipases has been studied (Wadsater et al. 2018).

pH-Switchable Lipid Bulk Phases and Lipid Nanoparticles

An attractive way to release cargo from a lipid phase or nanoparticle is to employ pH. By the addition of a pH-sensitive polymer to monoolein cubosomes of spacegroup Im3m, it was shown that the cubic structure is strongly perturbed upon lowering the pH from 7.5 to 5.5 (Kluzek et al. 2017). Sensitivity to pH can also be built in by incorporating an ionizable lipid component that will change its charge state over a certain pH range, perturbing the stability of the phase. It was shown that incorporating 10 mol% of oleic acid into monoolein induced a transformation from a cubosome (Pn3m) to a hexosome (H_{II}) upon lowering the pH from 7 to approximately 5.3 (Nakano et al. 2002). Similarly, by adding 3 wt% linoleic acid to monoolein, the lipid Im3m cubic phase reversibly transformed to an H_{II} phase upon lowering the pH from pH 7 to pH 2 (Negrini and Mezzenga 2011). This was shown to reduce the rate of release of the model hydrophilic drug phloroglucinol by a factor of four, implying that this system could help bypass the acidic environment of the stomach when trying to target drug release into the intestinal tract. Conversely, micellosomes (Fd3m) formed in a range of unsaturated fatty acid/monoolein mixtures at pH 4.9 transformed to hexosomes (H_{II}) upon raising the pH to 7.4 using phosphate buffered saline (Fong et al. 2020).

A system showing the opposite behavior was devised based upon incorporating into monolinolein a weak amphiphilic base, pyridinylmethyl linoleate, which protonates into a charged state on lowering the pH (Negrini et al. 2015). This system exhibited a pH-induced H_{II} – cubic (Pn3m) transition upon lowering the pH from 7.4 to 5.5. It was shown that the release of the anti-cancer drug doxorubicin increased by more than one order of magnitude after this phase transition, which is highly relevant for potential therapeutic use, given that the pH is lower in the region of tumors than in healthy tissue.

The idea of using an ionizable cationic lipid to induce a phase transition within lipid nanoparticles upon lowering the pH was further developed (see Figure 3.10)

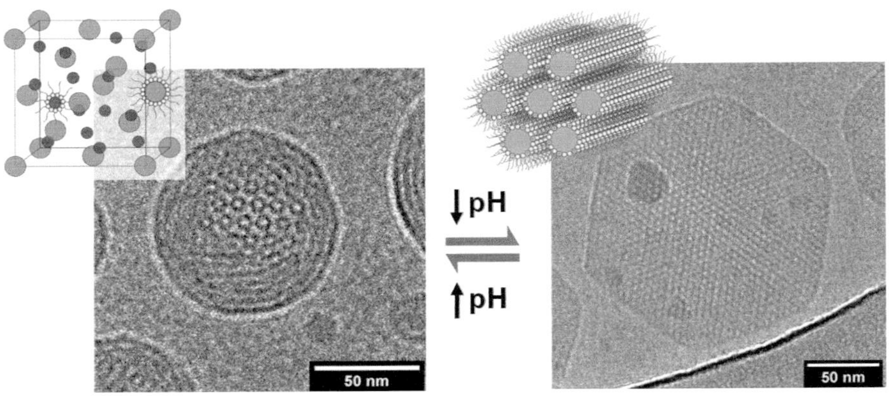

Figure 10. pH-switchable micellosomes. The incorporation of a small amount of a cationic ionizable lipid allows a phase transition to occur between a discontinuous inverse cubic internal structure of the micellosome and a more porous inverse hexagonal internal structure upon lowering the pH to below 6. Reproduced with permission from (Xu et al. 2021).

in a recent study of a monoolein-oleoyl alcohol mixture containing 3 mol% of the ionizable cationic lipid DOBAQ (Xu et al. 2021).

At pH 7.4, the system consists of micellosomes with an Fd3m cubic internal structure, consisting of isolated, discontinuous inverse micelles, which should release any water-soluble cargo molecules only very slowly. Upon lowering the pH to below pH 6 at both 25 and 37°C, a transition is triggered to an H_{II} phase, causing the internal structure to become porous, which should lead to a faster release of cargo molecules.

Hydrostatic pressure has significant effects both on the structure of non-lamellar phases and on non-lamellar phase transitions (reviewed in (Brooks et al. 2011)). Pressure sometimes swells non-lamellar phases but sometimes shrinks them (Barriga et al. 2019). The effect of hydrostatic pressure on cubosomes and hexosomes has also been investigated, and significant differences in the effects on bulk phases have been observed (Kulkarni et al. 2016).

In summary, the field of non-lamellar lipid phases and phase transitions remains an extremely active one, in part driven by the need for a better understanding at a fundamental level of the self-assembly behavior of lipids that underlies many biomedical applications such as vaccines and drug delivery.

References

Aleandri, S., D. Bandera, R. Mezzenga et al. 2015. Biotinylated Cubosomes: A versatile tool for active targeting and codelivery of paclitaxel and a fluorescein-based lipid dye. Langmuir 31(46): 12770–12776. https://doi.org/10.1021/acs.langmuir.5b03469.

Angelov, B., A. Angelova., M. Ollivon et al. 2003. Diamond-type lipid cubic phase with large water channels. J. Am. Chem. Soc., 125(24): 7188–7189. https://doi.org/10.1021/ja034578v.

Angelova, A., B. Angelov, V.M. Garamus et al. 2012. Small-angle X-ray scattering investigations of biomolecular confinement, loading, and release from liquid-crystalline nanochannel assemblies. J. Phys. Chem. Lett. 3(3): 445–457. https://doi.org/10.1021/jz2014727.

Angelova, A., B. Angelov, R. Mutafchieva et al. 2011. Self-assembled multicompartment liquid crystalline lipid carriers for protein, peptide, and nucleic acid drug delivery. Acc. Chem. Res. 44(2): 147–156. https://doi.org/10.1021/ar100120v.

Barriga, H.M., A.I. Tyler, N.L. McCarthy et al. 2015. Temperature and pressure tuneable swollen bicontinuous cubic phases approaching nature's length scales. Soft Matter 11(3): 600–607. https://doi.org/10.1039/c4sm02343a.

Barriga, H.M.G., O. Ces, R.V. Law et al. 2019. Engineering swollen cubosomes using cholesterol and anionic lipids. Langmuir 35(50): 16521–16527. https://doi.org/10.1021/acs.langmuir.9b02336.

Barriga, H.M.G., M.N. Holme and M.M. Stevens. 2019. Cubosomes: the next generation of smart lipid nanoparticles? Angew. Chem. Int. Ed. Engl. 58(10): 2958–2978. https://doi.org/10.1002/anie.201804067.

Belliveau, N.M., J. Huft, P.J.C. Lin et al 2012. Microfluidic synthesis of highly potent limit-size lipid nanoparticles for *in vivo* delivery of siRNA. Molecular Therapy-Nucleic Acids, 1 e37. https://doi.org/10.1038/mtna.2012.28.

Brooks, N.J., O. Ces, R.H. Templer et al. 2011. Pressure effects on lipid membrane structure and dynamics. Chem Phys Lipids 164(2): 89–98. https://doi.org/10.1016/j.chemphyslip.2010.12.002.

Bruinsma, R. 1992. Elasticity and excitations of minimal crystals. Journal De Physique Ii 2(3): 425–451. https//doi.org/10.1051/jp2:1992142.

Caffrey, M. 2015. A comprehensive review of the lipid cubic phase or in meso method for crystallizing membrane and soluble proteins and complexes. Acta Crystallogr. F Struct. Biol. Commun. 71(Pt 1): 3–18. https://doi.org/10.1107/S2053230X14026843.

Campelo, F., C. Arnarez, S.J. Marrink et al. 2014. Helfrich model of membrane bending: from Gibbs theory of liquid interfaces to membranes as thick anisotropic elastic layers. Adv Colloid Interface Sci. 208: 25–33. https://doi.org/10.1016/j.cis.2014.01.018.

Clulow, A.J., M. Salim, A. Awley et al. 2018. A closer look at the behaviour of milk lipids during digestion. Chem Phys Lipids 211: 107–116. https://doi.org/10.1016/j.chemphyslip.2017.10.009.

Conn, C.E. and Drummond. 2013. Nanostructured bicontinuous cubic lipid self-assembly materials as matrices for protein encapsulation. Soft Matter 9(13): 3449–3464. https://doi.org/10.1039/c3sm27743g.

Demurtas, D., P. Guichard, I. Martiel et al. 2015. Direct visualization of dispersed lipid bicontinuous cubic phases by cryo-electron tomography. Nat. Commun. 6: 8915. https://doi.org/10.1038/ncomms9915.

Downing, R., G. Volpe Bossa and S. May. 2020. Saddle-curvature instability of lipid bilayer induced by amphipathic peptides: a molecular model. Soft Matter 16(21): 5032–5043. https://doi.org/10.1039/d0sm00499e.

Duesing, P.M., R.H. Templer and J.M. Seddon. 1997. Quantifying packing frustration energy in inverse lyotropic mesophases. Langmuir 13(2): 351–359. https://doi.org/10.1021/la960602p.

Elani, Y., S. Purushothaman, P.J. Booth et al. 2015. Measurements of the effect of membrane asymmetry on the mechanical properties of lipid bilayers. Chemical Communications 51(32): 6976–6979. https://doi.org/10.1039/c5cc00712g.

Faizi, H.A., C.J. Reeves, V.N. Georgiev et al. 2020. Fluctuation spectroscopy of giant unilamellar vesicles using confocal and phase contrast microscopy. Soft Matter. https://doi.org/10.1039/d0sm00943a.

Faizi, H.A., J. Steinkuhler, S.L. Frey et al. 2019. Lipid charge increases the bending rigidity of bilayer membranes. Biophysical Journal 116(3): 507a–507a.

Foley, S.L., A. Hossein and M. Deserno. 2022. Fluid-gel coexistence in lipid membranes under differential stress. Biophysical Journal 121(16): 2997–3009. https://doi.org/10.1016/j.bpj.2022.07.021.

Fong, C., J. Zhai, C.J. Drummond et al. 2020. Micellar Fd3m cubosomes from monoolein—long chain unsaturated fatty acid mixtures: Stability on temperature and pH response. J. Colloid Interface Sci. 566: 98–106. https://doi.org/10.1016/j.jcis.2020.01.041.

Frewein, M.P.K., P. Piller, E.F. Semeraro et al. 2023. Distributing aminophospholipids asymmetrically across leaflets causes anomalous membrane stiffening. Biophys. J. 122(12): 2445–2455. https://doi.org/10.1016/j.bpj.2023.04.025.

Frewein, M.P.K., M. Rumetshofer and G. Pabst. 2019. Global small-angle scattering data analysis of inverted hexagonal phases. J. Appl. Crystallogr 52(Pt 2): 403–414. https://doi.org/10.1107/S1600576719002760.

Golani, G. and U.S. Schwarz. 2023. High curvature promotes fusion of lipid membranes: Predictions from continuum elastic theory. Biophys. J. 122(10): 1868–1882. https://doi.org/10.1016/j.bpj.2023.04.018.

Gruner, S.M. 1985. Intrinsic curvature hypothesis for biomembrane lipid composition: a role for nonbilayer lipids. Proc. Natl. Acad. Sci. U S A 82(11): 3665–3669. https://doi.org/10.1073/pnas.82.11.3665.

Gustafsson, J., H. LjusbergWahren, M. Almgren et al. 1996. Cubic lipid-water phase dispersed into submicron particles. Langmuir 12(20): 4611–4613. https://doi.org.

Harper, P.E., A.T. Cavazos, J.J. Kinnun et al. 2020. Vitamin E promotes the inverse hexagonal phase via a novel mechanism: implications for antioxidant role. Langmuir 36(18): 4908–4916. https://doi.org/10.1021/acs.langmuir.0c00176.

Helfrich, W. 1973. Elastic properties of lipid bilayers: theory and possible experiments. Z Naturforsch C 28(11): 693–703. https://doi.org/10.1515/znc-1973-11-1209.

Hossein, A. and M. Deserno. 2020. Spontaneous curvature, differential stress, and bending modulus of asymmetric lipid membranes. Biophys. J. 118(3): 624–642. https://doi.org/10.1016/j.bpj.2019.11.3398.

Hou, X., T. Zaks, R. Langer et al. 2021. Lipid nanoparticles for mRNA delivery. Nat. Rev. Mater 6(12): 1078–1094. https://doi.org/10.1038/s41578-021-00358-0.

Hu, M., J.J. Briguglio and M. Deserno. 2012. Determining the Gaussian curvature modulus of lipid membranes in simulations. Biophys. J. 102(6): 1403–1410. https://doi.org/10.1016/j.bpj.2012.02.013.

Hu, M., D.H. de Jong, S.J. Marrink and M. Deserno. 2013. Gaussian curvature elasticity determined from global shape transformations and local stress distributions: a comparative study using the MARTINI model. Faraday Discuss 161: 365–382; discussion 419–359. https://doi.org/10.1039/c2fd20087b.

Julicher, F. and R. Lipowsky. 1993. Domain-induced budding of vesicles. Phys Rev. Lett. 70(19): 2964–2967. https://doi.org/10.1103/PhysRevLett.70.2964.

Julicher, F. and R. Lipowsky. 1996. Shape transformations of vesicles with intramembrane domains. Phys. Rev E Stat. Phys Plasmas Fluids Relat Interdiscip Topics 53(3): 2670–2683. https://doi.org/10.1103/physreve.53.2670.

Kaltenegger, M., J. Kremser, M.P.K. Frewein et al. 2021. Intrinsic lipid curvatures of mammalian plasma membrane outer leaflet lipids and ceramides. Biochim Biophys Acta Biomembr 1863(11): 183–709. https://doi.org/10.1016/j.bbamem.2021.183709.

Kalutskii, M.A., T.R. Galimzyanov and K.V. Pinigin. 2023. Determination of elastic parameters of lipid membranes from simulation under varied external pressure. Phys. Rev. E. 107(2-1): 024414. https://doi.org/10.1103/PhysRevE.107.024414.

Kang, M. and C. Leal. 2016. Soft nanostructured films for actuated surface-based siRNA delivery. Advanced Functional Materials 26(31): 5610–5620. https://doi.org/10.1002/adfm.201600681.

Karamdad, K., R.V. Law, J.M. Seddon et al. 2016. Studying the effects of asymmetry on the bending rigidity of lipid membranes formed by microfluidics. Chemical Communications 52(30): 5277–5280. https://doi.org/10.1039/c5cc10307j.

Khelashvili, G., B. Kollmitzer, P. Heftberger et al. 2013. Calculating the bending modulus for multicomponent lipid membranes in different thermodynamic phases. J. Chem. Theory Comput. 9(9): 3866–3871. https://doi.org/10.1021/ct400492e.

Kim, H., J. Sung, Y. Chang et al. 2018. Microfluidics synthesis of gene silencing cubosomes. ACS Nano 12(9): 9196–9205. https://doi.org/10.1021/acsnano.8b03770.

Kluzek, M., A.I.I. Tyler, S. Wang et al. 2017. Influence of a pH-sensitive polymer on the structure of monoolein cubosomes. Soft Matter. 13(41): 7571–7577. https://doi.org/10.1039/c7sm01620d.

Kollmitzer, B., P. Heftberger, M. Rappolt et al. 2013. Monolayer spontaneous curvature of raft-forming membrane lipids. Soft Matter 9(45): 10877–10884. https://doi.org/10.1039/C3SM51829A.

Kulkarni, C.V., A. Yaghmur, M. Steinhart et al. 2016. Effects of high pressure on internally self-assembled lipid nanoparticles: a synchrotron small-angle X-ray scattering (SAXS) study. Langmuir 32(45): 11907–11917. https://doi.org/10.1021/acs.langmuir.6b03300.

Landau, E.M. and J.P. Rosenbusch. 1996. Lipidic cubic phases: a novel concept for the crystallization of membrane proteins. Proc. Natl. Acad. Sci. U S A 93(25): 14532–14535. https://doi.org/10.1073/pnas.93.25.14532.

Larsson, K. 2004. Bicontinuous cubic lipid-water particles and cubosomal dispersions. Mesoporous Crystals and Related Nano-Structured Materials 148: 41–51. eBook ISBN: 9780080536231.

Lee, W.B., R. Mezzenga and G.H. Fredrickson. 2008. Self-consistent field theory for lipid-based liquid crystals: hydrogen bonding effect. J. Chem. Phys. 128(7): 074504. https://doi.org/10.1063/1.2838624.

Leung, S.S.W. and C. Leal. 2019. The stabilization of primitive bicontinuous cubic phases with tunable swelling over a wide composition range. Soft Matter 15(6): 1269–1277. https://doi.org/10.1039/c8sm02059k.

Lu, L., W.J. Doak, J.W. Schertzer et al. 2016. Membrane mechanical properties of synthetic asymmetric phospholipid vesicles. Soft Matter 12(36): 7521–7528. https://doi.org/10.1039/c6sm01349j.

Luzzati, V., R. Vargas, A. Gulik et al. 1992. Lipid polymorphism: a correction. The structure of the cubic phase of extinction symbol Fd—consists of two types of disjointed reverse micelles embedded in a three-dimensional hydrocarbon matrix. Biochemistry 31(1): 279–285. https://doi.org/10.1021/bi00116a038.

Marsh, D. 2007. Lateral pressure profile, spontaneous curvature frustration, and the incorporation and conformation of proteins in membranes. Biophysical Journal 93(11): 3884–3899. https://doi.org/10.1529/biophysj.107.107938.

Martiel, I., L. Sagalowicz and R. Mezzenga. 2013. A reverse micellar mesophase of face-centered cubic Fm3m symmetry in phosphatidylcholine/water/organic solvent ternary systems. Langmuir 29(51): 15805–15812. https://doi.org/10.1021/la404307x.

Mazzoni, S., L.R. Barbosa, S.S. Funari et al. 2016. Cytochrome-c affects the monoolein polymorphism: consequences for stability and loadin efficiency of drug delivery systems. Langmuir 32(3): 873–881. https://doi.org/10.1021/acs.langmuir.5b03507.

Mezzenga, R., J.M. Seddon, C.J. Drummond et al. 2019. Nature-inspired design and application of lipidic lyotropic liquid crystals. Adv. Mater 31(35): e1900818. https://doi.org/10.1002/adma.201900818.

Minamikawa, H. and M. Hato. 1998. Reverse micellar cubic phase in a phytanyl-chained glucolipid/water system. Langmuir 14(16): 4503–4509. https://doi.org.

Mulet, X. B.J. Boyd and C.J. Drummond. 2013. Advances in drug delivery and medical imaging using colloidal lyotropic liquid crystalline dispersions. J. Colloid Interface Sci. 393: 1–20. https://doi.org/10.1016/j.jcis.2012.10.014.

Nakano, M., T. Teshigawara, A. Sugita et al. 2002. Dispersions of liquid crystalline phases of the monoolein/oleic acid/Pluronic F127 system. Langmuir 18(24): 9283–9288. https://doi.org/10.1021/la026297r.

Negrini, R., W.K. Fong, B.J. Boyd et al. 2015. pH-responsive lyotropic liquid crystals and their potential therapeutic role in cancer treatment. Chem. Commun. (Camb) 51(30): 6671–6674. https://doi.org/10.1039/c4cc10274f.

Negrini, R. and R. Mezzenga. 2011. pH-responsive lyotropic liquid crystals for controlled drug delivery. Langmuir 27(9): 5296–5303. https://doi.org/10.1021/la200591u.

Negrini, R. and R. Mezzenga. 2012. Diffusion, molecular separation, and drug delivery from lipid mesophases with tunable water channels. Langmuir 28(47): 16455–16462. https://doi.org/10.1021/la303833s.

Oka, T., N. Ohta and S. Hyde. 2018. Polar-nonpolar interfaces of inverse bicontinuous cubic phases in phytantriol/water system are parallel to triply periodic minimal surfaces. Langmuir 34(50): 15462–15469. https://doi.org/10.1021/acs.langmuir.8b03320.

Parsons, E., E. Mason, W. Góźdź et al. 2022. A square nano-array formed by surface aligned self-assembly of a bicontinuous lipid cubic phase. ChemRxiv, 27 June 2022, version 1; working paper. https://doi.org/10.26434/chemrxiv-2022-8s6nq.

Pilkington, C.P., J.M. Seddon and Y. Elani. 2021. Microfluidic technologies for the synthesis and manipulation of biomimetic membranous nano-assemblies. Phys. Chem. Chem. Phys. 23(6): 3693–3706. https://doi.org/10.1039/d0cp06226j.

Pramanik, A., Z. Xu, S.H. Shamsuddin et al. 2022. Affimer tagged cubosomes: targeting of carcinoembryonic antigen expressing colorectal cancer cells using *in vitro* and *in vivo* models. ACS Appl Mater Interfaces 14(9): 11078–11091. https://doi.org/10.1021/acsami.1c21655.

Rappolt, M., F. Cacho-Nerin, C. Morello et al. 2013. How the chain configuration governs the packing of inverted micelles in the cubic Fd3m-phase. Soft Matter 9(27): 6291–6300. https://doi.org/10.1039/c3sm50308a.

Rappolt, M., A. Hickel, F. Bringezu et al. 2003. Mechanism of the lamellar/inverse hexagonal phase transition examined by high resolution X-ray diffraction. Biophysical Journal 84(5): 3111–3122. https://doi.org.

Richardson, S.J., P.A. Staniec, G.E. Newby et al. 2014. Predicting the orientation of lipid cubic phase films. Langmuir 30(45): 13510–13515. https://doi.org/10.1021/la503313n.

Rittman, M., H. Amenitsch, M. Rappolt et al. 2013. Control and analysis of oriented thin films of lipid inverse bicontinuous cubic phases using grazing incidence small-angle X-ray scattering. Langmuir 29(31): 9874–9880. https://doi.org/10.1021/la401580y.

Salentinig, S., S. Phan, A. Hawley et al. 2015. Self-assembly structure formation during the digestion of human breast milk. Angew. Chem. Int. Ed. Engl. 54(5): 1600–1603. https://doi.org/10.1002/anie.201408320.

Saranathan, V., C.O. Osuji, S.G. Mochrie et al. 2010. Structure, function, and self-assembly of single network gyroid (I4132) photonic crystals in butterfly wing scales. Proc. Natl. Acad. Sci. U S A 107(26): 11676–11681. https://doi.org/10.1073/pnas.0909616107.

Saranathan, V., A.E. Seago, A. Sandy et al. 2015. Structural Diversity of Arthropod Biophotonic Nanostructures Spans Amphiphilic Phase-Space. Nano. Lett. 15(6): 3735–3742. https://doi.org/10.1021/acs.nanolett.5b00201.

Schwarz, U.S. and G. Gompper. 2000. Stability of inverse bicontinuous cubic phases in lipid-water mixtures. Phys. Rev. Lett. 85(7): 1472–1475. https://doi.org/10.1103/PhysRevLett.85.1472.

Seddon, J.M. 1990a. An inverse face-centered cubic phase formed by diacylglycerol-phosphatidylcholine mixtures. Biochemistry 29(34): 7997–8002. https://doi.org/10.1021/bi00486a031.

Seddon, J.M. 1990b. Structure of the inverted hexagonal (HII) phase, and non-lamellar phase transitions of lipids. Biochim. Biophys. Acta 1031(1): 1–69. https://doi.org/10.1016/0304-4157(90)90002-t.

Seddon, J.M. and R.H. Templer. 1993. Cubic phases of self-assembled amphiphilic aggregates. Philosophical Transactions of the Royal Society of London Series a-Mathematical Physical and Engineering Sciences 344(1672): 377–401. https://doi.org.

Seddon, J.M., N. Zeb, R.H. Templer et al. 1996. An Fd3m lyotropic cubic phase in a binary glycolipid/water system. Langmuir 12(22): 5250–5253. https://doi.org.

Seddon, J.M., J. Robins, T. Gulik-Krzywicki et al. 2000. Inverse micellar phases of phospholipids and glycolipids. Physical Chemistry Chemical Physics 2(20): 4485–4493. https://doi.org.

Shearman, G.C., O. Ces, R.H. Templer et al. 2006. Inverse lyotropic phases of lipids and membrane curvature. J. Phys. Condens Matter 18(28): S1105–1124. https://doi.org/10.1088/0953-8984/18/28/S01.

Shearman, G.C., A.I. Tyler, N.J. Brooks et al. 2009. A 3-D hexagonal inverse micellar lyotropic phase. J. Am. Chem. Soc. 131(5): 1678–1679. https://doi.org/10.1021/ja809280r.

Siegel, D.P. and M.M. Kozlov. 2004. The Gaussian curvature elastic modulus of N-monomethylated dioleoylphosphatidylethanolamine: Relevance to membrane fusion and lipid phase behavior. Biophysical. Journal 87(1): 366–374. https://doi.org/10.1529/biophysj.104.040782.

Templer, R.H., B.J. Khoo and J.M. Seddon. 1998. Gaussian curvature modulus of an amphiphilic monolayer. Langmuir 14(26): 7427–7434. https://doi.org.

Templer, R.H., J.M. Seddon, P.M. Duesing et al. 1998. Modeling the phase behavior of the inverse hexagonal and inverse bicontinuous cubic phases in 2: 1 fatty acid phosphatidylcholine mixtures. Journal of Physical Chemistry B 102(37): 7262–7271. https://doi.org.

Templer, R.H., J.M. Seddon, N.A. Warrender et al. 1998. Inverse bicontinuous cubic phases in 2: 1 fatty acid phosphatidylcholine mixtures. The effects of chain length, hydration, and temperature. Journal of Physical Chemistry B 102(37): 7251–7261. https://doi.org/DOI 10.1021/jp972835a.

Tyler, A.I., H.M. Barriga, E.S. Parsons et al. 2015. Electrostatic swelling of bicontinuous cubic lipid phases. Soft Matter 11(16): 3279–3286. https://doi.org/10.1039/c5sm00311c.

Wadsater, M., J. Barauskas, T. Nylander et al. 2014. Formation of highly structured cubic micellar lipid nanoparticles of soy phosphatidylcholine and glycerol dioleate and their degradation by triacylglycerol lipase. Acs Applied Materials & Interfaces 6(10): 7063–7069. https://doi.org/10.1021/am501489e.

Wadsater, M., J. Barauskas, F. Tiberg et al. 2018. The lipolytic degradation of highly structured cubic micellar nanoparticles of soy phosphatidylcholine and glycerol dioleate by phospholipase A(2) and triacylglycerol lipase. Chemistry and Physics of Lipids 211: 86–92. https://doi.org/10.1016/j.chemphyslip.2017.11.011.

Wilts, B.D., B. Apeleo Zubiri, M.A. Klatt et al. 2017. Butterfly gyroid nanostructures as a time-frozen glimpse of intracellular membrane development. Sci. Adv. 3(4): e1603119. https://doi.org/10.1126/sciadv 1603119.

Wilts, B.D., K. Michielsen, H. De Raedt et al. 2012. Iridescence and spectral filtering of the gyroid-type photonic crystals in Parides sesostris wing scales. Interface Focus 2(5): 681–687. https://doi.org/10.1098/rsfs.2011.0082.

Xu, Z.X., J.M. Seddon, P.A. Beales et al. 2021. Breaking isolation to form new networks: pH-triggered changes in connectivity inside lipid nanoparticles. Journal of the American Chemical Society 143(40): 16556–16565. https://doi.org/10.1021/jacs.1c06244.

Yaghmur, A. and O. Glatter. 2009. Characterization and potential applications of nanostructured aqueous dispersions. Advances in Colloid and Interface Science 147-48: 333–342. https://doi.org/10.1016/j.cis.2008.07.007.

Yaghmur, A. and I. Hamad. 2022. Microfluidic nanomaterial synthesis and *in situ* SAXS, WAXS, or SANS characterization: manipulation of size characteristics and online elucidation of dynamic structural transitions. Molecules 27(14). https://doi.org/10.3390/molecules27144602.

Zabara, A., J.T.Y. Chong, I. Martiel et al. 2018. Design of ultra-swollen lipidic mesophases for the crystallization of membrane proteins with large extracellular domains. Nat. Commun. 9(1): 544. https://doi.org/10.1038/s41467-018-02996-5.

Zelisko, M. F. Ahmadpoor, H. Gao et al. 2017. Determining the gaussian modulus and edge properties of 2D Materials: From Graphene to Lipid Bilayers. Phys. Rev. Lett. 119(6): 068002. https://doi.org/10.1103/PhysRevLett.119.068002.

Zhai, J., C. Fong, N. Tran et al. 2019. Non-lamellar lyotropic liquid crystalline lipid nanoparticles for the next generation of nanomedicine. ACS Nano 13(6): 6178–6206. https://doi.org/10.1021/acsnano.8b07961.

Zhen, G., R.T.M. Hinton, B.W. Muir et al. 2012 Glycerol-monooleate-based nanocarriers for siRNA delivery *in vitro*. Mol Pharmaceutics 9: 2450–2457. https://doi.org/10.1021/mp200662f.

4

Lipid Shape Matters

James Jennings, Enrico F. Semeraro and *Georg Pabst**

Introduction

Life within cellular membranes shares many things in common with our experiences of daily life: it is full of complicated interactions and involves compromises that can lead to stress and frustration. The molecules that make up membranes are structurally highly diverse but can be subsumed into two major classes: lipids and proteins. In general, lipids make up the structural matrix of membranes—a bilayer—into which proteins insert or adsorb to, and the function of membranes depends on intricate lipid-lipid, protein-protein, and lipid-protein interactions. These interactions include short-range specific interactions mediated by stereochemistry (covalent bonding, H-bonding, ion-bridges, etc.) and long-range interactions of nonspecific type (dispersion forces, solvation forces, Coulomb forces, entropic forces, etc.) (Israelachvili 2011).

Arguments based on the shape of 'membrane molecules' have been proven as simple yet powerful tools to gain significant insight into the physical principles of membrane organization and function (Frolov et al. 2011; Hyde et al. 1997; Zimmerberg and Kozlov 2006). For example, shape-based arguments readily allow understanding as to why conically-shaped amphiphiles, such as single-chain surfactants or lyso-lipids, preferably aggregate into micelles or why cylindrically-shaped amphiphiles prefer lamellar structures, and inverse-conically-shaped lipids non-lamellar structures (Israelachvili 2011). Yet shape-based arguments might also be misleading, for example, if a molecule's effective or apparent shape is unknown and mistakenly assumed to be identical to its steric shape. In other words, specific or nonspecific interactions may lead to significant deviations from the steric size of an amphiphile which can change in different lipid environments. Further, deviations of the above-named molecular shapes of lipids/surfactants (cone and cylinder) have also been proposed. For example, rotationally anisotropic shapes of amphiphiles have been invoked to stabilize specific membrane topologies, such as tubular structures

University of Graz, Institute of Molecular Biosciences, Biophysics Division, A-8010 Graz, Austria.
* Corresponding author: georg.pabst@uni-graz.at

or necks (Iglic and Kralj-Iglic 2003) or the disk-like shapes proposed for lipidoids with small polar headgroups that favor aggregation into cubic phases (Jennings and Pabst 2023).

In the absence of stress, monolayers of amphiphiles of the above-described shapes can display different topologies, which, in the most simple case of a 2-dimensional deviation from a flat surface, can be described by the so-called spontaneous *monolayer* curvature, i.e., the tendency to curve away or toward the aqueous phase (Seddon and Templer 1995). Unfortunately, the spontaneous monolayer curvature is often confused with the spontaneous *bilayer* curvature, i.e., the tendency of the full membrane to form an upward or downward bent surface. Unlike membrane curvature, spontaneous monolayer curvature cannot be observed directly in flat membranes but gives rise to elastic stress stored in membranes (elastic curvature stress) due to the deviation of each leaflet from its preferred spontaneous curvature, which has also been termed the frustrated state (Frolov et al. 2011; Seddon and Templer 1995). Importantly, this stored elastic energy contributes to the stability of a membrane, for example, toward membrane-active compounds, and also couples mechanically to membrane proteins, thus affecting their activity. Additionally, distinct spontaneous monolayer curvatures of lipid mixtures were theoretically shown to lead to the formation of nanoscopic domains (Allender and Schick 2022; Meinhardt et al. 2013); for details, see Chapter 2 of this book. Attempting to differentiate between spontaneous monolayer and bilayer curvatures clearly, the former is also often called *intrinsic lipid curvature*, emphasizing the role of the effective lipid shape. Yet, the connection between lipid shape and spontaneous monolayer curvature might be less straight forward than anticipated. Within this chapter, we will carefully address this issue after briefly recapping the connection between lipid shape and intrinsic lipid curvature and review recent advances in measuring this quantity (Section: Lipid curvatures). We then revisit models that show how intrinsic lipid curvature can non-specifically couple to protein activity (Section: Spontaneous Curvature-Mediated Lipid-Protein Interactions), including select experimental evidence, before we discuss the role of elastic curvature stress in its response to membrane-active compounds (Section: Inducing Spontaneous Membrane Curvature). Finally, we discuss potential future work that will either harness current knowledge or require additional insight from the effects of lipid shape (Section: Summary and Outlook).

Lipid Curvatures

Depending on structure and physicochemical properties, the effective shape of amphiphiles can vary considerably. Here, the effective shape refers to the geometric properties of a theoretically isolated lipid (or lipid-like) molecule located at some featureless polar/apolar interface (Figure 1A). The shape of such molecules can be characterized by the famous shape or packing parameter (Israelachvili 2011)

$$S = \frac{V_{\mathrm{HC}}}{A_0 D_{\mathrm{C}}}, \tag{1}$$

where V_{HC} is the hydrocarbon chain volume, A_0 is the optimum headgroup area at the polar/apolar interface, and D_{C} is the hydrocarbon chain length. $S < 1$ favors surfaces with positive curvatures (by convention for bending away from the aqueous phase), $S = 1$ planar surfaces and $S > 1$ negative curvatures (Israelachvili 2011; Seddon and Templer 1995).

Alternatively, the shape of an amphiphile can also be characterized in terms of the intrinsic lipid curvature:

$$C_0 = \frac{1}{R_0}, \tag{2}$$

where R_0 is the radius of curvature of a monolayer constructed by aligning the lipids like building blocks next to each other without imposing bending stress and in the absence of any specific lipid-lipid interactions. The radius of this ”unstressed” monolayer is measured at the *neutral plane*, which describes the position where molecular bending and lateral stretching modes are decoupled (Kozlov 2007; Leikin et al. 1996). Frequently, the neutral plane is confused with the *pivotal plane*, which occurs at the position where the bending of the lipid monolayer does not lead to any changes in the lateral area. Strictly speaking, neutral and pivotal planes concur only in the case of an infinitely large lateral compression modulus (Leikin et al. 1996).

Specific lipid-lipid interactions may modify the above-described monolayer curvature. Still requiring the absence of bending stress, one can differentiate between the two curvatures by introducing the so-called spontaneous monolayer curvtature, J_0. Assuming strictly additive interactions between lipid molecules, $J_0 = \langle C_0^i \rangle$, where the angle brackets refer to some averaging (linear or non-linear) over the individual intrinsic lipid;curvatures C_0 of a given lipid species i (Figure 1B). In general, however, $J_0 \neq \langle C_0^i \rangle$ (Figure 1C), as discussed further below. Analogously to the shape parameter $J_0 > 0$ indicates positive curvatures, $J_0 = 0$ planar surfaces, and $J_0 < 0$ negative curvatures. Draping the lipid molecules onto a cylindrical surface leads to a connection between the two parameters (Marsh 1996):

$$J_0 = \frac{2}{D_C}(1 - S). \tag{3}$$

For symmetric flat bilayers, the spontaneous curvatures of each monolayer cancel out, i.e., $J_{0b} = J_0^+ + J_0^- = 0$, because $J_0^+ = -J_0^-$, where the superscripts ‘$+/-$’ refer to the inner and outer membrane leaflets (monolayers). It follows from the Helfrich theory of membrane elasticity (Helfrich 1973), that the stored elastic energy for each leaflet:

$$E_m^s = \frac{\kappa_m J_0^2}{2} \tag{4}$$

where κ_m is the monolayer bending rigidity. Importantly, the spontaneous monolayer curvature enters quadratically in the equation above and thus dominates E_m^s. Moreover, the stored elastic energy of bilayers does not vanish ($E_b^s = E_m^{s,+} + E_m^{s,-} \neq 0$), as long as $J_0 \neq 0$. The bilayer spontaneous curvature of asymmetric bilayers can differ from zero, because of contributions from different leaflet curvatures ($J_0^+ \neq -J_0^-$), bending moduli, and lateral compressibilities (Hossein and Deserno 2020).

Measuring intrinsic lipid curvatures necessitates systems that allow the lipids to self-assemble into a monolayer of preferred spontaneous curvature. For $C_0 < 0$

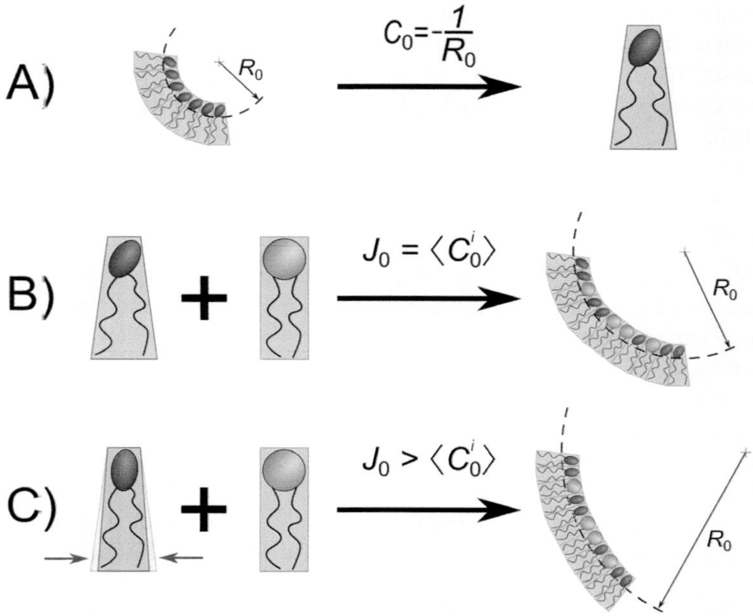

Figure 1. Differentiating between lipid shape, intrinsic lipid curvature, and monolayer spontaneous curvature. **A:** Schematic of the calculation of the (negative) intrinsic lipid curvature C_0 for unstressed, single-component monolayers [see, e.g., Frewein et al. (2019)]. **B:** Additive mixing of intrinsic lipid curvatures in case of ideal mixing. The monolayer spontaneous curvature is derived by averaging, i.e., $J_0 = \langle C_0^i \rangle$. **C:** Non-ideal mixing, leading to non-additive interactions, in which case the spontaneous monolayer curvature differs from the average intrinsic curvatures, $J_0 \neq \langle C_0^i \rangle$ [see, e.g., Lessen et al. (2022)].

systems, inverted hexagonal phases, H_{II}, provide the optimal structure from which C_0 can be inferred using SAXS analysis (Chen and Rand 1997; Gruner et al. 1986; Leikin et al. 1996; Rand et al. 1990; Tate and Gruner 1989); see Figure 2. In such experiments, free alkanes or alkenes are added at low amounts, which preferentially locate within the interstices and thus ensure a fully relaxed system (Chen and Rand 1997; Kirk and Gruner 1985; Rand et al. 1990; Vacklin et al. 2000).

The majority of membrane lipids do not form H_{II} phases. The intrinsic curvatures of these lipids can then be inferred experimentally by titration into to H_{II}-forming templates (e.g., Alley et al. 2008; Kooijman et al. 2005; Rand et al. 1990). DOPE displays a H_{II} phase over a broad range of temperatures and is thus the standard "host" lipid for such purposes.

However, there are some caveats within this approach that one needs to be aware of. Firstly, it necessitates homogeneous mixing of host and "guest" lipids, i.e., lipids should not form domains or phase separate. For example, guest lipids with much longer hydrocarbon tails than the hosts might accumulate within interstitial regions. Also, sufficiently large amounts of $C_0 \nless 0$ lipids may drive the system into a different structure, e.g., lamellar phase. Secondly, *non-linear mixing* effects need

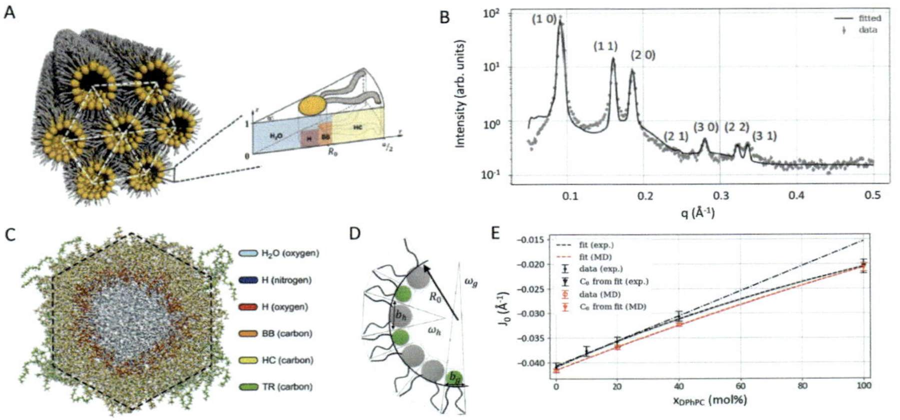

Figure 2. Determination of intrinsic lipid curvatures from H_{II} phases using SAXS. **A:** Schematic of a 2D hexagonal packing of inverted rod-like micelles mimicking stress-free lipid monolayers. The pie demonstrates the contributions of different quasi-molecular fragments in the compositional scattering model. **B:** H_{II} phase-SAXS pattern of DOPE at 35°C. Numbers in brackets give the Miller indices. The solid line gives the best global fit over all scattering vectors q. **C:** Snapshot of an all-atom molecular dynamics simulation of a DOPE unit cell. Note the distribution of tricosene (TR). **D:** Geometrical considerations for non-linear curvature mixing lipids with different shapes, characterized by angles ω_h and ω_g, respectively, and effective in-plane headgroup dimensions b_h and b_g, respectively. **E:** Dependence of J_0 on the concentration of DOPhPC in DOPE H_{II} phases as obtained from SAXS (black line and symbols) and MD simulations (red line and symbols). The dash-dotted line shows a linear extrapolation of experimental J_0 data. Figure adapted from (Frewein et al. 2019) with permission of IUCrJ Journals (CC-BY 4.0 license), copyright 2019; and from (Kaltenegger et al. 2021) with permission of Elsevier (CC- BY 4.0 license), copyright 2021.

to be considered in case of significant differences in shape between host and guest lipids (Kaltenegger et al. 2021). In this case:

$$J_0 = \frac{(1-x)C_0^h + x\xi_0 C_0^g}{1 - x + x\xi_0}, \tag{5}$$

where x is the mole fraction of guest lipid, $C_0^{h,g}$ are the intrinsic curvatures of host and guest lipids, and $\xi_0 = b_g/b_h$ is the ratio of the transverse-plane effective sizes of the lipids at the neutral plane (Figure 2D). Note that Equation (5) becomes linear for $\xi_0 = 1$.

The SAXS data analysis of H_{II} phases for C_0 has evolved from pure geometry-based considerations (Rand et al. 1990) to a global (full q-range) analysis of the scattering patterns (Frewein et al. 2019). The latter technique applies *compositional modeling*, i.e., parsing the lipids into quasimolecular fragments with volume constraints, in order to reach the highest structural fidelity. The neutral plane is assumed to coincide with the lipid's back- bone, in agreement with MD simulations (Sodt and Pastor 2013). Table 1 lists all intrinsic curvature values determined using this technique. A full list of lipid curvature values using different techniques and definitions can be found in Dymond (2021). Note also that lipids with $C_0 < 0$ typically show a pronounced temperature dependence, i.e., their intrinsic curvature becomes

Table 1. List of C0's determined by global analysis of HII phases using SAXS. Numbers in brackets give the uncertainties for the last digits.

lipid	C_0 (Å^{-1})	lipid	C_0 (Å^{-1})
DLPC[a,c]	0.011 (5)	C2:0 Cer[a,c]	–0.03 (2)
DMPC[a,c]	0.007 (2)	C6:0 Cer[a,c]	–0.09 (2)
DPPC[a,c]	0.005 (5)	C16:0 Cer[a,c]	–0.11 (2)
DSPC[a,c]	0.008 (3)	C24:0 Cer[a,c]	–0.15 (3)
DOPC[a,c]	–0.004 (4)	DOPE[a,c]	–0.041 (1)
POPC[a,c]	0.001 (4)	POPE[b,c]	–0.0317 (7)
DOPhPC[a,c]	–0.021 (1)	C16:1PE[a,c]	–0.037 (1)
pSM[a,d]	0.010 (6)	DMPE[b,e]	–0.0314 (8)

[a] Kaltenegger et al. 2021
[b] from Frewein et al. 2019
[c] 35°C
[d] 50°C
[e] 80°C

more negative at elevated temperatures, while lipids with $C_0 \approx 0$ do not change with temperature (Frewein et al. 2019; Kaltenegger et al. 2021).

Derivation of C_0, as described above assumes ideal mixing between phospholipid species. However, specific interactions arising from, for example, H-bonding or Coulombic forces in the headgroup regime may significantly alter the lipids' native C_0 values (Figure 1C). Such enthalpic contributions are known as *non-ideal mixing* and lead to *non-additive* curvature interactions that have been reported both from experiment and theory (Kaltenegger et al. 2021; Lessen et al. 2022; Sodt et al. 2016). The abundance of such interactions in lipid membranes containing amino phospholipids, sphingolipids, or cholesterol strongly implies that one needs to be very cautious with far-reaching conclusions based on simple shape-based arguments—in particular for complex lipid mixtures mimicking natural membranes—as $J_0 \neq \langle C_0^i \rangle$, i.e., there is no simple relation between spontaneous monolayer curvature and intrinsic lipid curvature. For example, we recently reported the effect of non-additive mixing for interactions of detergent with binary and ternary lipid mixtures (Angerer et al. 2023). More research in such directions will be required to obtain further understanding.

Spontaneous Curvature-Mediated Lipid-Protein Interactions

In spite of the above-discussed challenges of non-additive mixing of intrinsic lipid curvatures, J_0 is known to be of fundamental importance to the function of a cellular membrane. In particular, J_0 is involved in the mechanical coupling of bulk membrane properties to protein function (Brown 2017). These bulk interactions, which simultaneously occur with specific lipid-protein or first-shell interactions, also include hydrophobic matching, i.e., energetic penalties resulting from differences in the thickness of the lipid bilayer and the length of the transmembrane domain of the protein. For a discussion of these effects, we refer to previous extensive reviews

(Brown 2017; Marsh 2008); see also Chapter 3 of this book. Here, we focus on a mechanical coupling between collective membrane properties and proteins mediated by spontaneous monolayer curvature.

Cantor was the first to discuss the coupling between the lateral pressure profile and the function of a transmembrane protein (Cantor 1997a,b, 1999); see also Figure 3. In brief, any conformational change of a transmembrane protein that requires a change of shape is associated with work against the lateral pressure of its surrounding lipid matrix. A change of this lateral pressure, e.g., via a change of lipid composition or insertion of a membrane-active compound, consequently affects the conformational equilibrium of the protein. This may, for example, shift the opening probability of an ion channel (Pabst 2013), or the partitioning and aggregation of integral proteins in a given lipid environment/domain (Frewein et al. 2016). For example, convex-shaped proteins favor self-aggregation in liquid-ordered domains (Figure 3B) and asymmetrically shaped proteins prefer an asymmetric lipid membrane (Figure 3C). Dan and Safran (1998) further extended the lateral pressure concept and showed that lipids with $C_0 \neq 0$ exert a line tension on transmembrane proteins, γ_B, which can reach values of up to 100 kT/protein. In particular, $\gamma_B < 0$ would lead to a lateral expansion of the proteins, while $\gamma_B > 0$ were predicted to promote lateral protein compression.

In addition to these lateral pressure effects on protein activity, the insertion of a membrane protein also depends on spontaneous monolayer curvature. Spontaneous insertion of membrane proteins is a rather rare case in cells, where such processes are regulated by the translocon (White and Wimley 1999). However, partitioning and translocating membrane-active compounds, such as cell-penetrating or antimicrobial peptides, can be a proxy for protein insertion. Such processes were found to depend not only on—as frequently discussed—a delicate balance of electrostatic and hydrophobic/solvation interaction but also on J_0. For example, the insertion of the antimicrobial peptide PGLa was demonstrated to be facile in bilayers with $J_0 > 0$, but energetically highly costly for bilayers with $J_0 < 0$ (Leber et al. 2018). Notably, the initial free energy barrier for entering the hydrocarbon core is highly modulated by J_0 (Figure 4). In particular, monolayers with significant negative spontaneous curvature lead to high lateral pressures at the polar-apolar interface. This alludes to the need to choose the lipid composition of membrane mimics properly to ensure that the obtained results are relevant to any biological activity of the studied compound in cells, i.e., not only the charge balance needs to be taken into account, but also the hydrocarbon chain composition and the resulting J_0.

A completely new level of complexity is attained when considering lipid asymmetry, a hallmark of all cellular plasma membranes (Boon and Smith 2002; Lorent et al. 2020). In eukaryotic plasma membranes, phosphatidylcholines and sphingomyelin dominate the outer leaflet composition, while phosphatidylethanolamine and phosphatdiylserine are enriched in the cytoplasmatic membrane leaflet (Lorent et al. 2020; Verkleij et al. 1973). Such a specific leaflet composition can lead to unique structural and elastic membrane properties. For example, we recently showed that hydrocarbon chain interdigitation can, depending on the penetration depth, either induce order or disorder in the opposing lipid leaflet (Frewein et al. 2022). Intriguingly, asymmetric lipid bilayers were reported to produce

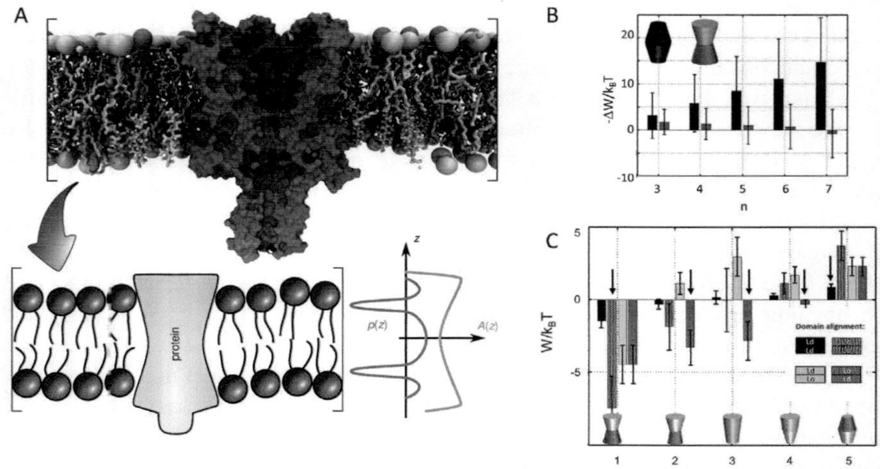

Figure 3. Membrane-mediated effects on integral proteins by coupling protein and lipid shapes. **A**: Schematic of the lateral pressure framework. An integral protein is translated into a geometric shape represented by the cross-sectional area $A(z)$ along the vertical coordinate z. Collective membrane properties result in a lateral pressure profile $p(z)$, against which the protein needs to perform work when changing conformation, or when translocating into a different lipid environment. **B**: Self-aggregation of convex- and concave shaped proteins in liquid disordered (LD) domains shown in terms of the energetic gain, ΔW, as a function of protein aggregation number, n. **C**: Stored elastic energy, W, for proteins of different shapes in symmetric or asymmetric membranes composed of correlated or anticorrelated LD and liquid-ordered (LO) phases. Arrows indicate energetically most favorite lipid environment for a given protein. Figures reproduced from (Frewein et al. 2016) with permission of The Royal Society of Chemistry (CC-BY 3.0 license), copyright 2016.

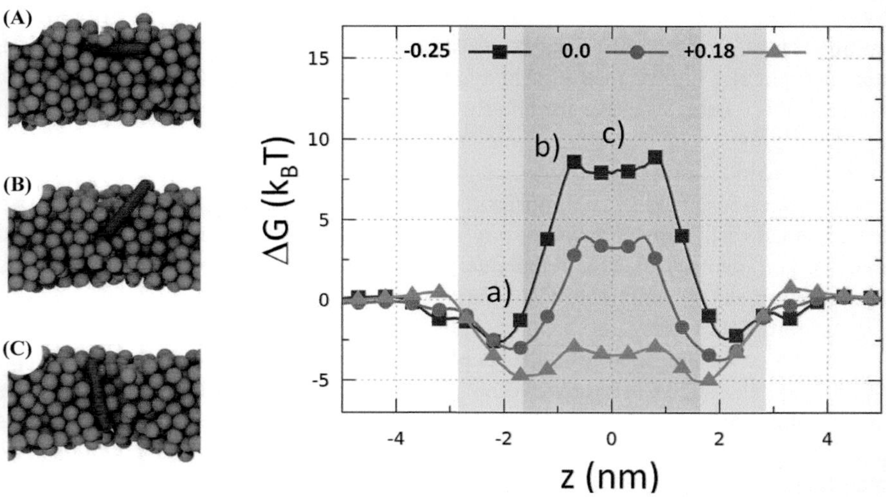

Figure 4. Dependence of the translocation-free energy pathway of an amphiphatic peptide on the spontaneous monolayer curvature as calculated from coarse-grained simulations; numbers in the legend correspond to J_0 in nm^{-1}. The left-hand side shows simulation screenshots of the most important steps of peptide translocation: (A) surface adsorption, (B) peptide tilting, and (C) peptide insertion. Figure reproduced from (Leber et al. 2018) with permission of Elsevier (CC-BY 4.0 license), copyright 2018.

much stiffer membranes than expected from the bending elasticity measured for their symmetric controls (Elani et al. 2015; Frewein et al. 2023; Karamdad et al. 2016; Lu et al. 2016; Rickeard et al. 2020). Recent theoretical developments demonstrate that such stiffening can occur either due to differing spontaneous monolayer curvatures in each leaflet as a result of their chemical composition or simply due to over- or under-crowding of lipids in a given leaflet, also termed "number asymmetry" (Hossein and Deserno 2020, 2021). Regarding the coupling of membrane asymmetry to membrane protein function, simple calculations based on the lateral pressure mechanism described earlier indicate that proteins with asymmetric shapes energetically prefer to be located within asymmetric membranes (Frewein et al. 2016), see Figure 3C. Moreover, the most recent data from our lab shows that the activity of an integral enzyme is mediated by the differential stress in asymmetric membranes (Piller et al. 2023). This indicates that the different J_0's of the membrane leaflets build up elastic energy stress [Equation (4)] that can be harnessed for mediating integral enzyme function. It is likely also the other membrane proteins are also susceptible to differential membrane stress.

Inducing Spontaneous Membrane Curvature

Above, we have already described the spontaneous monolayer curvature dependence of peptide/protein insertion and translocation across lipid bilayers. But what about the response of the bilayer to the presence of such a molecule? How does the induced structural remodeling of the membrane depend on the shapes of its constituting lipids or the peptide/protein itself? In general, membrane insertion of any compound (i.e., membrane-active compound) will affect J_0 and lead to a nonzero J_{0b}, in particular, if the compound remains located within one leaflet. Additionally, membrane-active compounds may induce topological changes, such as, transmembrane pores, or fusion necks, which can be characterized in terms of the Gaussian curvature (Kabelka et al. 2022; Semeraro et al. 2022b); see also Chapters 7 and 8 of this book.

Membrane-Active Peptides

The antimicrobial peptide PGLa provides an interesting case. PGLa adopts an amphiphatic linear α-helical secondary structure upon membrane insertion and can be approximately described by a rod-like shape with a hydrophilic and a hydrophobic face (Figure 5C). Depending on peptide/lipid concentration and intrinsic lipid curvature, linear amphiphatic peptides such as, e.g., PGLa may adopt either a surface-aligned topology ($C_0 < 0$), or tilt into the membrane at various angles ($C_0 \geq 0$) (Harmouche and Bechinger 2018; Perrin and Pastor 2016; Strandberg et al. 2013). Electrostatic interactions between the cationic amino acid residues of the peptides and anionic lipid headgroups further stabilize surface-aligned peptide topologies.

The peptide-induced spontaneous monolayer curvature has been calculated for POPE/POPG (3:1 mol/mol) mixtures using molecular dynamics simulations (Leber et al. 2018). Because $C_0 < 0$ for POPE (Table 1) and $C_0 \sim 0$ for POPG (Leber et al. 2018), we expect a net negative J_0 for the POPE/POPG (3:1 mol/mol) monolayer. Inserting PGLa in surface-aligned topology into the lipid's headgroup regime, we found small but significant differences in the lateral pressure profile (Figure 5A,B).

Interestingly, these changes not only occurred at the location of the peptides within the membrane but also propagated deeper into the interior of the lipid bilayer. The monolayer spontaneous curvature of the POPE/POPG mixture calculated from the MD simulations is ~ -0.032 Å^{-1} (Semeraro et al. 2022b). The presence of PGLa leads to a net increase of J_0 by about 0.003 Å^{-1}. Increasing the "footprint" of PGLa within the bilayer by simulating the presence of PGLa dimers led to a similar effect with ΔJ_0 ~ $+0.004$ Å^{-1}. In summary, surface-aligned PGLa induced positive, spontaneous monolayer curvatures within POPE/POPG mixtures (Figure 5D).

Inducing positive spontaneous monolayer curvature upon peptide insertion eventually destabilizes the bilayer when the peptide concentration is increased.

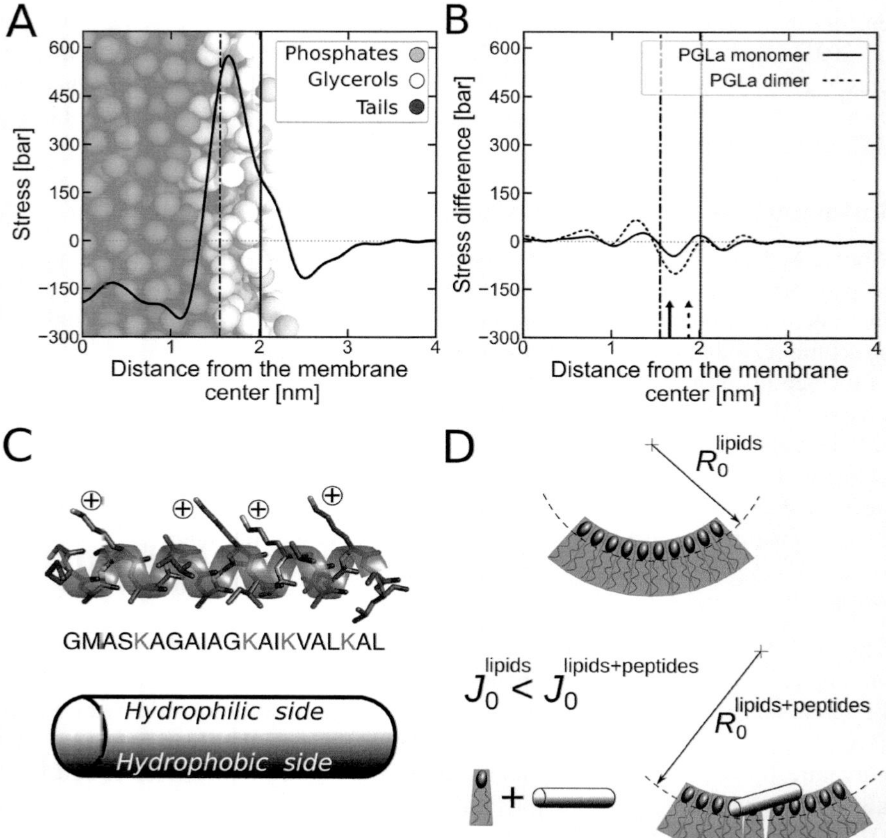

Figure 5. **(A-B)** Calculated lateral pressure profiles for POPE/POPG (3:1 mol/mol) bilayers with and without surface-aligned peptides. **(A)** Reference lateral pressure profile of the lipid-only membrane. Lateral pressure data are overlaid by a simulation snapshot (water beads are omitted for clarity). **(B)** Changes in the stress profiles induced by the presence of PGLa monomers and dimers. The positions of glycerol (dotted-dashed) and phosphate (solid) groups are indicated by vertical lines. The average positions of peptides are indicated by arrows. Panels A and B were adapted from (Semeraro et al. 2022b) with permission (CC-BY-NC-ND). **(C)** Ideal helix conformation of PGLa peptide and its equivalent rod-like representation (structure generated by Swiss-PdbViewer, Guex and Peitsch (1997), http://www.expasy.org/spdbv/). **(D)** Schematic of the net increase of J_0 upon PGLa insertion.

Analogously, also inducing negative spontaneous monolayer curvature will break up lipid bilayers above some critical concentration of the membrane-active compound. For example, compounds discussed in Section (Membrane-Active Lipid-Like Amphiphiles) are candidates for such a scenario. Any type of destabilization of the physical barrier function of membranes (pore formation, micellization, fusion, etc.) can be observed experimentally for example via the "escape" of fluorescent dyes, stored in the lumen of lipid vesicles (leakage). It was shown, for example, that PGLa causes different degrees of leakage from vesicles—most likely due to pore formation—depending on their average spontaneous monolayer curvature (Leber et al. 2018). Specifically, experiments were designed to maintain the overall surface charge density of the lipid bilayer by fixing the content of POPG but varying J_0 by mixing POPE ($C_0 < 0$) and lyso PE ($C_0 > 0$). Additionally, POPG/POPC mixtures (both with $C_0 \sim 0$) were also studied. Significant leakage was observed at low PGLa concentrations only for $J_0 \sim 0$ systems, while effects were negligible for $J_0 < 0$ (Leber et al. 2018). These findings reflect the higher energy cost for peptide insertion in bilayers with $J_0 < 0$ (Figure 4). In the regime of high membrane-bound PGLa content, the observed leakage is convoluted with membrane topology changes, leading to a transformation of the initial large unilamellar vesicles into multilamellar aggregates with collapsed interstitial water layers or non-lamellar structures such as cubic or sponge phases (Kabelka et al. 2020; Marx et al. 2021b). A similar scenario was observed for the lactoferricin derivative LF11-215, which inserts into lipid bilayers close to the location of the lipid backbone but with a more wedge-like shape (Marx et al. 2021b; Zweytick et al. 2011). These studies also included mixtures with anionic cardiolipin that can adopt a significant negative intrinsic curvature in the presence of salts (Chen et al. 2015). Interestingly, an overall drop of leakage from vesicles was observed only for $J_0 < 0$ mixtures upon increasing LF11-215 concentration (Marx et al. 2021b), which is most likely associated with the concomitantly occurring membrane topological changes also observed for PGLa.

In cells, peptide-induced formation of non-lamellar phases or multilamellar aggregates is unlikely due to the significantly enriched compositional and structural complexity. However, the same peptide was found to induce vesicle budding from the outer membrane of Gram-negative bacteria, while an acylated analog of LF11-215 caused the formation of tubular structures (Semeraro et al. 2022a; Zweytick et al. 2011). While this shows different avenues of propagating changes in membrane stress in natural systems, it is clear that this is also a result of matching the physicochemical properties of the membrane-active compound (including its molecular shape) with that of the membrane constituents, i.e., lipids and proteins. Another important aspect in this context is asymmetry, which has also been reported for bacterial membranes (Bogdanov et al. 2020; Boon and Smith 2002). Here, studies on lipid-only mimics of asymmetric membranes revealed elevated leakage in the presence of PGLa, paired with an increased membrane partitioning of the peptide (Marx et al. 2021a). That is, the differential stress between membrane leaflets, which emerges from the different C_0 values of its constituting lipids, plays a significant role in the membrane's response to antimicrobial peptides and should be considered in future studies.

Membrane-Active Lipid-Like Amphiphiles

Synthetic amphiphilic molecules with readily tunable chemical structures provide a route to explore the influence of molecular shape and intrinsic curvatures on membrane activity against cells and model bilayer systems. Lipid-like molecules ("lipidoids") are one such class of material that can be prepared by reacting polyamines with acrylates or epoxides to impart the hydrophilic headgroup and hydrophobic tails into the structure, respectively. Cationic lipidoids are modular in terms of headgroup size and charge, tail length and hydrophobicity. By selecting the number of tails and headgroup size, compounds of a variety of curvatures can be designed toward mimicking specific biological lipid behaviours. For example, lipidoids with six tails were compared structurally to the highly hydrophobic bacterial lipid A, a multitailed lipid that exists in the outer leaflet of Gram-negative bacteria (Carter et al. 2015; Jennings et al. 2020). Lipid A is known to possess highly negative curvature and diverse shapes that may disrupt flat phospholipid bilayers upon interaction (Schromm et al. 2000).

Similar to lipid A, multi-tailed lipidoids have been found to possess diverse molecular conformations and distinct steric shapes, which are uncovered by studying their self-assembly into liquid crystal phases using SAXS analysis. In addition to conventional inverted conical shapes (Figure 1A), which furnish inverted micellar cubic or hexagonal phases, a host of unusual conformations are accessible to multi-tail lipidoids. In particular, chair-like conformations, in which individual molecules can bridge between two bilayers, have been observed within lamellar phases (Jennings et al. 2020; Jennings and Pabst 2023), and pseudo-discs that stack into either inverse cubic or hexagonal phases (Jennings and Pabst 2023). The preference for these distinct shapes strongly depends on an interplay between the arrangement of amines within the headgroup (i.e., linear, branched, or cyclic), the tail structure (i.e., length and bulkiness) and the charged state and counterion associated with the protonated amine headgroup; see Figure 6 for summary. In general, protonated lipidoids with five or more tails most commonly adopt chair-like conformations that pack into the lamellar phase. However, disk shapes appear to prevail in lipidoids with branched headgroups and bulky tails protonated with strong acids. Meanwhile, introducing permanent positive charges into lipidoids (by methylation) enables other lipidoids to adopt inverted conical shapes (Jennings et al. 2023b). Overall, the complex balance of intermolecular forces at play in lipidoid molecules, and how they mimic lipids, is only beginning to be understood.

Modular lipidoid structures provide a platform to uncover correlations between the shape of a synthetic amphiphile and its interaction with biological lipid membranes. We recently demonstrated these principles by testing the toxicity of the compounds to prokaryotic (Gram-negative and Gram-positive bacteria) and eukaryotic (erythrocytes) cells as a proxy for membrane activity (Jennings et al. 2023a,b). This recent study revealed a particularly strong correlation between the propensity for a methylated lipidoid to adopt inverted conical shapes (that pack into inverse hexagonal phases), and their antimicrobial activity and membrane permeabilizing ability. Conversely, the lipidoids with a propensity for disk conformations only showed antibacterial activity at much higher concentrations.

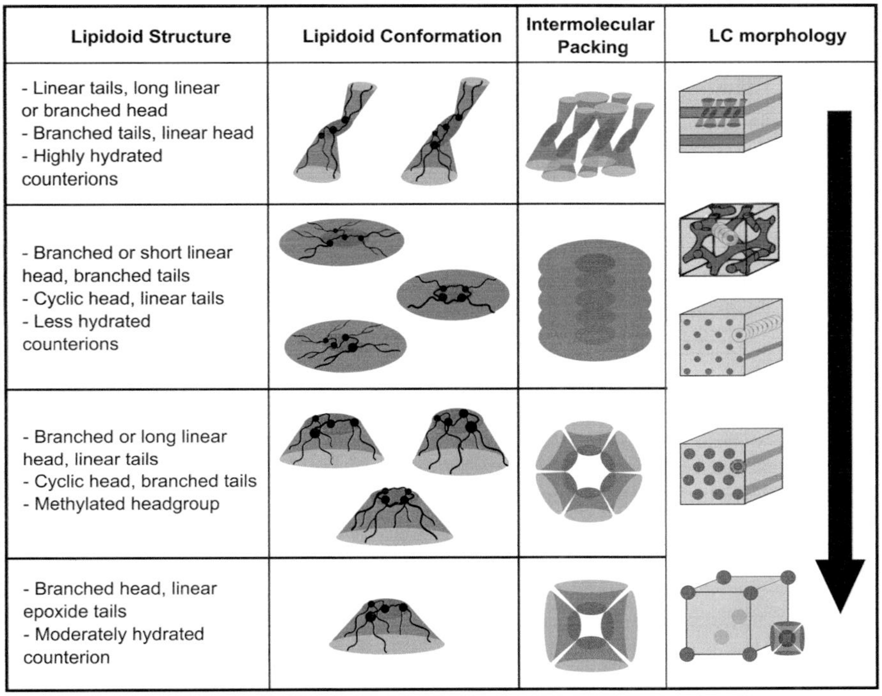

Figure 6. Table summarizing the current understanding of how lipidoid structure influences its preferred liquid crystalline (L_C) morphology and what we can infer from this about the underlying molecular conformations and intermolecular packing (Jennings and Pabst 2023). The compounds most active against biomembranes, as quantified by their toxicity against bacteria (Jennings et al. 2023b), are found to be those able to adopt inverse conical conformations.

Inverted conical lipidoids also showed high hemolytic activity, but the study revealed a number of structures that had higher selectivity for bacteria. Both Gram-positive bacteria and erythrocyte envelopes comprise a single cell membrane but differ both in the headgroups (more negatively charged species in bacteria) and variety of lipid tails, as well as the membrane asymmetry. Such differences imply that differently-shaped molecules would be required to cause disruption. In addition to informing the design of potential membrane-targeting antibacterials, combinatorial studies of lipidoids could provide design rules to inform the design of new membrane disrupting ingredients for other applications, e.g., within lipid nanoparticles that enable endosomal escape for drug delivery (Love et al. 2010).

Summary and Outlook

In this chapter, we provided a broad overview of the role of lipid shape in diverse scenarios of membranes, bestowing the lipid bilayer with specific values of stored elastic energy, discussed here mainly in terms of spontaneous monolayer curvature. A given spontaneous monolayer curvature is a collective property that can be quite different for each membrane leaflet, in particular for asymmetric lipid bilayers.

The caveat of this concept is that it is not always clear how collective properties emerge from the property of a single constituent due to possible enthalpic, i.e., non-additive contributions to mixing lipids. In the case of lipid shapes and the related intrinsic lipid curvatures, this may lead to spontaneous monolayer curvatures that are seemingly unrelated to the properties of the individual components. In particular interactions in the headgroup regime of electrostatic origin (including H-bonding) may significantly modify the projected area of the lipid headgroup and hence the lipid shape (Equation 1). Yet the relation between lipid shape and spontaneous monolayer curvature is not trivial because in general $J_0 \neq \langle C_0 \rangle$. Nevertheless, recent experimental and computational developments (Kaltenegger et al. 2021; Lessen et al. 2022) show promising paths toward in-depth insight.

Despite the need to gain further understanding of the intricate relation between J_0 and C_0, there is ample evidence for the role of monolayer spontaneous curvature in coupling to protein function or membrane remodeling by membrane-active compounds. Here, we particularly reviewed work from our laboratory, demonstrating that integral proteins respond to changes in J_0 by adapting their activity or partitioning within a given membrane environment. Such considerations are of specific importance in the case of membrane domains or transbilayer asymmetry. Additionally, the specific J_0 may be tipped into a certain direction by inserting membrane-active compounds into the lipid bilayer, leading to topological changes. In addition to the well-studied effects of antimicrobial peptides, a new class of lipid-like molecules (lipidoids) show promising effects in disrupting membranes due to their unusual molecular shapes.

From the few examples given, although highly biased by the authors' research interest and own work, it is clear that lipid shape really matters. The need to predict the J_0 from the C_0's of a given lipid mixture is currently the greatest challenge in the field and calls for further studies to increase our understanding. Such studies need also to consider thoroughly the effects of ions dissolved in the aqueous phase. The conformation of lipid headgroups has been shown to be highly sensitive to the presence of salts (Seelig et al. 1987), which again may change the effective headgroup size and, hence, the lipid shape. Indeed, Chen et al. (2015) reported significant changes of C_0 for charged lipids in the presence of calcium ions. This may lead to completely new insights on the physiological role of ions considering the coupling of J_0 to membrane protein function, reviewed above. A related but very different potential application of the concepts laid out within this chapter would be polymer-based nanodiscs, which have become popular recently for studying the structure and dynamics of membrane proteins (Glueck et al. 2022). Analogously to antimicrobial peptides, these polymers are also amphipathic. Their insertion into membranes and capability to extract proteins from their native environment are thus likely to be strongly influenced by balancing lipid shapes and polymer conformations, as discussed in this chapter.

References

Allender, D.W. and M. Schick. 2022. A theoretical basis for nanodomains. J. Membrane Biol. 255(4): 451–460.

Alley, S.H., O. Ces, M. Barahona et al. 2008. X-ray diffraction measurement of the monolayer spontaneous curvature of dioleoylphosphatidylglycerol. Chem. Phys. Lipids 154(1): 64–67.

Angerer, N., P. Piller, E.F. Semeraro et al. 2023. Interaction of detergent with complex mimics of bacterial membranes. Biophys. Chem. 296: 107002.

Bogdanov, M., K. Pyrshev, S. Yesylevskyy et al. 2020. Phospholipid distribution in the cytoplasmic membrane of gram-negative bacteria is highly asymmetric, dynamic, and cell shape dependent. Sci. Adv. 6(23).

Boon, J.M. and B.D. Smith. 2002. Chemical control of phospholipid distribution across bilayer membranes. Med. Res. Rev. 22(3): 251–281.

Brown, M.F. 2017. Soft matter in lipid-protein interactions. Annu. Rev. Biophys. 46: 379–410.

Cantor, R.S. 1997a. The lateral pressure profile in membranes: a physical mechanism of general anesthesia. Biochemistry 36(9): 2339–2344.

Cantor, R.S. 1997b. Lateral pressures in cell membranes: A mechanism for modulation of protein function. J. Phys. Chem. B 101(10): 1723–1725.

Cantor, R.S. 1999. Lipid composition and the lateral pressure profile in bilayers. Biophys. J. 76(5): 2625–2639.

Carter, M.C.D., D.S. Miller and J. Jennings. 2015. Synthetic mimics of bacterial lipid a trigger optical transitions in liquid crystal microdroplets at ultralow picogram-per-milliliter concentrations. Langmuir 31(47): 12850–12855.

Chen, Y.-F., K.-Y. Tsang, W.-F. Chang et al. 2015. Differential dependencies on ca2+ and temperature of the monolayer spontaneous curvatures of dope, dopa and cardiolipin: Effects of modulating the strength of the inter-headgroup repulsion. Soft Matter 11(20): 4041–4053.

Chen, Z. and R.P. Rand. 1997. The influence of cholesterol on phospholipid membrane curvature and bending elasticity. Biophys. J. 73(1): 267–276.

Dan, N. and S.A. Safran. 1998. Effect of lipid characteristics on the structure of transmembrane proteins. Biophys. J. 75(3): 1410–1414.

Dymond, M.K. 2021. Lipid monolayer spontaneous curvatures: A collection of published values. Chem. Phys. Lipids 239: 105117.

Elani, Y., S. Purushothaman, P.J. Booth et al. 2015. Measurements of the effect of membrane asymmetry on the mechanical properties of lipid bilayers. Chem. Commun. 51(32): 6976–6979.

Frewein, M., B. Kollmitzer, P. Heftberger et al. 2016. Lateral pressure-mediated protein partitioning into liquid-ordered/liquid disordered domains. Soft Matter 12(13): 3189–3195.

Frewein, M.P., P. Piller, E.F. Semeraro et al. 2023. Distributing aminophospholipids asymmetrically across leaflets causes anomalous membrane stiffening. Biophys. J.

Frewein, M.P.K., P. Piller, E.F. Semeraro et al. 2022. Interdigitation induced order and disorder in asymmetric membranes. J. Membrane Biol., pp. 1–15.

Frewein, M.P.K., M. Rumetshofer and G. Pabst. 2019. Global small-angle scattering data analysis of inverted hexagonal phases. J. Appl. Crystallogr. 52(Pt 2): 403–414.

Frolov, V.A., A.V. Shnyrova and J. Zimmerberg. 2011. Lipid polymorphisms and membrane shape. Cold Spring Harb. Perspect. Biol. 3(11): a004747.

Glueck, D., A. Grethen, M. Das et al. 2022. Electroneutral polymer nanodiscs enable interference free probing of membrane proteins in a lipid-bilayer environment. Small 18(47): e2202492.

Gruner, S.M., V.A. Parsegian and R.P. Rand. 1986. Directly measured deformation energy of phospholipid HII hexagonal phases. Faraday Discuss. Chem. Soc. (81): 29–37.

Guex, N. and M.C. Peitsch. 1997. Swiss-model and the swiss-pdb viewer: an environment for comparative protein modeling. Electrophoresis 18(15): 2714–2723.

Harmouche, N. and B. Bechinger. 2018. Lipid-mediated interactions between the antimicrobial peptides magainin 2 and pgla in bilayers. Biophys. J. 115(6): 1033–1044.

Helfrich, W. 1973. Elastic properties of lipid bilayers: theory and possible experiments. Z. Naturforsch. C. 28(11): 693–703.

Hossein, A. and M. Deserno. 2020. Spontaneous curvature, differential stress, and bending modulus of asymmetric lipid membranes. Biophys. J. 118(3): 624–642.

Hossein, A. and M. Deserno. 2021. Stiffening transition in asymmetric lipid bilayers: The role of highly ordered domains and the effect of temperature and size. J. Chem. Phys. 154(1): 014704.

Hyde, S.T., S. Andersson, K. Larsson et al. 1997. The Language of Shape. Elsevier, Amsterdam, Netherlands.

Iglic, A. and V. Kralj-Iglic. 2003. Chapter 4—effect of anisotropic properties of membrane constituents on stable shapes of membrane bilayer structure. *In*: Tien, H. and A. Ottova-Leitmannova (eds.). Membrane Science and Technology: Planar Lipid Bilayers (BLMs) and Their Applications. 7: 143–172. Elsevier.

Israelachvili, J.N. 2011. Intermolecular and surface forces. Academic Press, Burlington, MA, 3rd ed edition.

Jennings, J., D. Ašćerić, N. Malanovic et al. 2023a. Structure–activity relationships of cationic lipidoids against *Escherichia coli*. Antibiotics 12(8): 1300.

Jennings, J., Ašćerić, D., E.F. Semeraro et al. 2023b. Combinatorial screening of cationic lipidoids reveals how molecular conformation affects membrane-targeting antimicrobial activity. ACS Appl. Mater Interfaces 15(34): 40178–40190.

Jennings, J., M.C.D. Carter, C.Y. Son et al. 2020. Protonation-driven aqueous lyotropic self assembly of synthetic six-tail lipidoids. Langmuir 36(28): 8240–8252.

Jennings, J. and G. Pabst. 2023. Multiple routes to bicontinuous cubic liquid crystal phases discovered by high-throughput self-assembly screening of multi-tail lipidoids. Small 19(28): 2206747.

Kabelka, I., V. Georgiev, L. Marx et al. 2022. Magainin 2 and pgla in bacterial membrane mimics iii: Membrane fusion and disruption. Biophysical Journal 121: 852–861.

Kabelka, I., M. Pachler, S. Prévost et al. 2020. Magainin 2 and pgla in bacterial membrane mimics ii: Membrane fusion and sponge phase formation. Biophysical. Journal 118: 612–623.

Kaltenegger, M., J. Kremser, M.P. Frewein et al. 2021. Intrinsic lipid curvatures of mammalian plasma membrane outer leaflet lipids and ceramides. Biochim. Biophys. Acta, 1863(11): 183709.

Karamdad, K., R.V. Law, J.M. Seddon et al. 2016. Studying the effects of asymmetry on the bending rigidity of lipid membranes formed by microfluidics. Chem. Commun. 52(30): 5277–5280.

Kirk, G.L. and S.M. Gruner. 1985. Lyotropic effects of alkanes and headgroup composition on the lα-hii lipid liquid crystal phase transition: hydrocarbon packing versus intrinsic curvature. J. Physique 46: 761–769.

Kooijman, E.E., V. Chupin, N.L. Fuller et al. 2005. Spontaneous curvature of phosphatidic acid and lysophosphatidic acid. Biochemistry 44(6): 2097–2102.

Kozlov, M.M. 2007. Determination of lipid spontaneous curvature from xray examinations of inverted hexagonal phases. pp. 355–366. *In*: Dopico, A.M. (ed.). Methods in Membrane Lipids. Humana Press, Totowa, NJ.

Leber, R., M. Pachler, I. Kabelka et al. 2018. Synergism of antimicrobial frog peptides couples to membrane intrinsic curvature strain. Biophys. J. 114(8): 1945–1954.

Leikin, S., M.M. Kozlov, N.L. Fuller et al. 1996. Measured effects of diacylglycerol on structural and elastic properties of phospholipid membranes. Biophys. J. 71(5): 2623–2632.

Lessen, H.J., K.C. Sapp, A.H. Beaven et al. 2022. Molecular mechanisms of spontaneous curvature and softening in complex lipid bilayer mixtures. Biophys. J. 121(17): 3188–3199.

Lorent, J.H., K.R. Levental, L. Ganesan et al. 2020. The mammalian plasma membrane is defined by transmembrane asymmetries in lipid unsaturation, leaflet packing, and protein shape. Nat. Chem. Biol. 16: 644–653.

Love, K.T., K.P. Mahon, C.G. Levins et al. 2010. Lipid-like materials for low-dose, *in vivo* gene silencing. Proceedings of the National Academy of Sciences 107(5): 1864–1869.

Lu, L., W.J. Doak, J.W. Schertzer et al. 2016. Membrane mechanical properties of synthetic asymmetric phospholipid vesicles. Soft Matter 12(36): 7521–7528.

Marsh, D. 1996. Intrinsic curvature in normal and inverted lipid structures and in membranes. Biophys. J. 70(5): 2248–2255.

Marsh, D. 2008. Energetics of hydrophobic matching in lipid-protein interactions. Biophys. J. 94(10): 3996–4013.

Marx, L., M.P.K. Frewein, E.F. Semeraro et al. 2021a. Antimicrobial peptide activity in asymmetric bacterial membrane mimics. Faraday Discussions 232: 435–447.

Marx, L., E.F. Semeraro, J. Mandl et al. 2021b. Bridging the antimicrobial activity of two lactoferricin derivatives in *E. coli* and lipid-only membranes. Frontiers in Medical Technology 3: 625975.

Meinhardt, S., R.L.C. Vink and F. Schmid. 2013. Monolayer curvature stabilizes nanoscale raft domains in mixed lipid bilayers. Proc. Natl. Acad. Sci. USA 110(12): 4476–4481.

Pabst, G. 2013. Coupling membrane elasticity and structure to protein function. *In*: Iglič, A. and C.V. Kulkarni (eds.). Advances in Planar Lipid Bilayers and Liposomes. 18: 81–109. Academic Press, Burlington, MA.

Perrin, B.S. and R.W. Pastor. 2016. Simulations of membrane-disrupting peptides i: Alamethicin pore stability and spontaneous insertion. Biophys. J. 111(6): 1248–1257.

Piller, P., E.F. Semeraro, G.N. Rechberger et al. 2023. Allosteric modulation of integral protein activity by differential stress in asymmetric membranes. PNAS Nexus. 2(5).

Rand, R.P., N.L. Fuller, S.M. Gruner et al. 1990. Membrane curvature, lipid segregation, and structural transitions for phospholipids under dual-solvent stress. Biochemistry 29(1): 76–87.

Rickeard, B.W., M.H.L. Nguyen, M. DiPasquale et al. 2020. Transverse Lipid Organization Dictates Bending Fluctuations in Model Plasma Membranes 12(3): 1438–1447.

Schromm, A.B., K. Brandenburg, H. Loppnow et al. 2000. Biological activities of lipopolysaccharides are determined by the shape of their lipida portion. European Journal of Biochemistry 267(7): 2008–2013.

Seddon, J.M. and R.H. Templer. 1995. Polymorphism of lipid water systems. pp. 97–160. *In*: Lipowsky, R. and E. Sackmann (eds.). Structure and dynamics of membranes. North-Holland, Amsterdam.

Seelig, J., P.M. Macdonald and P.G. Scherer. 1987. Phospholipid head groups as sensors of electric charge in membranes. Biochemistry 26(24): 7535–7541.

Semeraro, E.F., L. Marx, J. Mandl et al. 2022a. Lactoferricins impair the cytosolic membrane of *Escherichia coli* within a few seconds and accumulate inside the cell. eLife, 11.

Semeraro, E.F., P. Pajtinka, L. Marx et al. 2022b. Magainin 2 and pgla in bacterial membrane mimics iv: Membrane curvature and partitioning. Biophysical Journal 121: 4689–4701.

Sodt, A.J. and R.W. Pastor. 2013. Bending free energy from simulation: correspondence of planar and inverse hexagonal lipid phases. Biophys. J. 104(10): 2202–2211.

Sodt, A.J., R.M. Venable, E. Lyman et al. 2016. Nonadditive compositional curvature energetics of lipid bilayers. Phys. Rev. Lett. 117(13): 138104.

Strandberg, E., J. Zerweck, P. Wadhwani et al. 2013. Synergistic insertion of antimicrobial magainin-family peptides in membranes depends on the lipid spontaneous curvature. Biophys. J. 104(6): L9–11.

Tate, M. and S.M. Gruner. 1989. Temperature dependence of the structural dimensions of the inverted hexagonal (hii) phase of phosphatidylethanolamine-containing membranes. Biochemistry 28: 4245–4253.

Vacklin, H.P., B.J. Khoo, K.H. Madan et al. 2000. The bending elasticity of 1-monoolein upon relief of packing stress. Langmuir 16(10): 4741–4748.

Verkleij, A.J., R.F. Zwaal, B. Roelofsen et al. 1973. The asymmetric distribution of phospholipids in the human red cell membrane. A combined study using phospholipases and freeze-etch electron microscopy. Biochim. Biophys. Acta 323(2): 178–193.

White, S.H. and W.C. Wimley. 1999. Membrane Protein Folding and Stability: Physical Principles 28: 319–365.

Zimmerberg, J. and M.M. Kozlov. 2006. How proteins produce cellular membrane curvature. Nat. Rev. Mol. Cell Biol. 7(1): 9–19.

Zweytick, D., G. Deutsch, J. Andrä et al. 2011. Studies on lactoferricin-derived *Escherichia coli* membrane-active peptides reveal differences in the mechanism of nacylated versus nonacylated peptides. Journal of Biological Chemistry 286: 21266–21276.

5

Continuum Modeling of Lipid Bilayers for Curvature Generation in Cellular Processes

P. Rangamani

Introduction

Lipid bilayers are the fundamental components of biological membranes in cells and intracellular organelles, and play a critical role in compartmentalization and cellular organization (Alberts et al. 1985). These membranes are extremely heterogeneous in composition, consisting of many different types of lipids, transmembrane and peripheral proteins, and ion channels (Bassereau et al. 2018; Harayama and Riezman 2018) (Figure 1A). Membrane-protein interactions often result in curvature generation, which is critical for trafficking in cells (Gruenberg 2001; Mukherjee and Maxfield 2000). The curvature generation capability of cellular membranes has been a focus of extensive experimental (Farsad et al. 2001; Ford et al. 2002; McIlhinney et al. 2005; McMahon and Gallop 2005) and theoretical studies (Hassinger et al. 2017; Helfrich 1973; Jülicher and Lipowsky 1993; Lipowsky 1991; Sackmann et al. 1986; Zimmerberg and Kozlov 2006). The shapes observed experimentally include tubes, pearls, and buds, which are observed both in cells and in reconstituted systems (Campelo and Herńandez-Machado 2007; Stachowiak et al. 2010; Yuan et al. 2021). In certain organelles, minimal surfaces (so named because they minimize the surface energy) such as helicoids and catenoids are also observed (Chabanon and Rangamani 2018, 2019; Mendel-sohn et al. 2021; Nixon-Abell et al. 2016; Terasaki et al. 2013; Wu et al. 2017). An additional layer of complexity is provided by the feedback between curvature generation and curvature sensing action of proteins (Antonny 2011). Beyond the observation of static shapes, dynamic modulation of curvature generation can arise from in-plane protein diffusion, protein

Department of Mechanical and Aerospace Engineering, University of California San Diego, La Jolla, CA, USA; Email: prangamani@ucsd.edu

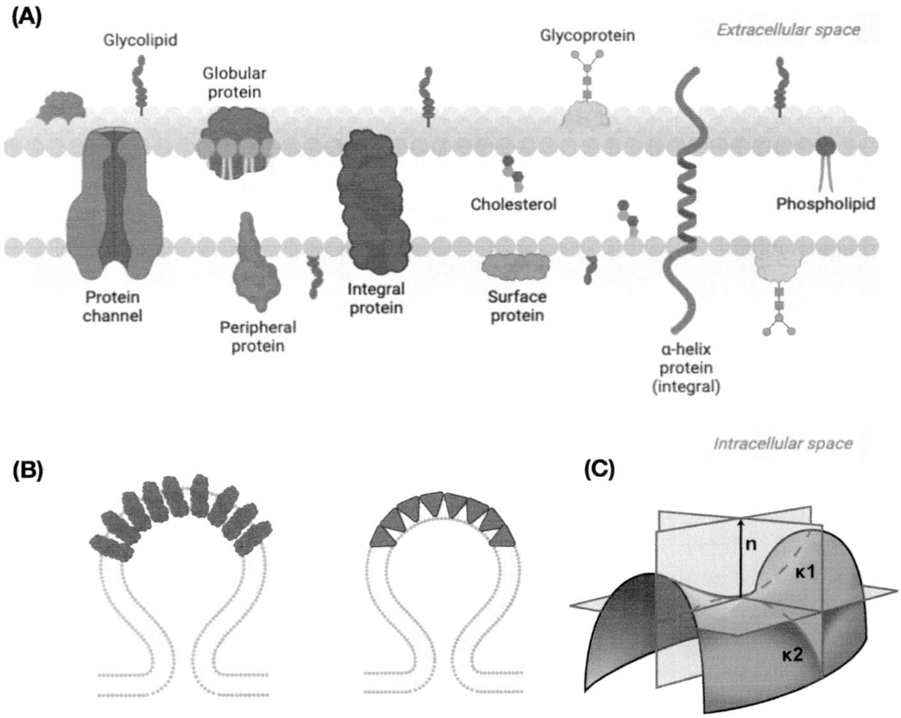

Figure 1. From biological complexity to mechanical simplicity. (A) A schematic depicting the compositional complexity of a plasma membrane. (B) Proteins of different structures can induce a curvature through aggregation and area differences in the leaflets. The schematic represents a small subset of possible spontaneous curvature-inducing mechanisms. (C) The membrane can be represented as a thin elastic shell, and continuum modeling uses the two principal curvatures of the surface to describe the strain energy. Panels (A) and (B) were created using Biorender.com. (C) Curvature schematic was modified from Wikipedia, credited to Eric Gaba (CC BY-SA 3.0).

aggregation (Gov 2008; Kahya et al. 2004; Mahapatra et al. 2021a; Mahapatra et al. 2021b; Reynwar et al. 2007; Saffman and Delbrück 1975), and advection caused by viscous flow of the lipids (Mahapatra et al. 2020; Noguchi and Gompper 2004; Rahimi and Arroyo 2012; Tran-Son-Tay et al. 1984). These dynamic rearrangements of cellular membranes are critical to cellular function, including membrane repair, trafficking, organelle dynamics, and even cell death (Zhang et al. 2018). From a biochemical standpoint, there are countless numbers of protein complexes that can induce curvature through different mechanisms (Alimohamadi and Rangamani 2018; McMahon and Gallop 2005) (Figure 1B). However, from a biophysical standpoint, the impact of these different molecular species can often be summarized as net stresses and bending moments acting on the membrane to generate out-of-plane curvature at different length scales (Deserno 2015; Jenkins 1977; Mahapatra et al. 2020; Steigmann 1999). As a result, continuum modeling lends itself well to the study of membrane curvatures greater than the thickness of the bilayer (Figure 1C).

The history of this field is rich going back to observations of the shape of the red blood cell (Canham 1970; Evans and Skalak 1980; Helfrich 1973) and has been used over the decades to model different cellular and subcellular membrane organization features. In this chapter, I will briefly summarize the details of the Helfrich model for continuum descriptions of bilayer bending. Then, using multiple illustrative examples, I will highlight how continuum mechanics models can be used to give leading-order insight into complex experimental observations. I will end with a brief discussion of future directions and how such models can integrate additional biological details.

Continuum Modeling of Lipid Bilayers

The origins of continuum modeling for lipid bilayers have their roots in liquid crystal theory (Collings 2002; Ericksen 1979; Naghdi 1972) and the theory of thin shell elasticity (Helfrich 1973; Jenkins 1977; Steigmann 1999). Lipid bilayers are roughly 4–5 nm thick (Case et al. 1974; Zhang et al. 1996) and are made of constituent lipid molecules. These amphiphilic lipids self-assemble into bilayers to minimize the exposure of the hydrophobic tails to water (Kučerka et al. 2006; Nagle and Tristram-Nagle 2000; White and Wiener 2017). More often than not, these lipids are organized roughly normal to the midplane of the bilayer (Leibler and Andelman 1987; Rand 1981; Seifert 1993). In many cases, the lipids can tilt against certain proteins-induced boundary conditions or in certain phases (Lenz and Schmid 2007; May 2000; Rangamani et al. 2014a; Rangamani et al. 2014b). The energetic cost of lipid tilt and associated deformations have been studied elsewhere, and we will not be considering them further in this chapter (Lubensky and MacKintosh 1993; Ou-Yang et al. 1999; Rangamani et al. 2014a; Rangamani et al. 2014b; Rangamani et al. 2014c). Our focus is on curvature generation at length scales far larger than the thickness of the bilayer. Therefore, we approximate the lipid bilayer to a thin elastic sheet and focus on surface deformation mechanics (Deserno 2015; Steigmann 1999). To capture the curvature generation capabilities of the membrane, we can use thin shell theory to ascribe a bending elasticity to the surface (Jenkins 1977; Steigmann 1999). The lipid membrane can deform due to external forces or intrinsic curvature generation by membrane proteins and lipids. Additionally, experiments have shown that the lipid bilayers have little capability to undergo stretch; they can withstand up to 4% area strain beyond which the membrane will rupture. In terms of in-plane transport, we note that the lipid bilayer is fluid in-plane despite being elastic in bending (Evans and Hochmuth 1978; Hochmuth et al. 1996; Hochmuth and Waugh 1987). Proteins can diffuse in the plane of the membrane (Kahya et al. 2004; Tran-Son-Tay et al. 1984) and can form aggregates on the membrane (Reynwar et al. 2007; Weber et al. 2019). In what follows, we will systematically explain how these different physics can be described in the continuum framework.

Membrane Bending Energy

A commonly used continuum model for capturing bending is the Canham-Helfrich-Evans (Canham 1970; Evans et al. 1980; Helfrich 1973) model. First appearing in the biophysics literature about 50 years ago, this model described the elastic behavior of the surface using a strain energy. This strain energy is written as a function of the principal curvatures, κ_1 and κ_2, of the surface (Figure 1C). The most commonly used form of this strain energy is given by the Helfrich Hamiltonian (Helfrich 1973).

$$W_{bending} = \kappa[H - C]^2 + \bar{\kappa}K. \quad (1)$$

Here, $H = \dfrac{\kappa_1 + \kappa_2}{2}$ and $K = \kappa_1\kappa_2$ are mean and Gaussian curvatures of the membrane, respectively (Figure 1C). κ and $\bar{\kappa}$ are the bending and Gaussian moduli, respectively. C, the spontaneous curvature, represents any bilayer asymmetry between the leaflets (see Figure 1B for some representations). Note that this energy functional differs from the standard Helfrich energy by a factor of 2, with the net effect being that our value for the bending modulus, κ, is twice that of the standard bending modulus typically encountered in the literature. We will revisit the spontaneous curvature in later sections. Deserno (2015) discusses other possible strain energy functions and their applicability. In this chapter, we will restrict our analysis of the membrane bending energy to Equation (1). In order to obtain the shape of the membrane from the Helfrich energy, the total energy, which is the energy density integrated over the surface area considered ($E = \int_\Omega W dA$), has to be minimized in the presence of additional constraints since the minimum energy state corresponds to mechanical equilibrium.

We systematically explore the different energy contributions, constraints, and free parameters that appear in the Helfrich energy and discuss how they can be related to experimental observations.

Spontaneous Curvature

The spontaneous curvature in Equation (1) captures the asymmetry of the two leaflets (Helfrich 1973). It is the preferred curvature of a free membrane and the configuration in which the membrane bending energy is minimized. Note that the spontaneous curvature is a scalar and generally refers to the isotropic curvature generation of the membrane. The spontaneous curvature has been the subject of extensive analysis to identify how different conditions can give rise to spontaneous curvature in the membrane. Spontaneous curvature can be affected by multiple factors, including the charge of the lipids, asymmetry of lipid composition, and protein-induced spontaneous curvature. In (Hirama et al. 2017), the authors calculate the spontaneous curvature of lipid monolayers and bilayers as a function of charge distribution of phospholipids. For individual proteins, the spontaneous curvature can be calculated using carefully calibrated molecular dynamic simulations. The composition difference between the two leaflets of the bilayer, particularly in cellular membranes, gives rise to an area difference between the two leaflets (Lipowsky 2013). This can lead to a net spontaneous curvature on the membrane, which is proportional to the area difference (Dobereiner et al. 1997; Miao et al. 1994; Mukhopadhyay et al. 2002).

While the spontaneous curvature for certain lipids and proteins has been measured or estimated, translating these values to the net spontaneous curvature on the membrane requires an additional assumption. If the spontaneous curvature induced due to individual proteins is known, then in the simplest case, we can propose that the spontaneous curvature is directly proportional to the density of the proteins on the membrane (Mahapatra et al. 2020). In this linear regime, the spontaneous curvature can be assumed to depend linearly on protein density σ (Agrawal and Steigmann 2011; Mahapatra et al. 2020) as:

$$C(\sigma) = \ell\sigma, \tag{2}$$

where the proportionality constant, ℓ, has units of length. For a complete coverage relationship with protein size and adsorption, the Langmuir adsorption isotherm can be used to obtain the protein surface coverage as a function of the solution concentration (Lipowsky 2013). When proteins crowd the membrane surface, they are known to induce curvature due to steric pressure (Stachowiak et al. 2010). Liese et al., estimated the spontaneous curvature as a function of protein crowding (Liese and Carlson 2021). In Derganc and Copic (2016), the authors examine the interplay between the lateral confinement of proteins on the membrane and protein crowding. Thus, multiple model formulations can be used to connect the protein concentration or lipid charge to the spontaneous curvature of the membrane. Additionally, the heterogeneity of the membrane can also be captured by controlling the area over which this spontaneous curvature is applied (Hassinger et al. 2017; Yuan et al. 2021). Such area dependence on spontaneous curvature is often an experimentally-inspired phenomenological function to match the final vesicle or tube size with the experimental observations (Yuan et al. 2021). We discuss the free parameters from a simulations perspective in the numerical modeling section.

Protein Diffusion

In the plane of the membrane, transmembrane proteins are known to diffuse (Kahya et al. 2004) even as they induce curvature in the bilayer (McMahon and Gallop 2005) (Figure 2). The diffusion of proteins on the membrane surface can be modeled using the principle of entropy maximization (Safran 2018). The entropy S of q proteins on n binding sites can be found from the number of combinations, $\Omega = {}^{n}C_{q}$, and is given by (Mahapatra et al. 2020; Mahapatra et al. 2021a; Mahapatra et al. 2021b):

$$S = k_B \log \Omega, \tag{3}$$

where k_B is the Boltzmann constant (Laurendeau 2005). For sufficiently large values of q and n, the entropic component of the free energy per binding site can be represented as a function of area fraction $\phi = q/n$ (Safran 2018):

$$\frac{W_{entropy}}{n} = -\frac{TS}{n} = k_B T\left[\phi \log \phi + (1-\phi)\log(1-\phi)\right], \tag{4}$$

where T is the temperature of system. The area fraction ϕ can also be represented as the ratio of the local protein density, σ, and the saturation density of proteins on the surface, σ_s. The free energy density per unit area of the membrane is obtained by multiplying the free energy density per binding site ($W_{entropy}/n$) with the saturation density of the proteins (σ_s). Note that the entropic component of the free energy

$W_{entropy}$ is minimized when the entropy S is maximum, corresponding to a uniform distribution of proteins in the domain.

The diffusivity of a spherical particle in an unbounded fluid is given by Stokes-Einstein relation, $D = k_B T/f$ (Guazzelli and Morris 2011). Saffman and Delbrück (1975) derive the expression for the mobility of a protein in a three-dimensional viscous medium surrounded by a three dimensional viscous fluid as:

$$\frac{1}{f} = \frac{1}{4\pi\mu h}\left(\log\frac{\mu h}{\mu' a} - 0.5772\right). \tag{5}$$

Here, h is the height of the membrane, μ and μ' are the viscosities of the membrane and surrounding fluid, and 0.5772 is the Euler constant. One of the advantages of this approach is that the effect of bulk fluid drag is incorporated as a modified viscosity in the two-dimensional medium and therefore, diffusion of proteins in the plane of the membrane can be studied without explicit considerations of the bulk fluid hydrodynamics. The diffusion coefficients of different lipids and proteins have been measured using experimental techniques and can be directly used in simulations for the diffusivity of the protein.

The interaction between in-plane protein diffusion and membrane bending has been modeled for different conditions (Gozdz 2011; Kralj-Iglič et al. 2005; Ramaswamy et al. 2005; Reynwar et al. 2007; Veksler and Gov 2007). Specifically, (Agrawal and Steigmann 2011) proposed a framework that included the chemical potential energy of membrane-protein interactions and membrane bending and demonstrated the interaction between membrane bending and diffusion.

Protein Aggregation

Since many biological membranes are densely packed with proteins on their surface, these proteins can form aggregates either through protein-specific interactions (Douglas and Cyr 2010) or due to curvature-mediated interactions (Reynwar et al. 2007) or both (Figure 2). Interactions between proteins can also lead to the formation of protein microdomains depending on the strength of interaction forces (Reynwar et al. 2007; Weber et al. 2019). The ability of these proteins to induce curvature, coupled with the ability of curvature to influence the lateral diffusion-aggregation dynamics, can result in a feedback loop between membrane curvature and protein density on the surface (Gera and Salac 2017; Klaus et al. 2016; Mahapatra et al. 2021a; Mahapatra et al. 2021b; Nitschke et al. 2012). The aggregation of proteins can be modeled using the interaction enthalpy of particles in a binary system. With the help of mean-field theory, a continuum representation of the aggregation free energy per binding site can be derived as (Givli et al. 2012; Safran 2018; Veksler and Gov 2007)

$$\frac{W_{aggregation}}{n} = \frac{\gamma}{2}\phi(1-\phi) + \frac{\gamma}{4\sigma_s}|\nabla\phi|^2, \tag{6}$$

where γ is the net effective interaction energy of the proteins. This term represents protein-protein attraction when $\gamma > 0$ and protein-protein repulsion when $\gamma < 0$, effectively capturing the energy of forming an interface (line tension).

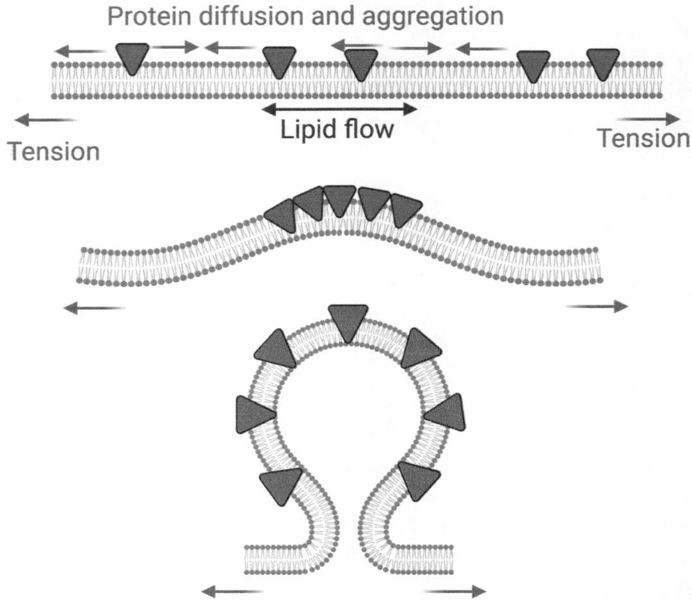

Figure 2. Transmembrane proteins can diffuse in the plane of the membrane while also interacting with one another and inducing curvature in the membrane. These interactions continue as the membrane bends (not shown in the curved states). As a result, these proteins can form aggregates on the membrane and reinforce the curvature generation. Lipid membranes have an in-plane viscosity that can affect in-plane transport. Membrane tension can counteract the bending. Models that can account for these coupled interactions are explained in this chapter (also see (Mahapatra 2022; Mahapatra et al. 2020)). Figure created using Biorender.com.

Area Incompressibility, Membrane Tension, and Surface Pressure

Lipid bilayers have very little capacity to withstand in-plane stretch (about 3–4%). As a result, they are mostly treated as incompressible surfaces. To incorporate this incompressibility, in continuum models, a Lagrange multiplier is used to enforce this constraint and is commonly interpreted as membrane tension. This reasonably straightforward interpretation of the Lagrange multiplier as membrane tension works well for inviscid membranes (membranes with negligible viscosity). In the past decade, many groups have proposed the addition of viscous effects in addition to membrane bending (Arroyo and DeSimone 2009; Rahimi and Arroyo 2012; Rangamani et al. 2013; Tozzi et al. 2019) building on the ideas presented by (Scriven 1960). More recently, it has been shown that the effect of intra-surface viscosity, in addition to membrane bending, allows for the calculation of local membrane tension in the presence of protein-induced spontaneous curvature (Rangamani et al. 2014a; Rangamani et al. 2014b) and the analysis of flow fields on minimal surfaces (Bahmani et al. 2016) (Figure 2; see also Section: Modeling Organelle Membranes). In addition to protein aggregation, in-plane viscous flow of the lipid molecules has been found to dominate some of the phase-transition kinetics of vesicle shapes (Noguchi and Gompper 2004). A series of studies by Arroyo and coworkers also developed a comprehensive framework for incorporating membrane-protein interactions using Onsager's variational principles (Arroyo et al. 2018; Torres-Sanchez et al. 2019; Tozzi et al. 2019). The viscous stresses in the plane of the membrane and their dependence on the rate of strain tensor were derived (Rahimi and Arroyo 2012; Rangamani et al. 2013). When in-plane viscosity is considered, the incompressibility condition for a deformable interface gives a relationship between the divergence of the velocity field and the mean curvature of the surface. When the viscous stresses are included in the governing equations, the interpretation of the Lagrange multiplier for area incompressibility as a surface pressure becomes clear (Bahmani et al. 2016; Rangamani et al. 2014a; Rangamani et al. 2014b). As we and others have shown in a series of theoretical works, the Lagrange multiplier is a surface pressure in the sense of a 2D pressure on the surface of a fluid rather than a membrane tension. Surface pressure can be understood as a force per unit length (analogous to pressure in three dimensions, which is force per unit area) and has the opposite sign of tension since they act in opposite directions. The flow field on the surface of the membrane responds to a surface pressure gradient, and analogies from interfacial fluid mechanics help build our intuition, particularly for heterogeneous membranes with protein domains, flow, and diffusion (also see supplementary material of (Shi et al. 2018)). Therefore, these two views of the Lagrange multiplier as surface pressure or membrane tension are not inconsistent with one another mathematically; they just carry opposite signs.

The heterogeneity of membrane tension, particularly in the case of a membrane with a heterogeneous composition, has been derived rigorously (Mahapatra et al. 2020; Mahapatra et al. 2021a; Mahapatra et al. 2021b; Rangamani et al. 2014a; Rangamani et al. 2014b). Lipowsky explains this concept quite elegantly by building intuition on how spontaneous curvature induced by proteins can induce a proportional spontaneous tension (Lipowsky 2013). This idea of spontaneous tension is extremely useful in terms of thinking about modeling complex cellular processes. From the energy terms given above, the spontaneous tension is roughly κC^2 for a flat membrane (Lipowsky 2013) and the effective tension of the membrane is $\lambda + \kappa C^2$ (Alimohamadi et al. 2021). For example, we recently showed that an "effective tension" that accounts for contributions from the Lagrange multiplier and the spontaneous tension can be used to model the shape of dendritic spines (Alimohamadi et al. 2021). Similarly, the impact of cytoskeleton on membrane tension can be interrogated in the context of this continuum model (Alimohamadi et al. 2020).

Equilibrium Description of the Free Energy

We obtain the total free energy density of the membrane, in terms of protein area fraction ϕ, by combining Equations (1), (4) and (6) as:

$$W = \underbrace{k_B T \sigma_s \left[\phi \log \phi + (1-\phi) \log (1-\phi)\right]}_{\text{entropic contribution}} + \underbrace{\frac{\gamma \sigma_s}{2}\phi(1-\phi) + \frac{\gamma}{4}|\nabla \phi|^2}_{\text{aggregation}} + \underbrace{\kappa(H - \ell\sigma)^2 + \bar{\kappa} K}_{\text{bending}}. \tag{7}$$

When this free energy is minimized over a prescribed area, we obtain the governing equations for the membrane shape.

Note: Those readers who wish to see these models applied to select biological phenomena can safely skip the next section and go to Section (Modeling Clathrin-Mediated Endocytosis) without loss of information.

Governing Equations for the Helfrich Model

Here, we briefly summarize the governing equations that result from energy minimization for the Helfrich model. We use the approach by Steigmann (1999) and Jenkins (1977) to derive the partial differential equations. We refer the reader to these and related papers for complete derivations for these equations and only summarize the equations here for brevity. In what follows Δ is the Laplace-Beltrami operator on the surface.

Bending Elasticity with Homogeneous Distribution of Spontaneous Curvature for Inviscid Membranes

When only the membrane bending energy and tension are considered for a homogeneous membrane with uniform spontaneous curvature, the partial differential equation that prescribes the shape of the membrane is given by:

$$\kappa[\Delta(H - C) + 2(H - C)(H^2 + HC - K)] - 2\lambda H = p. \tag{8}$$

Here, p is the pressure difference across the membrane.

Bending Elasticity with Heterogeneous Distribution of Spontaneous Curvature for Inviscid Membranes

When the spontaneous curvature C is prescribed as a function of the coordinates to capture any heterogeneity on the membrane, the Lagrange multiplier, correspondingly also changes along the surface, and these equations are given by:

$$\kappa[\Delta(H - C) + 2(H - C)(H^2 + HC - K)] - 2\lambda H = p, \ \nabla\lambda = 2\kappa(H - C)\nabla C. \tag{9}$$

Bending Elasticity and in-plane Viscous Flow for Homogeneous Membranes with no Spontaneous Curvature

We now introduce membrane viscosity, v, and the rate of strain tensor, **d**. Here, v_α are the covariant components of the velocity vector, w is the normal component, $()_{;\alpha}$ is the covariant derivative and $\Gamma^{\phi}_{\alpha\beta} = z_{\phi} z{,}\alpha\beta/a$ are the Christoffel symbols (Sokolnikoff 1951). (See Rangamani et al. 2013 for a complete derivation). We first define:

$$d_{\alpha\beta} = \frac{1}{2}(v_{\alpha;\beta} + v_{\beta;\mu}), \tag{10}$$

$$v_{\alpha;\beta} = v\alpha{,}\beta - v\phi\Gamma^{\phi}_{\alpha\beta}. \tag{11}$$

Then, the tangential balance equation and the incompressibility condition are given by:

$$\nabla\lambda - 4\nu w \nabla H + 2\nu(a^{\alpha\mu} d_{\gamma\mu;\alpha} - w_{,\alpha} b^{\alpha}_{\gamma}) = 0,\ \nabla \cdot \mathbf{v} = 2wH. \tag{12}$$

Here, **b** is the curvature tensor. The normal force balance equation is:

$$\underbrace{\kappa[\Delta H + 2H(H^2 - K)] - 2\lambda H}_{\text{elastic contribution}} + \underbrace{2\nu[b^{\alpha\beta} d_{\alpha\beta} - w(4H^2 - 2K)]}_{\text{Viscous contribution}} = p. \tag{13}$$

Bending Elasticity and in-plane Viscous Flow for Heterogeneous Membranes

When we include coordinate-dependent spontaneous curvature, the coupling between membrane bending and the flow field results in the following normal force balance.

$$\underbrace{\kappa[\Delta(H - C) + 2(H - C)(H^2 + HC - K)] - 2\lambda H}_{\text{Elastic contribution}} + \underbrace{2\nu[b^{\alpha\beta} d_{\alpha\beta} - w(4H^2 - 2K)]}_{\text{Viscous contribution}} = p. \tag{14}$$

The equations of motion in the tangent plane are:

$$\lambda_{,\gamma} - 4\nu w H_{,\gamma} + 2\nu(a^{\alpha\mu} d_{\gamma\mu;\alpha} - w_{,\alpha} b^{\alpha}_{\gamma}) = 2\kappa(H - C)\frac{\partial C}{\partial x^{\gamma}}. \tag{15}$$

Bending Elasticity, Protein Diffusion, Aggregation, and in-Plane Flow

Finally, we summarize the governing equations for the coupled dynamics of the system. Please see (Mahapatra et al. 2021a; Mahapatra et al. 2021b) for details. σ is the density of proteins on the membrane as before and ℓ is the length scale (see Section: Spontaneous Curvature). The tangential force balance becomes (Mahapatra et al. 2020; Rangamani et al. 2013; Steigmann 1999):

$$\nabla\lambda + \underbrace{2\nu(\nabla \cdot \boldsymbol{d} - \nabla w \cdot \boldsymbol{b}) - 4\nu w \nabla H}_{\text{viscous}} = - \nabla\sigma\left[\underbrace{k_B T \log\frac{\phi}{1-\phi}}_{\text{entropic}} - \underbrace{2\kappa\ell(H - \ell\sigma_s\phi)}_{\text{bending}} - \underbrace{\left(\frac{\gamma}{2}(2\phi - 1) + \frac{\gamma}{2\sigma_s}\Delta\phi\right)}_{\text{aggregation}}\right]. \tag{16}$$

Along with the surface incompressibility condition:

$$\nabla \cdot \boldsymbol{v} = 2wH, \tag{17}$$

Equation (16) describes how the surface pressure gradient is balanced by the tangential contributions of lipid flow, membrane bending, and membrane-protein interactions. Equations (16) and (17) constitute the governing equations for the velocity field and tension on the evolving surface of the membrane. The shape of the surface is obtained by the normal force balance and is given by:

$$\underbrace{\kappa\Delta(H-\ell\sigma_s\phi)+2\kappa(H-\ell\sigma_s\phi)(2H^2-K)-2H(\kappa(H-\ell\sigma_s\phi)^2+\bar{\kappa}K)}_{\text{bending}}-\underbrace{2\nu\left[\boldsymbol{b}:\boldsymbol{d}-w(4H^2-2K)\right]}_{\text{viscous}}$$
$$-2H\left[\underbrace{k_BT\sigma_s\{\phi\log\phi+(1-\phi)\log(1-\phi)\}}_{\text{entropic}}+\underbrace{\frac{\gamma\sigma_s}{2}\phi(1-\phi)+\frac{\gamma}{4}|\nabla\phi|^2}_{\text{aggregation}}\right]=\underbrace{p+2\lambda H}_{\text{capillary}}. \tag{18}$$

While this equation is complex and contains many terms, it can be understood intuitively by making the following observations. In the absence of all other stresses (bending, viscous, entropic, and aggregation), Equation (18) simply reduces to the Young-Laplace law. When the viscous, entropic, and aggregation terms are removed, we recover the so-called "shape equation" (Equation 8). The additional terms capture the non-trivial coupling between protein density, aggregation, lipid flow, and membrane bending. Equation (16) and Equation (18) both involve the area fraction of proteins $\phi = \sigma/\sigma_s$, which evolves according to the mass conservation equation given by

$$\phi_t+\nabla\cdot(\boldsymbol{v}\phi)=\frac{1}{f}\Delta\phi\left[\frac{k_BT}{1-\phi}+2\kappa\ell^2\sigma_s\phi-\gamma\phi\right]-\frac{1}{f}\phi\left[2\kappa\ell\Delta H+\frac{\gamma}{2\sigma_s}\Delta^2\phi\right]$$
$$+\frac{1}{f}\nabla\phi\cdot\left[\nabla\phi\left(\frac{k_BT}{(1-\phi)^2}+2\kappa\ell^2\sigma_s-\gamma\right)-2\kappa\ell\nabla H-\frac{\gamma}{2\sigma_s}\nabla(\Delta\phi)\right], \tag{19}$$

where ϕ_t denotes the time derivative $\frac{\partial\phi}{\partial t}$. Note that, in the absence of flow and protein-induced spontaneous curvature, Equation (19) reduces to the Cahn-Hilliard equation for aggregation-diffusion (Mahapatra et al. 2021a; Mahapatra et al. 2021b). Additionally, if we eliminate protein aggregation ($\gamma = 0$), in the limit of dilute concentration of proteins ($\phi \ll 1$), we recover the classical equation for Fickian diffusion.

Other Energy Considerations and Alternate Models

In addition to the physics described above, one can include additional terms to the energy to account for the effect of the fourth-order energy (Fournier and Galatola 1997), adhesion (Long et al. 2012), protein binding-unbinding (Lipowsky 2013), grafted brush-like polymers (Shurer et al. 2019), conforming cytoskeleton (Hendrickson et al. 2020), and any localized forces (Alimohamadi et al. 2020; Derényi et al. 2002; Vasan et al. 2020). We note that this list is non-exhaustive but opens avenues for the modeling community to incorporate experimentally relevant observations into the mechanical framework.

We briefly discuss one alternate formulation of the Helfrich model and the modified bending energy prevalent in the literature. This formulation can be quite

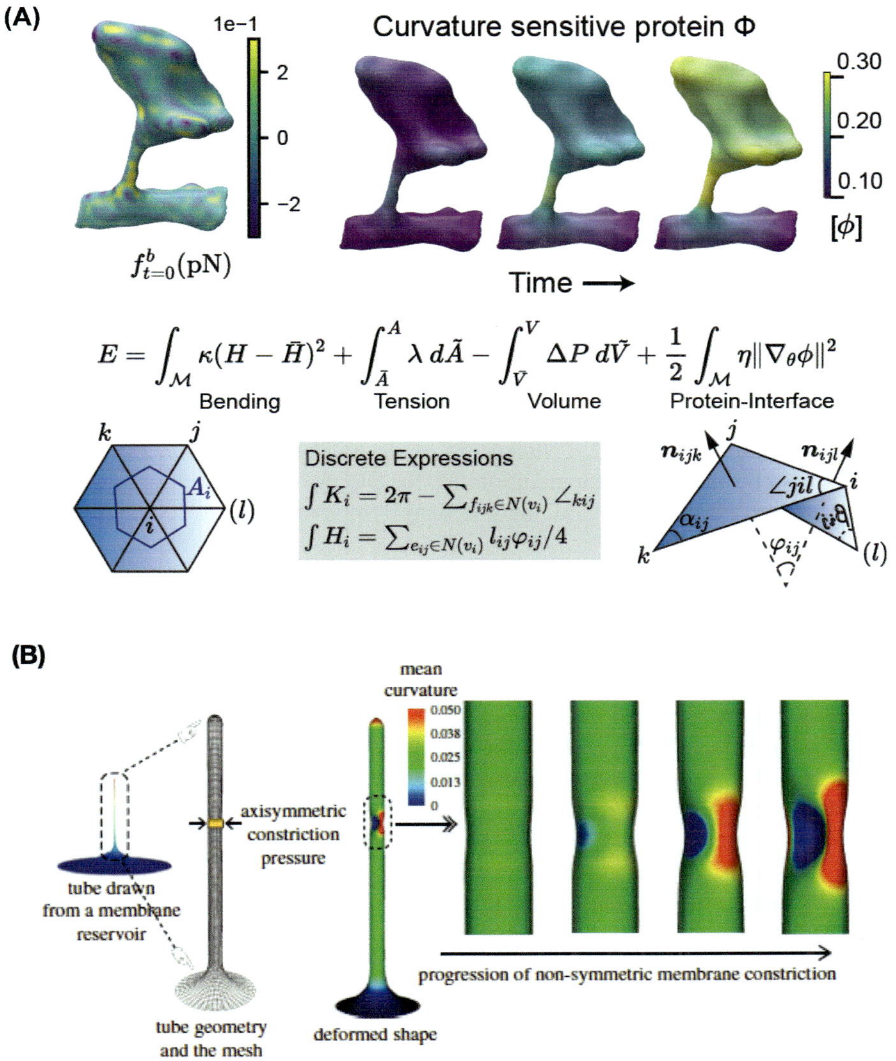

Figure 3. Different computational schemes for solving the governing equations for membrane bending. (A) Overview of Mem3DG: a discrete differential geometry framework (Zhu et al. 2022), figure courtesy, Dr. Chris Lee. (B) Nonsymmetric deformation modes of a cylinder observed using finite element analysis; figure reproduced from (Auddya et al. 2021).

useful for obtaining certain geometries. In our description of the spontaneous curvature thus far, we only account for isotropic curvatures, i.e., the two principal curvatures are affected equally by the spontaneous curvature. However, there are many proteins that induce an anisotropic curvature, including BAR-domain proteins. In such cases, the Helfrich free energy can be modified to account for the two different curvatures induced on the membrane (Kralj-Iglič et al. 2005). Therefore, in addition to the mean curvature, H, we introduce a deviatoric curvature $D = \frac{\kappa_1 + \kappa_2}{2}$ and write a strain energy as (Iglič et al. 2005; Mahapatra et al. 2021a; Mahapatra et al. 2021b.; Mahapatra et al. 2021; Walani et al. 2015):

$$W = \kappa (H - C_0)^2 + \kappa_d (D - D_0)^2 . \quad (20)$$

The first term describes the isotropic component of the strain energy, and the second term describes the anisotropic component. D_0 is the spontaneous anisotropic curvature, sometimes as spontaneous deviatoric curvature. κ and κ_d are the corresponding bending rigidities of the membrane. The governing equations for the energy using the curvature deviator are given in (Mahapatra and Rangamani 2023). This formulation has been used extensively to study the mechanics of protein-mediated tube formation in membranes (Iglič et al. 2005; Mahapatra and Rangamani 2023; Mahapatra et al. 2021a; Mahapatra et al. 2021b). Beyond the bilayer itself, models have been proposed for incorporating the energy of a conforming cytoskeleton (Hendrickson et al. 2020) and the mechanics of a red blood cell cytoskeleton (Alimohamadi et al. 2020).

Computational Methods to Solve the Governing Equations

The equations described in Section (Governing Equations for the Helfrich Model) are a system of nonlinear coupled partial differential equations. A chapter on continuum modeling is incomplete without a discussion of how these equations can be solved numerically. While a detailed discussion of the computational methods required to obtain solutions to these equations is beyond the scope of the current chapter, we provide a brief summary here for completeness. Solving these partial differential equations requires often deciding on surface parametrization (axisymmetric surface, Monge parametrization, etc.). For the most part, to solve the equations described in Section (Membrane Bending Energy), one can make an assumption of axisymmetry and cast the partial differential equations into a system of ordinary differential equations with boundary values. A detailed algorithm was presented in (Luke 1982) back in the 1980s and subsequently, boundary value problem solvers have become robust and easily available in MATLAB and Python. Sometimes, if one is studying small deformation from a plane, using a Monge parametrization of the surface, then the resulting partial differential equations have Laplacian operators in them and standard finite element methods (custom code, COMSOL, Fenics) can be employed with some success (Mahapatra et al. 2020; Mahapatra et al. 2021a; Mahapatra et al. 2021b; Rangamani et al. 2014a; Rangamani et al. 2014b; Sahu et al. 2017).

When additional physics is included, such as in-plane flow, diffusion, and aggregation, even for these relatively simple coordinates, the computational schemes can get quite complex, particularly when one aims to solve the coupled deformation and transport problems. This is because the computational scheme has to satisfy mechanical equilibrium for the membrane at all times but step forward in time to keep track of the evolution of the protein density or the velocity field. In (Mahapatra et al. 2021a; Mahapatra et al. 2021b), we solved the dimensionless governing equations of the equation of motion and the mass conservation of proteins in the linear Monge regime numerically inside a square domain and for axisymmetric geometry for large deformation. We developed a finite difference scheme to solve the transport equations for the protein density, membrane shape, and velocity. The velocity and the shape equation were solved using a Fourier spectral method (Canuto et al. 2006; Hasimoto 1959). In all cases, a semi-implicit scheme was used for the time marching for the protein density ϕ with a time step Δt, where the nonlinear terms involving velocity and curvature were treated explicitly. In contrast, the nonlinear aggregation-diffusion terms were treated with linear implicit terms.

Finally, when one wishes to relax the assumption of axisymmetry or Monge parametrization and solve for the complete deformation of the membrane, advanced computational schemes are needed (Auddya et al. 2021; Sahu et al. 2020; Sauer et al. 2017; Vasan et al. 2020; Zhu et al. 2022). To fully capture the physics involved in transport and bending, we also developed a discrete differential geometry framework that retains all the relevant physics in discrete terms (Figure 3A) (Zhu et al. 2022). Recently, in a series of collaborative efforts, we used isogeometric analysis to simulate membrane bending elasticity without any assumptions of axisymmetry (Auddya et al. 2021) (Figure 3B).

Free Parameters and Sensitivity Analysis

In order to solve these governing equations, a number of parameters must be provided. These include the area of the membrane, the bending moduli, membrane tension, spontaneous curvature, the area over which the spontaneous curvature is applied, diffusion coefficient, aggregation/repulsion strength, and viscosity. Simulations are often conducted using dimensionless parameters to minimize the number of free parameters and find physically meaningful parameter groups. However, if one wishes to vary parameters one at a time to mimic experimental conditions, then grounding these parameters in their physiological range is important. There are some ongoing attempts to infer the free parameters from the shape of the membrane, but they need more development. Additionally, the numerical simulations of the governing equations can be extremely sensitive to these parameters (Rangamani et al. 2021). Unfortunately, these technical challenges with simulating membrane deformation are not discussed as often as they should be in the modeling literature.

Computational Methods to Calculate Membrane Curvature in 3D Reconstructions

Although not the main focus of this chapter, for completeness, we also touch upon how one can calculate the principal curvatures of a surface. This is especially relevant for 3D electron micrograph reconstructions that allow for the visualization of membrane architectures *in situ* (Mendelsohn et al. 2021; Terasaki et al. 2013; Wu et al. 2017). Calculating the curvatures on these reconstructed surfaces can give insight into how far from idealization membrane structures are. The 3D reconstructions generate surfaces that often have irregular mesh structures. Using geometry-preserving adaptive meshing, one can turn these irregular meshes into smooth, regular meshes. Then, using standard formulas from discrete differential geometry, one can calculate the two principal curvatures on these surfaces. This approach not only tells us how curved in membrane structures are in reality but also generates meshes that can be used in simulations for mechanical deformation that can be used in computation (Figure 3A). An example of the curvature distribution on dendritic spines is shown in Figure 4. The original 3D reconstructions were obtained from

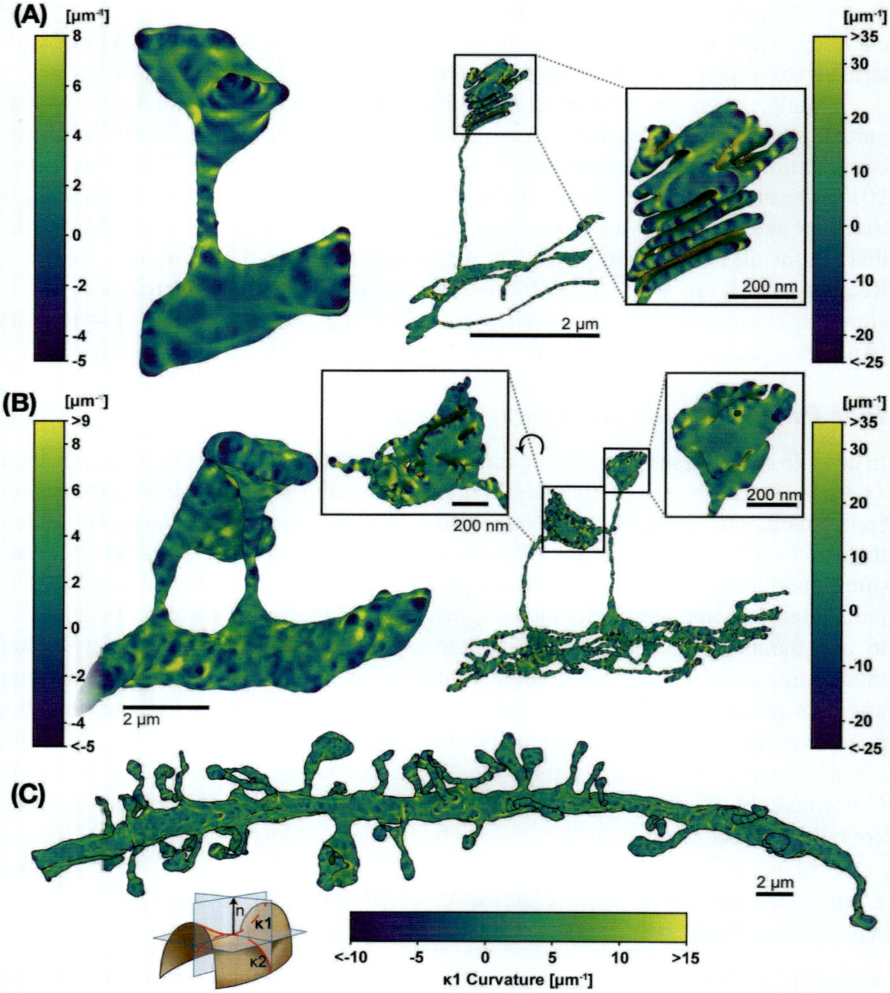

Figure 4. The signed first principal curvature, corresponding to the maximum curvature at each mesh point, is estimated using GAMer 2. Color bars correspond to curvature values with units of 1/μm. We have adopted the sign convention where negative curvature values refer to convex regions. Geometries are (A) single spine model, left: plasma membrane, right: endoplasmic reticulum; (B) two spine model, left: plasma membrane, right: endoplasmic reticulum; and (C) plasma membrane of the dendritic branch model. Scale bars: full geometries 2 μm, inlays: 200 nm. Original spine geometries from (Wu et al. 2017) and this figure reproduced with permission from (Lee et al. 2020). Curvature schematic modified from Wikipedia, credited to Eric Gaba (CC BY-SA 3.0).

Wu et al. (2017), and then using the meshing pipeline described in (Lee et al. 2020), we generated high-quality meshes that were amenable to the calculation of the principal curvatures. Using this framework, we computed the principal curvatures along the plasma membranes and the endoplasmic reticulum of two different dendritic spines (Figure 4A, B). We also computed the curvature distribution along the entire dendrite (Figure 4C) demonstrating that these calculations can be scaled to different cellular and subcellular membranes without loss of accuracy.

Modeling Clathrin-mediated Endocytosis

One of the areas of application of the Helfrich energy for membrane curvature generation has been in the study of vesicle formation during endocytosis. Endocytosis is a process by which the plasma membrane curves in response to a stimulus (specific cargo loading), and the resulting curved vesicle pinches off and subsequently undergoes other downstream processes (Figure 5A). There are different types of endocytosis, including clathrin-mediated and clathrin-independent endocytosis. Here, we use clathrin-mediated endocytosis (CME) as an example to illustrate how mechanical modeling has shed insight into the underlying mechanisms of a complex biological process. CME is a fairly complex biological process involving a complex and intricate orchestration of vast molecular machinery that comprises more than 60 proteins. Experimentalists have worked toward painstakingly dissecting the kinetics and the coupling between the protein machinery and membrane deformation (Kaksonen et al. 2003; Kaksonen et al. 2005). As a result, this is a field that is amenable to biophysical modeling that is rooted in experimental observations. There have been many studies that have focused on such modeling, often focusing on specific questions inspired by experimental observations (Dmitrieff and Nédélec 2015; Lemière et al. 2021;

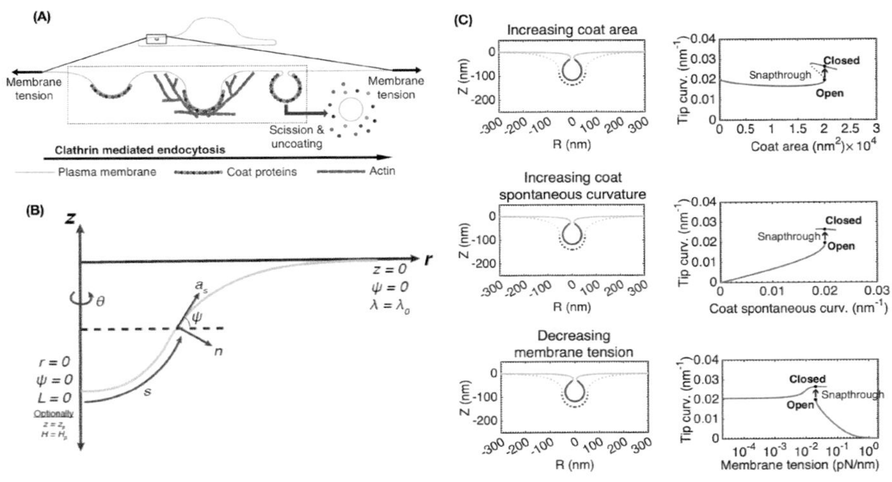

Figure 5. (A) Schematic describing the different steps in clathrin-mediated endocytosis. (B) The axisymmetric parametrization of the membrane, which boundary conditions. (C) Simulation results demonstrate the dependence of vesiculation on different parameters in the Helfrich model. Figure reproduced with permission from (Hassinger et al. 2017).

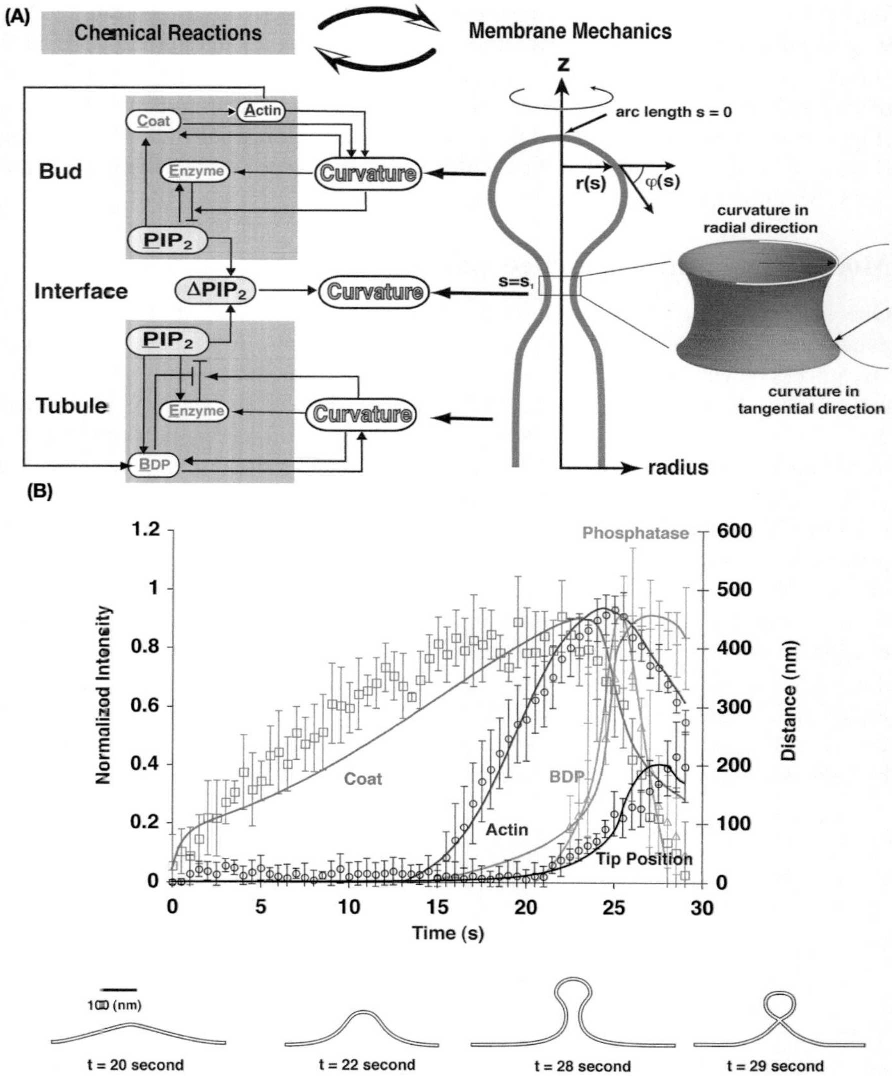

Figure 6. Mechanochemical feedback mechanism for endocytosis in budding yeast proposing curvature as an organizing principle for BAR-domain protein assembly and PIP2 hydrolysis (see Figure 2 in (Liu et al. 2009) for details). (B) Fitting of the results calculated from the model to experimental results—solid lines are model outcomes and discontinuous lines are experimental measurements (see Figure 3 in (Liu et al. 2009) for details). Figures reproduced from (Liu et al. 2009). BDP stands for BAR-domain proteins.

Liu et al. 2006; Liu et al. 2009; Ma and Berro 2021; Walani et al. 2015), a few of which will be discussed below. CME can be described as a series of steps in which the coat proteins assemble on the membrane to induce a spontaneous curvature (Figure 5A). As the effect of this curvature increases, either through increasing spontaneous curvature itself or due to an increase in coat area, the membrane bends to form a vesicle. Actin can exert forces to support vesicle internalization (Akamatsu et al. 2020; Serwas et al. 2022) and subsequently, the vesicle undergoes scission. Note that continuum modeling using the Helfrich framework cannot capture the scission process. Using the Helfrich energy given in Equation (1) in axisymmetric coordinates (Figure 5B), one can systematically explore the effect of a number of free parameters, including coat curvature, coat area, and membrane tension (prescribed as a boundary condition, Figure 5B). The details of the calculations are presented in (Hassinger et al. 2017). The formation of a bud depends on these different parameters, as shown in Figure 5C.

From this model of CME, we predicted that membrane tension can act as a barrier to the formation of vesicles during endocytosis (Hassinger et al. 2017). This was consistent with previous experimental observations (Boulant et al. 2011; Bucher et al. 2018). We further explained that such tension dependence occurs as a "snap-through instability" wherein a small amount of work done on the system (through coat area addition or decreasing tension) can "snap" the vesicle from a U-shape (open neck) to a Ω-shape (closed neck) configuration. A similar instability was described for CME in yeast where the membrane forms tubular invaginations (Walani et al. 2014) and in tube formation in general (Mahapatra and Rangamani 2023; Mahapatra et al. 2021a; Mahapatra et al. 2021b).

One of the key questions in CME is: What are the forces required for a vesicle to undergo scission? In order to incorporate the role of lipid domains and the resulting line tension, the Helfrich energy in Equation (1) was modified to incorporate the line tension (Liu et al. 2006). Using this framework, the authors showed that the line tension resulting from $PI(4,5)P_2$ domain formation possibly provided sufficient force to induce vesicle scission. This was a particularly important prediction in yeast endocytosis, where constricting proteins like dynamin are not involved in endocytosis. On the other hand, when constricting molecules like dynamin are involved in endocytosis, the force generated by these molecules can be represented by localized "collars" resulting in constriction (Vasan et al. 2020) (also see Figure 3B for tube constriction). Similarly, actin-mediated forces can be modeled using localized forces on the membrane for CME (Hassinger et al. 2017; Liu et al. 2006, 2009; Walani et al. 2014).

The mechanical models described thus far do not account for the kinetics of the 60-odd proteins involved in endocytosis. In Liu et al. (2009), the authors modeled a kinetic description of the key players involved in CME in yeast. The model focused on curvature as an organizing principle of the kinetic events such that a dynamic feedback loop was established between the membrane curvature and the kinetics of proteins. The model was constructed in a modular fashion to account for the role of lipids, coat proteins, the BAR-domain proteins that accumulate in the tube region, actin-mediated force generation, and enzymatic hydrolysis of PIP2 (Figure 6A).

First and foremost, this model captured the dynamics of the different endocytic proteins as observed in experiments (Figure 6B), providing an important validation of the model against experiments. We note that this is a non-trivial effort requiring optimizing multiple free parameters for a nonlinear model. Next, the model went on to propose the idea that positive feedback between the curvature sensing and curvature deforming activities of the BAR-domain proteins plays an important role in deforming the membrane. Additionally, a second positive feedback loop results in the dynamic evolution of interfacial forces. Regions of the membrane covered in the BAR domain proteins are protected from the action of enzymes that hydrolyze PIP2. Therefore, the regions that are not covered by BAR-domain proteins are susceptible to PI(4,5)P_2 hydrolysis, creating an interfacial force or a line tension. As noted before, line tension can promote the curvature at the interface (Liu et al. 2006) and curvature enhances the phosphatase activity. Experiments showed that synaptojanin, a phosphoinositide dephosphatase hydrolyzed PI(4,5)P_2 in a curvature-dependent manner (Chang-Ileto et al. 2011), suggesting geometric control of enzyme activity. The coupling between these two mechanochemical feedback loops results in successful endocytosis. Endocytosis in yeast has also been modeled using continuum approaches, where, in addition to the different factors discussed above, the role of turgor pressure was considered extensively (Dmitrieff and Nédélec 2015). Thus, modeling CME using continuum approaches has provided substantial insights into the different steps involved in membrane curvature generation.

Modeling Organelle Membranes

Beyond curvature generation for tubes and buds, one of the attractive features of the Helfrich model is that it can not only capture energy minimization when a force or a curvature-inducing mechanism is applied but also for minimal surfaces (Figure 7). Minimal surfaces are surfaces with zero mean curvature. Flat planes, catenoids (Figure 7A), and helicoids (Figure 7B) are all minimal surfaces. Flat planes have both mean and Gaussian curvatures equal to zero everywhere because both the principal curvatures are zero. Helicoids and catenoids have zero mean curvature and negative Gaussian curvature everywhere (Kreyszig 1959). That is, the two principal curvatures for helicoids and catenoids are equal and opposite in sign. Helicoids and catenoids are also isomorphic surfaces, which means geometrically, we can map a helicoid to a catenoid (Figure 7B, top row).

By inspection, it is obvious that when the spontaneous curvature is zero, flat planes minimize the membrane bending energy. Electron microscopy reconstructions of the endoplasmic reticulum membrane showed that they can be arranged as helicoids in certain locations (Terasaki et al. 2013). The most dramatic visualization of such a helicoid-like structure can be visualized in the spine apparatus, a specialized endoplasmic reticulum compartment in dendritic spines (Wu et al. 2017). As explained elegantly by Marshall (2013), helicoidal structures in endoplasmic reticulum membranes can minimize the free energy of bending. Similarly, mitochondrial membranes are known to form triply periodic minimal surfaces under certain conditions. Triply periodic minimal surfaces have a lattice-like structure, and their mean curvature is zero everywhere. Catenoids appear in the neck regions in both buds and tubes prior to constriction and scission (Chabanon and Rangamani 2018). The cristae junctions connecting the mitochondrial inner membrane and the cristae membrane also showed catenoid-like organization (Mendelsohn et al. 2021). The analysis of catenoids as energy-minimizing surfaces and their role in connecting the necks was done by Goldstein and Powers (Powers et al. 2002).

Why is this discussion of minimal surfaces important for membrane curvature and the Helfrich energy? The Helfrich energy can also be used to solve inverse problems, i.e., if we know the geometry of a surface, can we calculate non-zero spontaneous curvatures that these surfaces accommodate? We used the Helfrich model to answer these questions in a few separate studies (Bahmani et al. 2016; Chabanon and Rangamani 2018, 2019). First, using the Helfrich model with the flow but no spontaneous curvature, we showed that the flow field on minimal surfaces could be derived analytically and satisfies the properties of a Killing vector field (Bahmani et al. 2016). When solving for the spontaneous curvature field for catenoids, we showed that as the neck of the catenoid decreases, the sign of the spontaneous curvature changes from positive to negative (Chabanon and Rangamani 2018) (Figure 7A). The switch in the sign of the spontaneous curvature as a function of neck radius may suggest that more than one class of proteins may be required to minimize the energy of catenoids with smaller necks. This switch in the sign of the

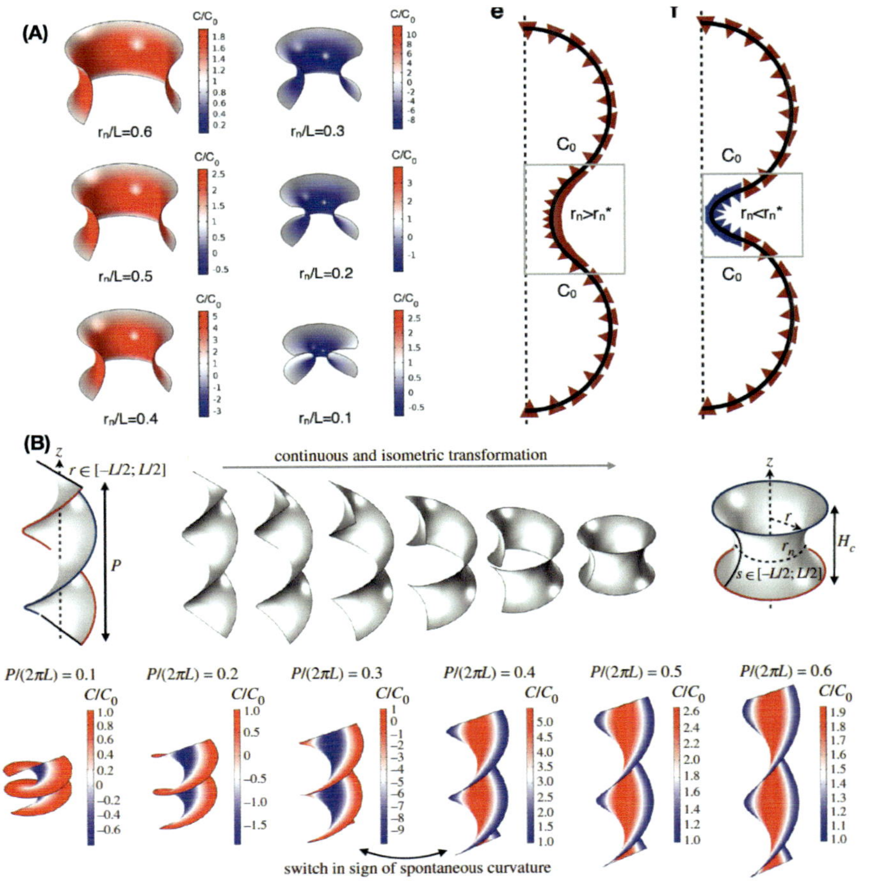

Figure 7. Solving inverse problems using the Helfrich model on minimal surfaces. (A) Distribution of spontaneous curvatures on catenoids of different neck radii shows a switch in the sign of spontaneous curvature as the neck radius decreases below a critical size. Reproduced from Figure 3 of (Chabanon and Rangamani 2018), see the paper for details. (B) A similar switch in the sign of the spontaneous curvature is observed in helicoids as their pitch is changed. Reproduced from Figure 2 of (Chabanon and Rangamani 2019). The top row shows the continuous and isometric transformations between catenoids and helicoids.

spontaneous curvature is associated with an energy barrier. The passage from one side of the energy barrier to the other will require additional mechanisms such as relaxation of the boundary conditions or external forces.

A similar switch in the sign of spontaneous curvature and as an associated energy barrier is seen in helicoids (Chabanon and Rangamani 2019). In this case, the energy barrier arises as the pitch of the helicoid changes (Figure 7B). These results, consistent with our previous results obtained for catenoids of varying neck radius (Chabanon and Rangamani 2018), suggest that at least two distinct curvature-inducing mechanisms are required to admit helicoids below the critical pitch as energy minimizers and solutions to the equation. Additionally, the existence of an energy barrier at a specific pitch can be interpreted as a possible biophysical mechanism preventing the collapse of a helicoidal membrane below a critical pitch and therefore providing a geometrical control of the membrane structure through the Gaussian curvature.

To go beyond the simplifications of the simple surfaces we described above, the Helfrich model has also been used to explore how the semicylindrical edges of organelle membranes, including the representation of endoplasmic reticulum, shaping proteins by including reticulons (Hu et al. 2008). Even though these are mathematically relatively simple models, they highlight how mechanical models can help give insight into structures of organelles.

Modeling Pearled Structures

Beyond the formation of spherical vesicles, biological membranes also form pearled structures. Pearled structures in membranes can occur due to polymer binding on the membrane (Campelo and Hernández-Machado 2007), the glycocalyx (Shurer et al. 2019), or in Golgi vesicle trafficking (Raote et al. 2020a; Raote et al. 2020b; Raote et al. 2018). Most recently, they have been observed in experiments where phase-separating proteins such as FUS-LC (N-terminal low-complexity domain of fused in sarcoma) can drive membrane curvature with a "pearls-on-a-string" morphology (Yuan et al. 2021) and protein chimeras can drive both inward and outward curvatures (Yuan et al. 2022). Pearling of lipid bilayers is reminiscent of the well-known Rayleigh-Plateau instability, where surface tension plays a critical role (Campelo et al. 2007).

The Helfrich model has also been used to model pearled structures under different conditions. In the case of protein-driven pearling, other factors beyond membrane tension may play a role. This morphology is typically associated with an area mismatch between the two membrane leaflets (Sanborn et al. 2013; Tsafrir et al. 2001). The observed set of shapes—pearls, unduloids, and cylinders—can be classified as Delaunay surfaces (Kreyszig 1959), which have a constant, non- zero mean curvature. More importantly, Delaunay surfaces, particularly unduloids and their variants, are known to minimize the Helfrich energy for membrane bending (Figure 8), (Naito et al. 1995). The radius and shape of the unduloids depend on a single dimensionless parameter $\alpha = \sqrt{(\lambda/(2\kappa C_0^2) + 1/4)}$, where λ is the membrane tension, κ is the bending modulus, and C_0 is the spontaneous curvature. When $\alpha = 0.75$, the membrane resembles a cylinder, and for $\alpha > 0.75$, the membrane

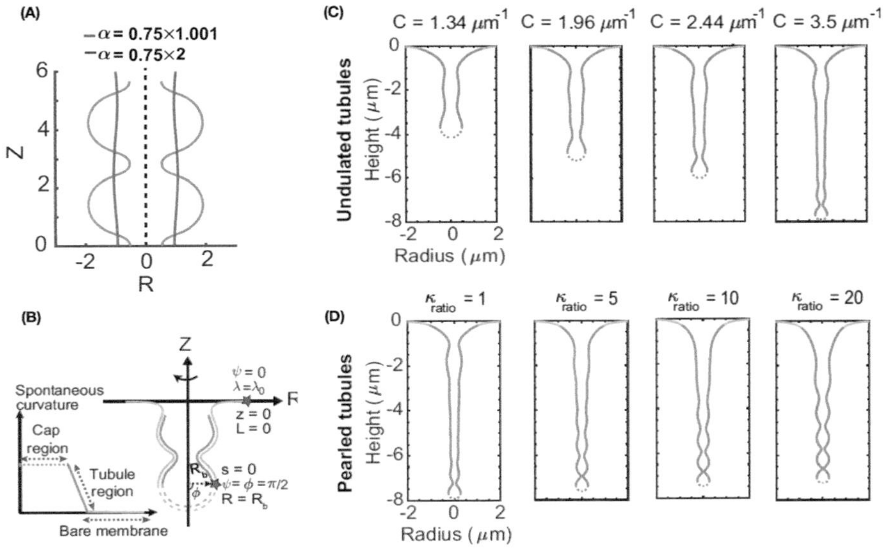

Figure 8. Modeling the formation of pearled tubes due to protein-induced spontaneous curvature. (A) Conditions in which unduloids form. (B) Axisymmetric geometry, boundary conditions, and different regions in the simulation. (C) Formation of undulated tubules as a function of spontaneous curvature and (D) formation of pearled tubules as a function of the ratio of bending moduli and spontaneous curvature. Figure modified with permission from (Yuan et al. 2021).

resembles an unduloid (Figure 8A). Numerical simulations using the axisymmetric geometry and boundary conditions shown in (Figure 8B) showed that increasing the spontaneous curvature in the protein-rich phase resulted in the formation of undulating tubules (Figure 8C) similar to those observed in experiments (Yuan et al. 2021). Our simulations also predicted that the tubule diameter would increase with the square root of the bending modulus in line with the results of our experiments (see (Yuan et al. 2021) for model comparison with the experiment). Finally, we investigated how heterogeneous membrane bending moduli (represented by κ_{ratio} in Figure 8D) might affect the membrane morphology. Here, κ_{ratio} is the ratio of the bending modulus of the membrane-protein composite to the bending modulus of the bare membrane. It represents the fact that the protein-coated region tends to have higher rigidity than the uncoated region (Yuan et al. 2021). When the heterogeneous modulus is considered, the pearls along the tube are better resolved (Figure 8D). A discussion of the parameters and the associated sensitivity can be found in (Rangamani et al. 2021).

Dynamic Modeling of Bulky Cargo Export: Integrating Multiple Physics

As a final example of integrating the different physics discussed throughout this chapter and to illustrate their applicability to a biological example, I discuss the transport of bulky cargo such as procollagen from the endoplasmic reticulum exit

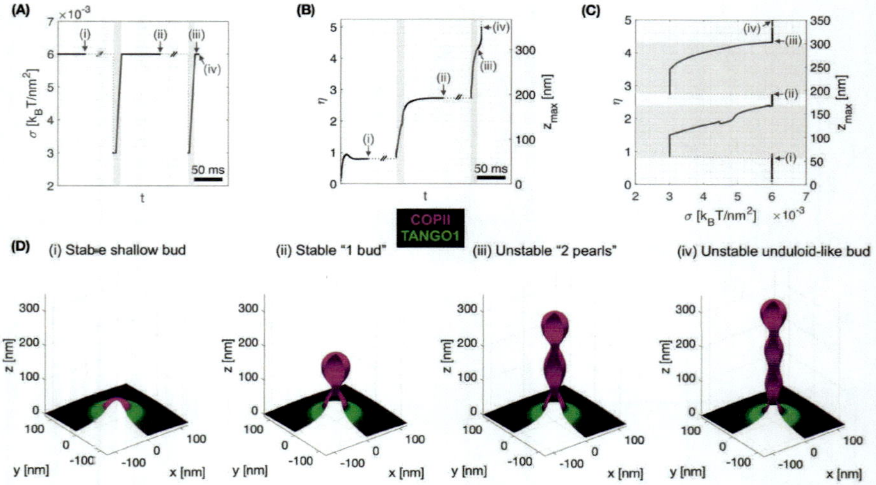

Figure 9. Simulations from a physical model of TANGO1 stabilized buds for collagen export. (A) The transient reduction in tension protocol used in the model. (B) Shape parameter, η, is defined effectively as the height over the radius of the bud and maximum height and *zmax* as a function of time in response to the stimulus applied in (A) and (C) as a function of tension. (D) Shape evolution of the transport intermediate from the simulations. Figure reproduced from Figure 3 of (Raote et al. 2020a; Raote et al. 2020b); see the original study for details of simulations and parameters.

sites (ERES). Briefly, collagen export from the endoplasmic reticulum requires the concerted action of TANGO1, a protein that was shown to form a ring in concert with the COPII molecules to corral the coat proteins (Raote et al. 2018). Furthermore, TANGO1 couples membrane flow with cargo transport. The membrane helix organization of TANGO1 was shown to result in the formation of a diffusion barrier in curved membranes. This barrier led to a reduction of membrane flow to minimize lipid mixing during the cargo transfer process (Raote et al. 2020a; Raote et al. 2020b).

With this detailed set of experimental observations, a physical model was developed to capture these different features (Raote et al. 2020a; Raote et al. 2020b). This model included the contributions of the membrane bending energy, the elastic energy of the TANGO1 ring, the entropic free energy of mixing/demixing of TANGO1 and COPII, the interface energy, and affinities for different proteins. Additionally, in-plane transport such as shear stress and protein diffusion were included on an incompressible membrane. Using this thermodynamically consistent model and numerical simulations in axisymmetry, the authors of this study showed that the model can capture the TANGO1 ring self-organization and how this ring can stabilize the transporters (Figure 9). Furthermore, the model predicted that the formation of a large transport intermediate is facilitated by a transient reduction in membrane tension (Figure 9A). The reduction in transient membrane tension is justified as a likely scenario in which new membrane may fuse to the endoplasmic reticulum from other membrane compartments. This addition of membrane thus allows the buds to grow and change their shape from shallow invaginations to pearled structures that would be able to promote the encapsulation of procollagen molecules (Figure 9B,C,D). Finally, a simple analytically tractable model reveals that this transition can be explained as a ratchet-like mechanism wherein the tension-mediated carrier growth can be thought of in terms of transitions between energetically stable configurations.

Future Directions

In this chapter, I have introduced the standard continuum mechanics approach for modeling membrane mechanics, discussed the different physical phenomena that can be included, and the computational approaches to solve these problems. This work has its roots in red blood cell shape, liquid crystal theory, and thin shell elasticity and is an exemplar of how a seemingly simple strain energy can describe a variety of biological phenomena. Future directions in this field will require multiscale efforts to incorporate lipid composition and molecular details (see Ingólfsson et al. 2014) into continuum models to capture the compositional features into the mechanical models. Incorporation of thermal fluctuations, bulk fluid interactions, and representations of the cortical cytoskeleton are all critical next steps to leverage the impact of the Helfrich model on exciting current biological observations in the cellular context. This chapter is also an ode to the Helfrich-Canham-Evans model of lipid bilayers and how it has empowered us to develop many biophysical models of cellular membranes.

Acknowledgments

The author would like to thank her many current and former group members, Dr. Haleh Alimohamadi, Dr. Julian Hassinger, Dr. Arijit Mahapatra, Dr. Ritvik Vasan, Dr. Sage Malingen, Dr. Chris Lee, Mr. Cuncheng Zhu, and Dr. Morgan Chabanon for many years of collaborative work on membrane mechanics. Special thanks to Dr. Arijit Mahapatra for his careful reading of the manuscript and his detailed feedback. The author would also like to thank her collaborators, Dr. David Drubin, Dr. David Saintillan, Dr. Jeanne Stachowiak, and Dr. David Steigmann, for many years of engaging discussions modeling membranes and bringing the models closer to experiments. Last but not least, the author would also like to thank the many eminent scholars who contributed to this field and apologize for not being able to cite all their works due to space limitations.

References

Agrawal, A. and D.J. Steigmann. 2011. A model for surface diffusion of trans-membrane proteins on lipid bilayers. Zamm-Z Angew. Math. Me. 62(3): 549–563.

Akamatsu, M., R. Vasan, D. Serwas et al. 2020, January. Principles of self-organization and load adaptation by the actin cytoskeleton during clathrin-mediated endocytosis. Elife, 9.

Alberts, B., A. Johnson, J. Lewis et al. 1985. Molecular biology of the cell. Garland Science.

Alimohamadi, H., M.K. Bell, S. Halpain et al. 2021, June. Mechanical principles governing the shapes of dendritic spines. Front. Physiol. 12: 657074.

Alimohamadi, H. and P. Rangamani. 2018. Modeling membrane curvature generation due to membrane–protein interactions. Biomolecules 8: 120–147.

Alimohamadi, H., A.S. Smith, R.B. Nowak et al. 2020, May. Non-uniform distribution of myosin-mediated forces governs red blood cell membrane curvature through tension modulation. PLoS Comput. Biol. 16(5): e1007890.

Antonny, B. 2011. Mechanisms of membrane curvature sensing. Annu. Rev. Biochem. 80: 101–123.

Arroyo, M. and A. DeSimone. 2009. Relaxation dynamics of fluid membranes. Phys. Rev. E. 79(3): 031915.

Arroyo, M., N. Walani, A. Torres-Sánchez et al. 2018. Onsager's variational principle in soft matter: introduction and application to the dynamics of adsorption of proteins onto fluid membranes. In The role of mechanics in the study of lipid bilayers (pp. 287–332). Springer.

Auddya, D., X. Zhang, R. Gulati et al. 2021, November. Biomembranes undergo complex, non-axisymmetric deformations governed by Kirchhoff–Love kinematics and revealed by a three-dimensional computational framework. Proceedings of the Royal Society A: Mathematical, Physical and Engineering Sciences 477(2255): 20210246.

Bahmani, F., J. Christenson and P. Rangamani. 2016. Analysis of lipid flow on minimal surfaces. Continuum Mech. Thermodyn. 28: 503–513.

Bassereau, et al. 2018. The 2018 biomembrane curvature and remodeling roadmap. J. Phys. D. Appl. Phys. 51: 343001.

Boulant, S., C. Kural, J.-C. Zeeh et al. 2011, August. Actin dynamics counteract membrane tension during clathrin-mediated endocytosis. Nat. Cell Biol. 13(9): 1124–1131.

Bucher, D., F. Frey, K.A. Sochacki et al. 2018, March. Clathrin-adaptor ratio and membrane tension regulate the flat-to-curved transition of the clathrin coat during endocytosis. Nat. Commun. 9(1): 1109.

Campelo, F. and A. Hernández-Machado. 2007. Model for curvature-driven pearling instability in membranes. Phys. Rev. Lett. 99: 088101.

Canham, P.B. 1970. The minimum energy of bending as a possible explanation of the biconcave shape of the human red blood cell. J. Theoret. Biol. 26: 61–81.

Canuto, C., M.Y. Hussaini, A. Quarteroni et al. 2006. Spectral methods. Springer.

Case, G.D., J.M. Vanderkooi and A. Scarpa. 1974. Physical properties of biological membranes determined by the fluorescence of the calcium ionophore a23187. Arch. Biochem. Biophys. 162(1): 174–185.

Chabanon, M. and P. Rangamani. 2018. Gaussian curvature directs the distribution of spontaneous curvature on bilayer membrane necks. Soft Matter 12: 2281–2294.

Chabanon, M. and P. Rangamani. 2019. Geometric coupling of helicoidal ramps and curvature-inducing proteins in organelle membranes. J. R. Soc. Interface. 16: 20190354.

Chang-Ileto, B., S.G. Frere, R.B. Chan et al. 2011. Synaptojanin 1-mediated PI(4,5)P2 hydrolysis is modulated by membrane curvature and facilitates membrane fission. Developmental Cell 20(2): 206–218.

Collings, P.J. 2002. Liquid crystals: nature's delicate phase of matter. Princeton University Press.

Derényi, I., F. Jülicher and J. Prost. 2002 Formation and interaction of membrane tubes. Phys. Rev. Lett. 88(23): 238101.

Derganc, J. and A. Copic. 2016. Membrane bending by protein crowding is affected by protein lateral confinement. Biochimica et Biophysica Acta (BBA)-Biomembranes 1858(6): 1152–1159.

Deserno, M. 2015. Fluid lipid membranes: From differential geometry to curvature stress. Chem. Phys. Lipids 185: 11–45.

Dmitrieff, S. and F. Nédélec. 2015, October. Membrane mechanics of endocytosis in cells with turgor. PLoS Comput. Biol. 11(10): e1004538.

Dobereiner, H.-G., E. Evans, M. Evans et al. 1997. Mapping vesicle shapes into the phase diagram: A comparison of experiment and theory. Physical Review E 55(4): 4458.

Douglas, P.M. and D.M. Cyr. 2010. Interplay between protein homeostasis networks in protein aggregation and proteotoxicity. Biopolymers: Orig. Res. Biomol. 93(3): 229–236.

Ericksen, J. 1979. Theory of Cosserat surfaces and its applications to shells, interfaces and cell membranes. In Proc. int. symp. on recent developments in the theory and application of generalized and oriented media (pp. 27–39).

Evans, E.A. and R. Skalak. 1980. Mechanics and thermodynamics of bio-membranes. CRC Press, Boca Raton (FL).

Evans, E.A. and R.M. Hochmuth. 1978. Mechanochemical properties of membranes. In Current topics in membranes and transport. Elsevier. 10: 1–64.

Farsad, K., N. Ringstad, K. Takei et al. 2001. Generation of high curvature membranes mediated by direct endophilin bilayer interactions. J. Cell biol. 155(2): 193–200.

Ford, Marijn G.J., Ian G. Mills, Brian J. Peter et al. 2002. Curvature of clathrin-coated pits driven by epsin. Nature 419(6905): 361.

Fournier, J.B. and P. Galatola. 1997. Tubular vesicles and effective fourth-order membrane elastic theories. Europhysics. Lett. 39(2): 225–230.

Gera, P. and D. Salac. 2017. Cahn–Hilliard on surfaces: A numerical study. Appl. Math. Letters 73: 56–61.

Givli, S., H. Giang and K. Bhattacharya. 2012. Stability of multicomponent biological membranes. SIAM Journal on Applied Mathematics 72(2): 489–511.

Gov, N. 2018. Guided by curvature: shaping cells by coupling curved membrane proteins and cytoskeletal forces. Philosophical Transactions of the Royal Society of London. Series B, Biological Sciences 373(1747): 20170115.

Gozdz, W.T. 2011. Shape transformation of lipid vesicle induced by diffusing macro-molecules. J. Chem. Phys. 134: 371–379.

Gruenberg, J. 2001. The endocytic pathway: A mosaic of domains. Nat. Rev. Mol. Cell Biol. 2: 721–730.

Guazzelli, E. and J.F. Morris. 2011. A physical introduction to suspension dynamics (Vol. 45). Cambridge University Press.

Harayama, T. and H. Riezman. 2018. Understanding the diversity of membrane lipid composition. Nat. Rev. Mol. Cell Biol. Cell Biol. 19: 281–296.

Hasimoto, H. 1959. On the periodic fundamental solutions of the stokes equations and their application to viscous flow past a cubic array of spheres. J. Fluid Mech. 5(2): 317–328.

Hassinger, J.E., G. Oster, D.G. Drubin et al. 2017. Design principles for robust vesiculation in clathrin-mediated endocytosis. Proc. Natl. Acad. Sci. U.S.A. 114(7): E1118–E1127.

Helfrich, W. 1973. Elastic properties of lipid bilayers: theory and possible experiments. Z Naturforsch C 28(11-12): 693–703.

Hendrickson, B., M. Shirani and D.J. Steigmann. 2020, May. Equilibrium theory for a lipid bilayer with a conforming cytoskeletal membrane. Mathematics and Mechanics of Complex Systems 8(1): 69–99.

Hirama, T., S.M. Lu, J.G. Kay et al. 2017, December. Membrane curvature induced by proximity of anionic phospholipids can initiate endocytosis. Nat. Commun. 8(1): 1–14.

Hochmuth, F., J.-Y. Shao, J. Dai et al. 1996. Deformation and flow of membrane into tethers extracted from neuronal growth cones. Biophys. J. 70(1): 358–369.

Hochmuth, R.M. and R.E. Waugh. 1987, January. Erythrocyte membrane elasticity and viscosity. Annu. Rev. Physiol. 49(1): 209–219.

Hu, J., Y. Shibata, C. Voss et al. 2008. Membrane proteins of the endoplasmic reticulum induce high-curvature tubules. Science 319(5867): 1247–1250.

Iglǐc, A., B. Babnik, U. Gimsa et al. 2005. On the role of membrane anisotropy in the beading transition of undulated tubular membrane structures. Journal of Physics A: Mathematical and General 38(40): 8527.

Ingólfsson, H.I., M.N. Melo, F.J. Van Eerden et al. 2014. Lipid organization of the plasma membrane. J. Am. Chem. Soc. 136(41): 14554–14559.

Jenkins, J.T. 1977. The equations of mechanical equilibrium of a model membrane. SIAM J. Appl. Math. 32(4): 755–764.

Jülicher, F. and R. Lipowsky. 1993. Domain-induced budding of vesicles. Phys. Rev. Lett. 70: 2964–2967.

Kahya, N., D. Scherfeld, K. Bacia et al. 2004. Lipid domain formation and dynamics in giant unilamellar vesicles explored by fluorescence correlation spectroscopy. J. Struct. Biol. 147: 77–89.

Kaksonen, M., Y. Sun and D.G. Drubin. 2003, November. A pathway for association of receptors, adaptors, and actin during endocytic internalization. Cell 115(4): 475–487.

Kaksonen, M., C.P. Toret and D.G. Drubin. 2005, October. A modular design for the clathrin- and Actin-Mediated endocytosis machinery. Cell 123(2): 305–320.

Klaus, C.J.S., K. Raghunathan, E. DiBenedetto et al. 2016. Analysis of diffusion in curved surfaces and its application to tubular membranes. Mol. Biol. Cell 27(24): 3937–3946.

Kralj-Iglǐc, V., H. Hagerstand, P. Veranic et al. 2005. Amphiphile-induced tubular budding of the bilayer membrane. Eur. Biophys. J. 34: 1066–1070.

Kreyszig, E. 1959. Differential geometry (352 p). Toronto: University of Toronto Press.

Kučerka, N., S. Tristram-Nagle and J.F. Nagle. 2006. Structure of fully hydrated fluid phase lipid bilayers with monounsaturated chains. The Journal of Membrane Biology 208: 193–202.

Laurendeau, N.M. 2005. Statistical thermodynamics: fundamentals and applications. Cambridge University Press.

Lee, C.T., J.G. Laughlin, N.A. de La Beaumelle et al. 2020, April. 3D mesh processing using GAMer 2 to enable reaction-diffusion simulations in realistic cellular geometries. PLoS Comput. Biol. 16(4): e1007756.

Leibler, S. and D. Andelman. 1987. Ordered and curved meso-structures in membranes and amphiphilic films. J. Phys. France. 48: 2013.

Lemière, J., Y. Ren and J. Berro. 2021, May. Rapid adaptation of endocytosis, exocytosis, and eisosomes after an acute increase in membrane tension in yeast cells. Elife, 10.

Lenz, O. and F. Schmid. 2007. Structure of symmetric and asymmetric "ripple" phases in lipid bilayers. Physical Review Letters 98(5): 058104.

Liese, S. and A. Carlson. 2021, June. Membrane shape remodeling by protein crowding. Biophys. J. 120(12): 2482–2489.

Lipowsky, R. 1991. The conformation of membranes. Nature 349: 475–481.

Lipowsky, R. 2013. Spontaneous tubulation of membranes and vesicles reveals membrane tension generated by spontaneous curvature. Faraday Discuss 161: 305–331.

Liu, J., M. Kaksonen, D.G. Drubin et al. 2006, July. Endocytic vesicle scission by lipid phase boundary forces. Proc. Natl. Acad. Sci. USA 103(27): 10277–10282.

Liu, J., Y. Sun, D.G. Drubin et al. 2009, September. The mechanochemistry of endocytosis. PLoS Biol. 7(9): e1000204.

Long, R., C.-Y. Hui, A. Jagota et al. 2012, July. Adhesion energy can regulate vesicle fusion and stabilize partially fused states. J. R. Soc. Interface 9(72): 1555–1567.

Lubensky, T. and F. MacKintosh. 1993. Theory of "ripple" phases of lipid bilayers. Physical review Letters 71(10): 1565.

Luke, J. C. 1982, April. A method for the calculation of vesicle shapes. SIAM J. Appl. Math. 42(2): 333–345.

Ma, R. and J. Berro. 2021. Endocytosis against high turgor pressure is made easier by partial protein coating and a freely rotating base. Biophys. J.

Mahapatra, A. 2022. Transport phenomena in lipid bilayers. University of California, San Diego.

Mahapatra, A. and P. Rangamani. 2023, May. Formation of protein-mediated bilayer tubes is governed by a snap through transition. Soft Matter.

Mahapatra, A., D. Saintillan and P. Rangamani. 2020, December. Transport phenomena in fluid films with curvature elasticity. J. Fluid Mech. 905(A8).

Mahapatra, A., D. Saintillan and P. Rangamani. 2021. Curvature-driven feedback on aggregation-diffusion of proteins in lipid bilayers. Soft Matter.

Mahapatra, A., C. Uysalel and P. Rangamani. 2021. The mechanics and thermodynamics of tubule formation in biological membranes. J. Memb. Biol. 254(3): 273–291.

Marshall, W.F. 2013, July. Differential geometry meets the cell. Cell 154(2): 265–266.

May, S. 2000. Protein-induced bilayer deformations: the lipid tilt degree of freedom. European Biophysics Journal 29: 17–28.

McIlhinney, J., N. Hooper, J.L. Gallop et al. 2005. Bar domains and membrane curvature: bringing your curves to the bar. In Biochem. Soci. Sympo. 72: 223–231.

McMahon, H.T. and J.L. Gallop. 2005, December. Membrane curvature and mechanisms of dynamic cell membrane remodelling. Nature 438(7068): 590–596.

Mendelsohn, R., G.C. Garcia, T.M. Bartol et al. 2021, October. Morphological principles of neuronal mitochondria. J. Comp. Neurol.

Miao, L., U. Seifert, M. Wortis et al. 1994. Budding transitions of fluid-bilayer vesicles: the effect of area-difference elasticity. Physical Review E 49(6): 5389.

Mukherjee, S. and E.R. Maxfield. 2000. Role of membrane organization and membrane domains in endocytic lipid trafficking. Traffic. 1: 203–211.

Mukhopadhyay, R., H.W.G. Lim and M. Wortis. 2002, April. Echinocyte shapes: bending, stretching, and shear determine spicule shape and spacing. Biophys. J. 82(4): 1756–1772.

Naghdi, P.M. 1972. Theory of shells and plates. In Handbunch der physik (Vol. C, pp. 425–640). *In*: Truesdell.

Nagle, J.F. and S. Tristram-Nagle. 2000. Structure of lipid bilayers. Biochimica et Biophysica Acta (BBA)-Reviews on Biomembranes 1469(3): 159–195.
Naito, H., M. Okuda and O.Y. Zhong-can. 1995, May. New solutions to the helfrich variation problem for the shapes of lipid bilayer vesicles: Beyond delaunay's surfaces. Phys. Rev. Lett. 74(21): 4345–4348.
Nitschke, I., A. Voigt and J. Wensch. 2012. A finite element approach to incompressible two-phase flow on manifolds. J. Fluid Mech. 708: 418.
Nixon-Abell, J., C.J. Obara, A.V. Weigel et al. 2016. Increased spatiotemporal resolution reveals highly dynamic dense tubular matrices in the peripheral ER. Science 354(6311): aaf3928.
Noguchi, H. and G. Gompper. 2004. Fluid vesicles with viscous membranes in shear flow. Phys. Rev. Lett. 93(25): 258102.
Ou-Yang, Z.-C., J.-X. Liu, Y.-Z. Xie et al. 1999. Geometric methods in the elastic theory of membranes in liquid crystal phases (Vol. 2). World Scientific.
Powers, T.R., G. Huber and R.E. Goldstein. 2002. Fluid-membrane tethers: minimal surfaces and elastic boundary layers. Physical Review E 65(4): 041901.
Rahimi, M. and M. Arroyo. 2012. Shape dynamics, lipid hydrodynamics, and complex vis-coelasdticity of bilayer membranes. Phys. Rev. E 86: 011932.
Ramaswamy, S., J. Toner and J. Prost. 2005. Nonequilibrium fluctuations, travelling waves, and instabilities in active membranes. Phys. Rev. Lett. 84: 3494–3497.
Rand, R. 1981. Interacting phospholipid bilayers: measured forces and induced structural changes. Ann. Rev. Biophys. Bioeng. 10(1): 277–314.
Rangamani, Padmini, Ashutosh Agrawal, Kranthi K. Mandadapu et al. 2013. Interaction between surface shape and intra-surface viscous flow on lipid membranes. Biomechanics and Modeling in Mechanobiology 12(4): 833–45.
Rangamani, P., A. Behzadan and M. Holst. 2021, March. Local sensitivity analysis of the "membrane shape equation" derived from the helfrich energy. Math. Mech. Solids 26(3): 356–385.
Rangamani, P., A. Benjamini, A. Agrawal et al. 2014, August. Small scale membrane mechanics. Biomech. Model. Mechanobiol. 13(4): 697–711.
Rangamani, P., K.K. Mandadap and G. Oster. 2014. Protein-induced membrane curvature alters local membrane tension. Biophysical Journal 107(3): 751–762.
Rangamani, P. and D. Steigmann. 2014. Variable tilt on lipid membranes. Proceedings of the Royal Society A: Mathematical, Physical and Engineering Sciences 470(2172): 20140463.
Raote, I., M. Chabanon, N. Walani et al. 2020, November. A physical mechanism of TANGO1-mediated bulky cargo export. Elife, 9.
Raote, I., A.M. Ernst, F. Campelo et al. 2020, May. TANGO1 membrane helices create a lipid diffusion barrier at curved membranes. Elife, 9.
Raote, I., M. Ortega-Bellido, A.J. Santos et al. 2018, March. TANGO1 builds a machine for collagen export by recruiting and spatially organizing COPII, tethers and membranes. Elife, 7.
Reynwar, Benedict J., Gregoria Illya, Vagelis A. Harmandaris et al. 2007. Aggregation and vesiculation of membrane proteins by curvature-mediated interactions. Nature 447(7143): 461–64.
Sackmann, E., H.P. Duwe and H. Engelhardt. 1986. Membrane bending elasticity and its role for shape fluctuations and shape transformations of cells and vesicles. Faraday Discuss. Chem. Soc. 81: 281–290.
Saffman, P. and M. Delbrück. 1975. Brownian motion in biological membranes. Proc. Nat. Acad. of Sci. 72(8): 3111–3113.
Safran, S. 2018. Statistical thermodynamics of surfaces, interfaces, and membranes. CRC Press.
Sahu, A., Y.A.D. Omar, R.A. Sauer et al. 2020. Arbitrary Lagrangian–Eulerian finite element method for curved and deforming surfaces: I. General theory and application to fluid interfaces. J. Comput. Phys. 407: 109253.
Sahu, A., R.A. Sauer and K.K. Mandadapu. 2017. Irreversible thermodynamics of curved lipid membranes. Phys. Rev. E. 96: 042409.
Sanborn, J., K. Oglecka, R.S. Kraut et al. 2013. Transient pearling and vesiculation of membrane tubes under osmotic gradients. Faraday Discuss. 161: 167–176.
Sauer, R.A., T.X. Duong, K.K. Mandadapu et al. 2017. A stabilized finite element formulation for liquid shells and its application to lipid bilayers. J. Comput. Phys. 330: 436–466.
Scriven, L.E. 1960. Dynamics of a fluid surface. Chem. Eng. Sci. 12: 98–108.

Seifert, U. 1993. Curvature-induced lateral phase segregation in two-component vesicles. Phys. Rev. Lett. 70: 1335.

Serwas, D., M. Akamatsu, A. Moayed et al. 2022, May. Mechanistic insights into actin force generation during vesicle formation from cryo-electron tomography. Dev. Cell 57(9): 1132–1145.e5.

Shi, Z., Z.T. Graber, T. Baumgart et al. 2018, December. Cell membranes resist flow. Cell 175(7): 1769–1779.e13.

Shurer, C.R., J.C.-H. Kuo, L.M. Roberts et al. 2019, April. Physical principles of membrane shape regulation by the glycocalyx. Cell.

Sokolnikoff, I.S. 1951. Tensor analysis: Theory and applications. Wiley.

Stachowiak, J.C., C.C. Hayden and D.Y. Sasaki. 2010, April. Steric confinement of proteins on lipid membranes can drive curvature and tubulation. Proc. Natl. Acad. Sci. USA 107(17): 7781–7786.

Steigmann, D.J. 1999. Fluid films with curvature elasticity. Arch. Rat. Mech. 150: 257–272.

Terasaki, M., T. Shemesh, N. Kasthuri et al. 2013. Stacked endoplasmic reticulum sheets are connected by helicoidal membrane motifs. Cell 154(2): 285–296.

Torres-Sanchez, A., D. Millan and M. Arroyo. 2019. Modelling fluid deformable surfaces with an emphasis on biological interfaces. J. Fluid Mech. 872: 218–271.

Tozzi, C., N. Walani and M. Arroyo. 2019. Out-of-equilibrium mechanochemistry and self-organization of fluid membranes interacting with curved proteins. New J. Phys. 21: 093004.

Tran-Son-Tay, R., S.P. Sutera and P.R. Rao. 1984. Determination of red blood cell membrane viscosity from rheoscopic observations of tank-treading motion. Biophys. J. 46: 1335–1338.

Tsafrir, I., D. Sagi, T. Arzi et al. 2001. Pearling instabilities of membrane tubes with anchored polymers. Phys. Rev. Lett. 86(6): 1138–41.

Vasan, R., S. Rudraraju, M. Akamatsu et al. 2020. A mechanical model reveals that non-axisymmetric buckling lowers the energy barrier associated with membrane neck constriction. Soft Matter 16: 784–797.

Veksler, A. and N.S. Gov. 2007. Phase transitions of the coupled membrane-cytoskeleton modify cellular shape. Biophys. J. 93: 3798–3810.

Walani, N., J. Torres and A. Agrawal. 2014. Anisotropic spontaneous curvatures in lipid membranes. Phys. Rev. E 89(6): 062715.

Walani, N., J. Torres and A. Agrawal. 2015. Endocytic proteins drive vesicle growth via instability in high membrane tension environment. Proceedings of the National Academy of Sciences 112(12): E1423–E1432.

Weber, C., T. Michaels and L. Mahadevan. 2019. Spatial control of irreversible protein aggregation. Elife 8: e42315.

White, S.H. and M.C. Wiener. 2017. Determination of the structure of fluid lipid bilayer membranes. In Permeability and stability of lipid bilayers (pp. 1–19). CRC Press.

Wu, Y., C. Whiteus, C.S. Xu et al. 2017, June. Contacts between the endoplasmic reticulum and other membranes in neurons. Proceedings of the National Academy of Sciences 114(24): E4859–E4867.

Yuan, F., H. Alimohamadi, B. Bakka et al. 2021, March. Membrane bending by protein phase separation. Proc. Natl. Acad. Sci. USA 118(11).

Yuan, F., J. Houser, A. Sangani et al. 2022, December. The ins and outs of membrane bending by intrinsically disordered proteins.

Zhang, R., S. Tristram-Nagle, W. Sun et al. 1996. Small-angle x-ray scattering from lipid bilayers is well described by modified caillé theory but not by paracrystalline theory. Biophys. J. 70(1): 349–357.

Zhang, Y., X. Chen, C. Gueydan et al. 2018. Plasma membrane changes during programmed cell deaths. Cell Res. 28: 9–21.

Zhu, C., C.T. Lee and P. Rangamani. 2022, September. Mem3DG: Modeling membrane mechanochemical dynamics in 3D using discrete differential geometry. Biophysical Reports 2(3): 100062.

Zimmerberg, J. and M.M. Kozlov. 2006. How proteins produce cellular membrane curvature. Nat. Rev. Mol. Cell Biol. 7: 9–19.

6

Lipid Membrane Remodeling by (Bio)Polymers and Nanoparticles

Mechanistic Insights From Multi-Scale Simulations

Sanjoy Paul,[1] *Sayantan Mondal,*[1] *Zeke A. Piskulich,*[1]
Taraknath Mandal[2] and *Qiang Cui*[3,*]

Introduction

Lipid membranes constitute the boundaries between the cell and the environment, and also between the different compartments of the cell. For many cellular events, the key step involves remodeling the membrane (Jarsch et al. 2016). Prominent examples include membrane fusion during synaptic transmission and viral infection, and membrane budding for the generation of cellular vesicles and virus particles. For biomedical applications such as drug delivery, membrane remodeling for the internalization of the delivery vehicle is also a crucial consideration (Mitchell et al. 2021). Finally, for synthetic biology, generation of internal compartments in a controlled fashion remains a major challenge (Rideau et al. 2018). Therefore, it is evident that understanding the molecular mechanisms that govern the driving force and kinetics of membrane remodeling has broad implications.

Thanks to decades of experimental (Jarsch et al. 2016), theoretical (Kozlov and Taraska 2023) and computational (Bassereau et al. 2018) studies, many mechanisms for membrane remodeling have been established and understood at least in an outline form. Nevertheless, new mechanisms continue to emerge, and one recent fascinating addition concerns the interaction between protein condensates and lipid membranes (Yuan et al. 2021). Considering the myriad sets of possible direct and

[1] Department of Chemistry, Boston University, 590 Commonwealth Avenue, Boston, MA 02215.
[2] Department of Physics, Indian Institute of Technology – Kanpur, Kanpur 208016, India.
[3] Departments of Chemistry, Physics and Biomedical Engineering, Boston University, 590 Commonwealth Avenue, Boston, MA 02215.
* Corresponding author: qiangcui@bu.edu

indirect contributions, the relative importance of different potential contributions to a specific problem is often not clear, calling for an in-depth analysis of the process at the molecular level. Gaining a molecular level of understanding is essential to our ability to rationally modulate the process of membrane remodeling for biomedical and biotechnological purposes.

Along this line, multi-scale computational studies can make major contributions to our molecular understanding of these processes (Bassereau et al. 2018). The requirement of "multi-scale" membrane remodeling is apparent, considering the broad length and time scales involved. Indeed, large-scale membrane remodeling often requires the action of a large number of proteins occurring at least millisecond time scales, which are out of the reach of a fully atomistic description. On the other extreme, continuum mechanics models have played a major role in establishing the fundamental principles behind many membrane remodeling processes, such as the stability of the fusion intermediates (Chernomordik and Kozlov 2008) and the importance of shallow membrane insertion (Campelo et al. 2008; Kozlov and Taraska 2023). Unfortunately, it is not straightforward to include molecular level of details at the continuum level, although some progress has recently been made in that respect (Zhu et al. 2016). Therefore, systematic and effective integration of computational models at different scales (Bassereau et al. 2018; Saunders and Voth 2013) remains a major challenge to the field.

In this book chapter, we first briefly summarize the three types of computational approaches commonly used in the analysis of membrane remodeling: atomistic simulations, coarse-grained models and continuum mechanics analysis. Then, we discuss several applications from our group to illustrate the values and limitations of the various approaches and, in some cases, how models at different resolutions (scales) can start to be connected to lead to new mechanistic insights. Finally, we conclude by summarizing a few major observations and highlighting several remaining technical and mechanistic questions.

General Computational Models

In this section, we briefly review the computational approaches at different length scales and comment on their current applicability.

Atomistic Simulations

The most detailed level of computational investigation involves atomistic molecular dynamics simulations, which capture the complex interactions between protein motifs, lipid molecules and interfacial water/ions that are key to determining membrane morphology. Thanks to recent developments of hardware and software, the size and complexity of biological systems that can be meaningfully studied with atomistic models have increased tremendously in recent years (Corradi et al. 2019; Marrink et al. 2019). It is rather "routine" to conduct atomistic simulations for systems that contain ~ millions of atoms on the time scale of hundreds of nanoseconds. For the analysis of membrane remodeling, however, such system sizes are still rather "modest" considering the cooperative interactions of many proteins required for large-scale membrane remodeling. Therefore, atomistic simulations are usually most effective at probing how specific interactions between a few proteins and the membrane develop local membrane deformations. It is also important to keep in mind that the periodic boundary conditions used in most atomistic simulations tend to suppress membrane deformation, thus, the finite size effects should be carefully investigated.

Along this line, a useful computational protocol for probing local membrane curvature development is the "membrane ribbon" approach (Mandal et al. 2020; Wu and Schulten 2014) (see Figures 2 and 3), in which the periodicity is disrupted along one lateral direction. It is also possible to conduct the "membrane bicelle" simulation (Bhaskara et al. 2019; Paul et al. 2023) in which periodicity is disrupted in both lateral directions, although to avoid membrane edge effects the required system size for such simulations can be very large. Another relevant technical question for the ribbon simulation concerns the time scale of relaxation, which can be slow due to inter-leaflet frictions (Anthony et al. 2022); thus, it is essential to conduct independent simulations to examine the reproducibility of the results carefully.

As an alternative to explicitly observing local membrane curvature development in the ribbon protocol, one could instead compute the lateral pressure profile, the first moment of which is related to the local spontaneous curvature (Sodt and Pastor 2014; Sodt et al. 2016). This calculation is often computationally expensive as a large number of snapshots (including both atomic positions and velocities) are required for smooth numerical results, especially in the presence of the peripheral protein.

Finally, it is worth noting that the adequate equilibration of atomistic simulations for membrane curvature generation can be challenging. For example, many proteins induce local membrane curvature through the relatively shallow insertion of amphiphatic motifs into the bilayer (Kozlov and Taraska 2023). Equilibrating the proper insertion depth, orientation and rearrangements of local lipid distributions can take time scales exceeding hundreds of nanoseconds. One possible solution is to use enhanced sampling techniques (Valsson et al. 2016); this require the identification of relevant collective variables (Mori et al. 2016), which may not be straightforward for complex protein/membrane interactions. An effective alternative is to use the Highly Mobile Membrane Mimetics (HMMM) model (Ohkubo et al. 2012), in which a majority of the lipid tails are replaced by an organic solvent (e.g., dichloro ethane) to enhance lipid diffusion and (de)mixing; the area per lipid is also slightly expanded to facilitate the equilibration of protein insertion. Once equilibrated, structures from the HMMM simulations can be used to reconstruct the full lipid models to analyze membrane curvature generation.

Coarse-grained Models

In order to overcome the limitations of the length (tens of nanometers) and time scales (10^{2-3} nanoseconds) of typical atomistic membrane simulations, coarse-grained simulations have become popular for the analysis of membrane-mediated processes, including membrane remodeling by proteins and nanomaterials. The most celebrated model in this context is the MARTINI model (Marrink et al. 2023), which reduces the resolution from an atomistic model based approximately on a 4 to 1 mapping of non-hydrogen atoms to coarse-grained beads. Such reduction in the number of particles also allows a substantially larger integration time steps (~ 10–20 fs), which together leads to an effective speedup of more than two orders of magnitude than atomistic simulations. Therefore, it is quite feasible to conduct MARTINI simulations for systems of ~ 10^2 nanometers at multiple to tens of microseconds. For even larger systems at longer time scales, it is necessary to eliminate the explicit treatment of solvent, which usually represents the dominant component of biomolecular simulations. There are several such "solvent-free" coarse-grained models for lipids and simple polymers (Arnarez et al. 2015; Cooke and Deserno 2005); models at a similar level of coarse-graining for proteins/peptides have emerged as well (Dignon et al. 2018; Ruff et al. 2015), although balancing highly structured and disordered regions have only recently received careful treatment (Latham and Zhang 2022).

Due to the approximate nature of coarse-grained models, it is important to calibrate them to suit the specific problem of interest. For example, one major issue of the popular MARTINI model concerns the treatment of electrostatics, which play major roles in membrane curvature generation since many relevant proteins are recruited to the membrane surface through electrostatic interactions with anionic lipids. Several variants of the standard MARTINI model have been developed over the years (Marrink et al. 2023), and they have shown different levels of success in the description of membrane curvature sensing and generation. For example, the BMW-MARTINI model, which explicitly considers electrostatic moments of water clusters, was shown to provide a more faithful description of interfacial electrostatics and therefore protein binding interface to the membrane (Wu et al. 2011). However, the model predicts the incorrect sign for the membrane curvature induced by the binding of the ESCRT-III complex due to a combination of over-prediction of protein insertion depth and inaccurate spontaneous curvature of anionic lipids (see below) (Mandal et al. 2021). Similarly, different binding behaviors of functionalized nanoparticles to a lipid bilayer were observed with three variants of the MARTINI model (Das et al. 2019). Therefore, calibration using relevant experimental or atomistic simulation data is critically important for coarse-grained simulations of membrane remodeling processes.

Continuum Models

Continuum mechanics models, such as the celebrated Helfrich model (Boal 2002), have been used extensively for the analysis of membrane remodeling by biomolecules (Kozlov and Taraska 2023; Weikl 2018) and nanomaterials (Bahrami et al. 2016). At the most basic level, the membrane morphology is dictated by the competition between membrane-free energy and protein/membrane (or nanomaterials/membrane) adhesion. In the simplest model, the membrane free energy consists of bending and stretching contributions, which implicate membrane curvature and membrane tension, respectively. More sophisticated models that include molecular details (e.g., lipid tilt contributions) have also been developed (Hamm and Kozlov 2000). There are several studies that aim to directly compare the prediction of particle-based models and continuum models for membrane deformation induced by either peripheral or transmembrane proteins. In many cases, the continuum model was surprisingly reliable, even at the length scale approaching that of a single lipid molecule; for example, atomistic, coarse-grained and continuum models gave consistent predictions for the shape of the membrane fusion pore (Yoo et al. 2013a; Yoo et al. 2013b) as well as the membrane shape induced by hydrophobic mismatch associated with a transmembrane protein (Sodt et al. 2017; Yoo and Cui 2013).

The applicability of the continuum mechanics model to membrane remodeling processes can be enhanced by a mesh representation of the membrane (Hansen et al. 2017; Pezeshkian et al. 2020; Zhu et al. 2022), which can be used to represent complex morphologies. While progress has been made to represent proteins at the continuum level, it is more common to represent the proteins "implicitly" by adjusting the local properties of the membrane mesh, such as local spontaneous curvature and bending moduli (Ayton and Voth 2010). In principle, these local properties can be obtained from atomistic simulations, representing one way that atomistic and continuum models can be integrated in a "serial multi-scale" fashion(Saunders and Voth 2013). Once parameterized, the continuum model can be used to describe complex and large-scale membrane remodeling for systems that reach the micron level, such as giant unilamellar vesicles (GUVs) and compare more explicitly with experimental (e.g., electron microscopy) observations (Simunovic et al. 2015).

Case Studies and Discussions

In this section, we discuss recent studies from our group to illustrate several fundamental mechanisms for membrane remodeling by proteins and nanomaterials. Through these examples, we hope to illustrate the value of using models of different resolutions, as well as the importance of calibrating computational models (especially coarse-grained ones) to avoid potential artifacts.

Membrane Curvature Generation by Shallow Insertion of Protein Motifs: ESCRT-III and Sar1

One well-established mechanism for membrane curvature generation is the insertion of amphiphatic motifs into the bilayer (Campelo et al. 2008; Kozlov and Taraska 2023). The amphipathic nature of the motif ensures that the insertion depth is relatively shallow, reaching the boundary between the polar and hydrophobic segments of lipids. Such shallow insertion is essential to the significant perturbation of local lipid packing and positive curvature generation; a deep insertion tends to lead to a smaller degree of perturbation or even negative curvature generation. However, it remains unclear whether insertion depth alone is sufficient for predicting the degree of curvature generation. Moreover, as discussed above, membrane remodeling usually involves a large number of proteins; how local membrane bending works collectively to generate different larger scale deformations remains to be explored more systematically.

In this section, we touch upon these issues using two systems, Sar1 and ESCRT- III, analyzed in our recent work (Mandal et al. 2020, 2021; Paul et al. 2023). Sar1 is a small GTPase involved in the coat protein complex II (COPII)-mediated protein transport pathway, and its GTP-bound state is believed to initiate the local positive membrane curvature at subdomains on the endoplasmic reticulum (ER) (Hanna et al. 2016). ESCRT-III is part of the protein machinery that carries out the membrane scission reactions required for many cellular processes (Vietri et al. 2020), and its filamental form is proposed to be involved in initiating local membrane curvature such that the spiral ESCRT-III structure lies in the inside of the membrane invagination (Pfitzner et al. 2021). Therefore, Sar1 and ESCRT-III appear to represent two systems that lead to membrane curvature generation in different directions or with distinct topologies. Since both Sar1 and ESCRT-III feature an amino-terminal amphipathic helix, the mechanism that leads to apparently different signs of membrane curvature at the larger scale remains unclear.

Local Effects: Insertion Depth vs. Volume and the Role of Lipid Spontaneous Curvature

We first examine the effect of Sar1 or ESCRT-III binding on the local membrane morphology. As shown in Figure 1, both Sar1 and ESCRT-III feature an amphiphathic helix at the amino terminus, which hints at similar behaviors at the local scale.

In the case of Sar1, the interesting feature is that the structural orientation of the amino-terminal helix depends on the nucleotide-binding state. In the GTP-bound state, the helix of 23 residues was not resolved in the crystal structure and assumed to be fully detached from the rest of the protein and available to interact with the membrane; in the GDP-bound state, part of the helix (residues 13 to 23) was observed in the corresponding crystal structure to retract into a surface pocket, leaving only the first 12 residues to engage with the membrane. This difference is expected to be relevant to the experimental observation (Hanna et al. 2016) that Sar1 in the GTP state can generate significant membrane curvature, while Sar1 in the GDP state binds more weakly to the membrane without significant curvature generation. However, since much of the amino-terminal helix is not resolved in the crystal structures of Sar1, the dependence of the membrane remodeling activity on the nucleotide state needs to be better understood.

Structural differences between GDP and GTP-bound Sar1s mainly appear at the amino-terminal region and thus demand a chemically detailed model to capture their dynamical dissimilarities in the presence of the membrane. Accordingly, we have conducted extensive atomistic simulations of Sar1 in different nucleotide states (Paul et al. 2023); a comparison was also made for Sar1s from different organisms

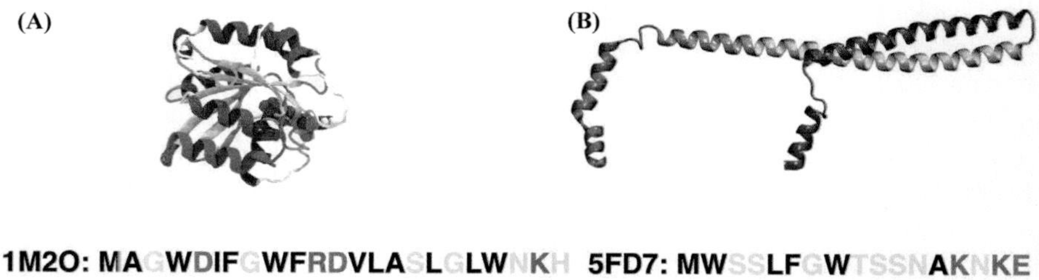

Figure 1. Structures for (A) Sar1 and (B) ESCRT-III based on the corresponding crystal structures (PDB codes: 1M2O and 5FD7); missing motifs, especially the amino-terminal helices, are modeled using Alpha fold and chimera, respectively (Mandal et al. 2020; Paul et al. 2023). Below each structure, the sequence of the amino-terminal amphipathic helix is shown.

and in different oligomeric states (monomer vs. dimer). As discussed above, the simulations started with equilibration of Sar1 using the Highly Mobile Membrane Mimetic (HMMM) model, which was followed by full-atomic-scale lipid-bilayer simulations for probing the insertion depth; finally, membrane ribbon simulations were used to probe the degree of local membrane curvature generation.

As shown in Figure 2A-D, due to its detached nature, the amino-terminal helix in the GTP-bound Sar1 is fully inserted into the bilayer. By contrast, only the first half of the amino-terminal helix is in the bilayer in the GDP-bound state. However, in terms of the depth of the inserted motif, the two states do not differ substantially (considering that both nucleotide states feature some level of variations

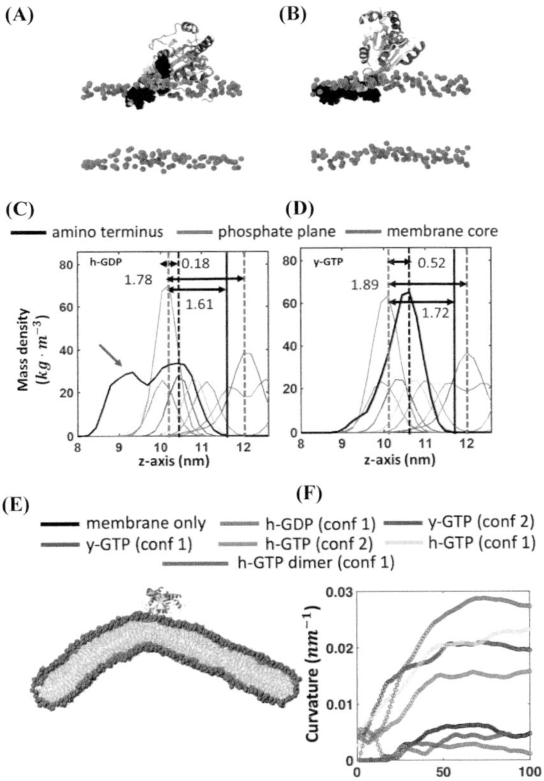

Figure 2. Key results for Sar1 and its membrane bending activity from atomistic simulations. (A-B) Snapshots of Sar1 in GDP- and GTP-bound states absorbed onto the fully atomistic membrane after 160 ns of simulation. The amino-terminal residues are shown in the van der Waals representation. Only phosphate atoms of the membrane are shown for clarity. (C-D) Mass density profile of various atoms of the DOPC lipid molecules and the amino-terminal region of Sar1 in different nucleotide states. The phosphate plane is considered as the membrane periphery and is highlighted by a light dotted line around $z \sim 10$. The dark dotted line represents the peak position of the amino-terminal anchor, and the light dotted line around $z \sim 12$ indicates the core of the membrane. The black solid line marks the region inside the membrane where the mass density of the amino-terminal anchor decays to 0. (E) Membrane ribbon simulation indicated significant membrane bending induced by the binding of a dimeric Sar1 in the GTP-bound state while no bending in the presence of the GDP-bound monomeric state. (F) Evolution of the magnitude of curvature with respect to the timescale of membrane ribbon simulation (in the unit of ns).

in independent simulations), with the deepest part reaching ~ 1.6–1.7 nm below the phosphate plane. The insertion depth is also similar to the dimeric form of Sar1 in the GTP state (Paul et al. 2023).

The membrane ribbon simulations, however, showed that the nucleotide and oligomeric states impacted the local membrane curvature significantly (Figure 2E-F). The dimeric form of the GTP-bound state led to a local mean curvature on the order of 0.03 nm^{-1}, while the monomeric GTP-bound Sar1 simulations led to smaller mean curvatures in the range of 0.015 to 0.02 nm^{-1}; the GDP-bound form did not induce any significant curvature. Therefore, the insertion depth alone was unable to explain the degree of local membrane curvature generation. Instead, we observed that for six independent simulations, the generated membrane curvature better correlated with the volume of protein motifs inserted into the bilayer. In other words, the orientation of the inserted protein motif, which was largely horizontal in the GTP-bound Sar1 but substantially tilted in the GDP-bound state (Figure 2A-B), also influenced the degree of membrane curvature generation. With a higher number of atoms inserted into a compact area of the bilayer, the dimeric Sar1-GTP form was able to generate a highly bent membrane with a radius of curvature of ~ 30 nm; this result was consistent with the experimental observation that Sar1s form a lattice assemblies on constricted membrane surfaces where the smallest repeating unit was a dimer. The observed curvatures in the simulations were also consistent with experimental studies that reported the formation of COPII-coated carriers of 25 to 50 nm radius (Hariri et al. 2014). The curvature values are comparable with that observed in the case of proteins from the Bin/Amphyphysin/Rvs (BAR) family (Arkhipov et al. 2009; Blood and Voth 2006; Yu and Schulten 2013). However, it is important to note that BAR domains produce membrane curvature through the scaffolding mechanism, where membrane sculpting can be continued even in the absence of the amphipathic segments (Yu and Schulten 2013). On the other hand, Sar1 in the GDP-bound form cannot remodel the membrane due to insufficient exposure to the amphipathic amino terminus.

With ESCRT-III, atomistic simulations yielded similar membrane binding and bending activities. With a trimer of ESCRT-III units, membrane ribbon simulations revealed a significant and positive local curvature with a radius of also ~ 30 nm, induced by the amino-terminal amphiphatic helix of 17 residues (Figure 3D). It is worth noting that the degree of membrane curvature generation for the same ESCRT-III trimer at the coarse-grained level was sensitive to the details of the MARTINI variants used (Mandal et al. 2021) (Figure 3A–C). With the standard MARTINI2.2 model, the size of the membrane-binding interface was underestimated due to the issue with electrostatics, as discussed above. Nevertheless, the model correctly predicted that ESCRT-III senses and generates a positive mean membrane curvature. With the POL-MARTINI model, the binding to the membrane is further reduced, leading to essentially no curvature generation or sensing activities. With the BMW-MARTINI model, the membrane-binding interface was well predicted, yet the model incorrectly predicted negative mean curvature generation and sensing activities.

To better understand the somewhat surprising observation for BMW-MARTINI, we constructed several related systems and analyzed their properties (Mandal et al. 2021). The results suggested that the incorrect membrane curvature predicted by BMW-MARTINI was due to the overestimated insertion depth of the amphipathic

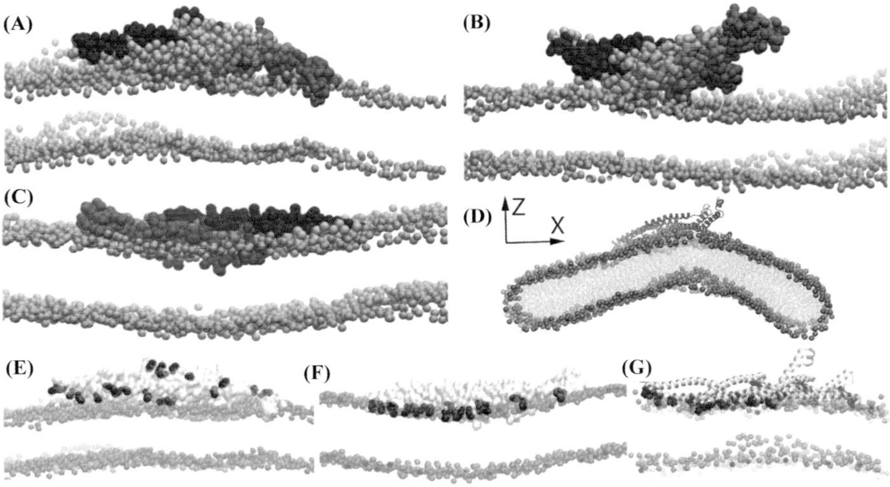

Figure 3. Equilibrated structures of the protein-membrane interface sampled by (A) regular MARTINI, (B) POL-MARTINI, (C) BMW-MARTINI, and (D) atomistic membrane ribbon models. Positively charged lysine residues located in the α_2 and α_3 helices are highlighted for the structures obtained from (E) regular MARTINI, (F) BMW-MARTINI, and (G) atomistic simulations.

helix and the incorrect sign for the spontaneous curvature of anionic lipids. On the one hand, this analysis highlighted the importance of calibrating coarse-grained models for membrane remodeling simulations and pointed out ways that such models can be improved. On the other hand, the comparison of closely related coarse-grained models suggests that by modulating local lipid composition and insertion depth of protein motifs, it is possible to qualitatively change the sign for membrane curvature sensing and generation by the same protein, leading to plasticity and regulation of membrane activity of peripheral proteins.

Co-operative Interactions of Local Curvatures

As alluded to earlier, while Sar1 and ESCRT-III are expected to lead to similar membrane bending at the local scale, their biological activities are rather different. While Sar1s are expected to initiate positive curvature generation, ESCRT-III is believed to induce membrane invagination such that their spiral filaments reside in the interior of the neck. How do we qualitatively explain such differences at a larger scale?

The key factor appears to lie in the different organizations of these proteins (Figure 4). The ESCRT-III proteins form a long linear filament through stable protein-protein interfaces; atomistic simulations found that the filament has a natural tendency to bend and twist (Mandal et al. 2020), which is consistent with the experimental observation (Chiaruttini et al. 2015; Pfitzner et al. 2021; Shen et al. 2014) that ESCRT-III filaments grow in spiral forms. We thus proposed a plausible mechanism for how ESCRT-III filament formation might drive membrane invagination (Figure 4A-C). As the filament grows on the membrane surface into a circular ring,

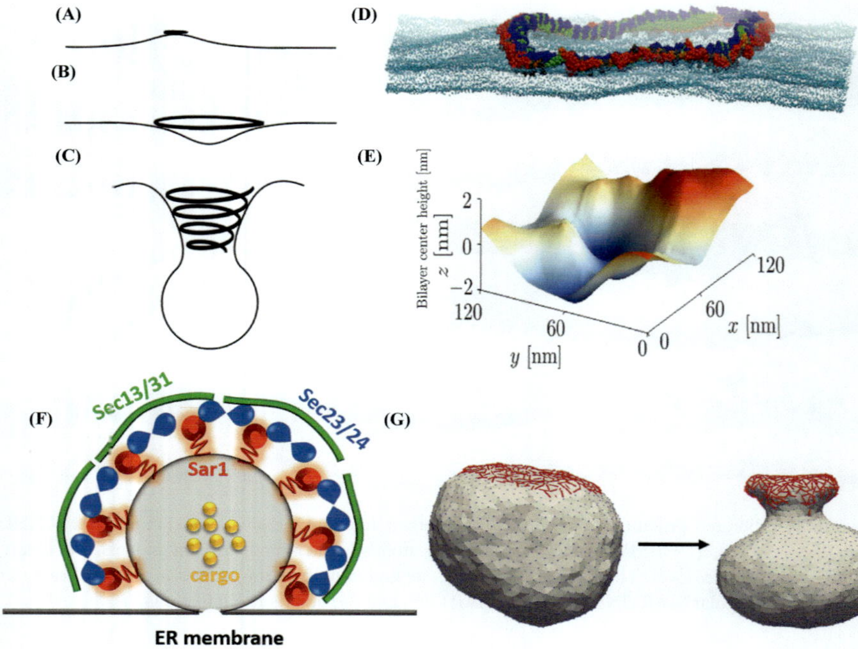

Figure 4. (A-C) A cartoon representation of a plausible mechanism for the membrane curvature development catalyzed by ESCRT-III filaments. (A) Initial adsorption of ESCRT-III induces local positive curvature because of the insertion of the amino-terminal helix; (B) adsorption of a ring of ESCRT-III polymers induces a negative curvature at the center of the circular ring; (C) as the ESCRT-III filament continues to elongate, bending and twisting deformations of the filament lead to the formation of a three-dimensional helical spiral that creates the neck of the invagination. (D) The configuration of a ring of ESCRT-III protein filament after 2.4 μs MD simulation (side view). (E) Three-dimensional color plots of the filament-induced membrane morphology shows significant negative curvature surrounding the protein filament. (F) Cartoon representation of a vesicular budding mediated by COPII on the subdomain of ER. Sar1 proteins together with Sec23/24 form the inner coat layer, whereas Sec13/31 proteins assemble to form the outer coat layer. (G) A continuum model (dynamical triangulated surface (Hansen et al. 2017)) based simulation of vesicular budding induced by a randomly oriented assembly of amhipathic helices (S. Paul, work in progress). Here, a closed membrane surface (white) is constructed in the form of the triangular where curvature-generating helices (red line) aggregate on 20 % surface area (left). After a million-step Monte Carlo simulation, the membrane surface constricts and the bud-neck forms (right).

the local positive curvature generated by the individual amino-terminal helices induces a negative curvature at the center of the ring. As the filament continues to grow, the intrinsic twist of the filament leads to the downward bending of the membrane to achieve a more energetically favorable membrane-filament interface. Due to the relatively low bending modulus of the filament and strong filament-membrane interaction, the filament prefers to grow in a 3D helical spiral, leading to the neck of the invagination.

We have tested this model with both simple continuum mechanics analysis (Mandal et al. 2020) and explicit coarse-grained simulations (Mandal et al. 2021). In the latter case, four ESCRT-III linear filaments were arranged in a square form on a large patch of lipid membrane (with a lateral dimension of 120 nm × 120 nm, Figure 4D) using the regular MARTINI2.2 model. After a few microseconds of simulations, the filaments spontaneously rearranged into an elliptical form with a considerable degree of negative membrane invagination enclosed in the center (Figure 4E), which supported our mechanistic model. For a recent summary of available models of ESCRT action, see Pfitzner (Pfitzner et al. 2021).

The situation is rather different in the case of Sar1. Electron microscopy studies observed that Sar1s form regular lattices together with Sec 23/24. The highly oriented nature of Sar1 proteins in such arrangements suggests strongly anisotropic curvature developments in the membrane, which is consistent with the tubule formation under *in vitro* condition (Hariri et al. 2014). In the case of procollagen export, multiple COPII buds can assemble in the presence of TANGO1, leading to a cylindrical protrusion of the membrane that can accommodate bulky procollagen cargoes (Raote et al. 2020). However, the canonical COPII coats are spherical in nature (Figure 4F), thus the arrangements of Sar1 under *in vivo* conditions are likely very different from the lattice structure revealed by the electron microscopy. Indeed, recent studies (Kasberg et al. 2021) suggested that Sec 23/24 undergo liquid-liquid phase separation (LLPS) and interact with Sar1 and the ER membrane in the form of protein condensates, similar to the case of clathrin-mediated endocytosis (Day et al. 2021). In this model, the arrangements of Sar1 proteins are likely more random, thus the local positive curvatures may add up to a more isotropic global curvature and therefore the development of a spherical vesicle rather than a cylindrical tubule. This mechanism is being tested explicitly with a continuum mechanics model with membrane meshes (Figure 4G).

Membrane Remodeling by Biopolymer Condensates and Nanoparticles

In this subsection, we'll further illustrate the value of coarse-grained models in exploring the mechanism of membrane remodeling by focusing on biopolymer condensates and functionalized nanoparticles. These systems are computationally expensive to analyze with full atomistic models, especially in terms of processes involving liposomes.

Biopolymer Condensate Interaction with Lipid Bilayers and Liposomes

Recent experimental studies have established that LLPS of proteins at the lipid membrane is involved in various cellular processes such as signal transduction (Chong and Forman-Kay 2016). It was observed that the protein condensates could induce significant membrane curvatures, vesicle biogenesis, multilamerllar structure formation, tubulation, and pore formation (Day et al. 2021; Snead and Gladfelter 2019; Yuan et al. 2021; Zhao and Zhang 2020). Therefore, LLPS at the membrane surface can serve as a novel mechanism for membrane remodeling. The molecular factors that govern the interaction between protein condensates and lipid membrane and the downstream effects of this interaction have not been well understood. For example, it remains unclear whether the main driving force for the membrane deformation is a favorable protein-membrane interaction or an enhanced protein-protein interaction and what are the governing factors that control the sign of the generated curvature (Kozlov and Taraska 2023; Yuan et al. 2021, 2023). Although there exist several theoretical/computational studies that focused on the LLPS of disordered proteins and polyelectrolytes in the bulk solution phase (Bremer et al. 2022; Danielsen et al. 2019; Das et al. 2020; Rumyantsev et al. 2019; Zheng et al., 2020), their behavior in the presence of a membrane/liposome has not been well understood from a microscopic perspective.

As an initial effort along this line of research, we have explored the interaction between polyelectrolyte coacervates and lipid membranes using classical molecular dynamics simulations by employing different coarse-grained models. In the first study (Mondal and Cui 2022), we used the MARTINI v3.0 model (Souza et al. 2021) to analyze the interactions between poly-lysine (K_{30}), poly-glutamate (E_{30}) and lipid bilayers of different lipid compositions (POPC and POPG mixtures with the fraction of PG ranging from 0 to 100%). We chose the poly-lysine/poly-glutamate system because it was well studied both experimentally (Dinic 2021; Priftis and Tirrell 2012) and computationally; in particular, the MARTINI 3.0 model was shown to give an adequate description of the coacervate in bulk solution when compared to available experimental data (Tsanai et al. 2021).

Under the conditions of our simulation, which corresponds to the polyelectrolyte concentration of approximately 5% (w/w), an equimolecular mixture of poly-lysine and poly-glutamate formed coacervate readily in solution and then was recruited to the membrane when the mole fraction of the anionic lipids was beyond 20% (Figure 5A and B). As the mole fraction of the anionic lipids increased, the coacervate wetted the bilayer to a larger degree and became less spherical and more spread out; with 100% PG, the coacervate completely wetted the membrane and formed a thin polyelectrolyte layer on top of the membrane.

Adsorption of the coacervate to the bilayer also led to major changes in the membrane. Although the polyelectrolytes did not penetrate into the bilayer, the coacervate led to significant negative membrane bending with the mean curvature in the range of 0.1 to 0.2 nm^{-1} (Figure 5B). The magnitude of the generated curvature decreases with decreasing concentration of anionic lipids. The results can be semi-quantitatively explained by a theoretical model for polymer-membrane interactions (Kim and Sung 2001), which predicts negative curvature generation due to strong polymer-membrane attractions. Analysis of the energy components before and after coacervate binding to the membrane in our simulations (Figure 5G) revealed that the underlying driving force for the coupled coacervate/membrane deformation was indeed the favorable poly-lysine/lipid interactions, which compensated the decrease in poly-lysine/poly-glutamate attraction as the coacervate became more spread out on the membrane surface; moreover, polyelectrolyte/counterion interactions also became more favorable as the coacervate deformed. The attractive interaction between poly-lysine and anionic lipids also led to a notable degree of local lipid demixing and a higher local concentration of anionic lipids beneath the coacervate (Figure 5E and F).

In addition to this, we have also explored the effect of the sequence in membrane-coacervate interaction. A mixture of E_{30} and K_{30} can be considered "blocky" sequences where a long stretch of positively/negatively charged residue is present. In Figure 5C, we show the effect of a polyampholyte coacervate that contains a less

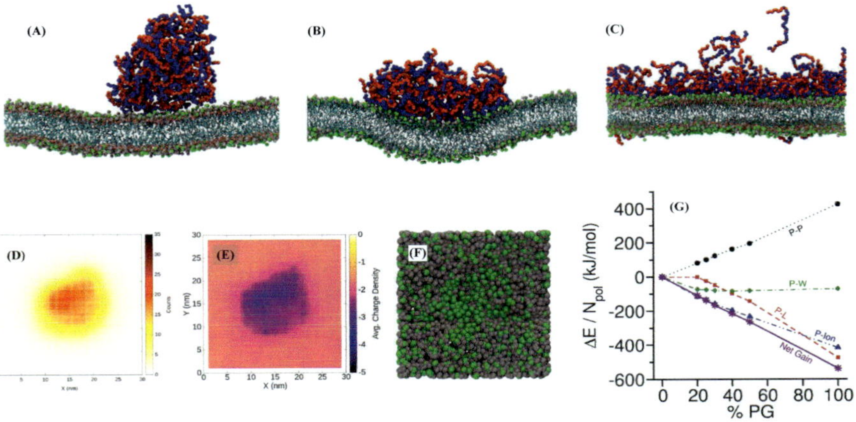

Figure 5. Interaction between polyelectrolyte/polyampholyte coacervates and lipid membranes of different compositions from MARTINI 3.0 simulations. Snapshots of (A) E_{30} and K_{30} polyelectrolyte coacervate adsorbed on PC : PG = 75 : 25 membrane. (B) E_{30} and K_{30} polyelectrolyte coacervate adsorbed on PC : PG = 50 : 50 membrane. (C) A polyampholyte coacervate adsorbed on PC : PG = 50 : 50 membrane. The polyampholyte is also composed of E and K residues but with a scrambled charge patterning. Color codes: the backbone beads of E and K residues are shown in red and blue, respectively. POPC head-groups are shown in gray, and those of POPG are shown in green. Lipid tails are shown in cyan. (D) Polyelectrolyte coacervate hotspot (density contour) on PC : PG = 75 : 25 membrane. (E) Wetting-induced demixing of lipids in the upper leaflet of PC : PG = 75 : 25 membranes upon interacting with the polyelectrolyte coacervate. (F) A snapshot (top view) of the upper leaflet that shows the POPG segregation at and around the protein hotspot shown in 'd'. (G) Component-wise interaction energies are functions of anionic lipid concentration in the bilayer. The energy values obtained from the pure POPC system are used as a reference. The self-interaction energy of polyelectrolytes increases with the negative charge concentration of the bilayer, which indicates a destabilizing effect. On the other hand, the polyelectrolyte-lipid and polyelectrolyte-ion interaction energies decrease with the concentration of anionic lipids. These two stabilizing interactions produce a net gain in the enthalpy.

"blocky" charge pattern on a 50% POPG membrane. It has been observed that the coacervate completely loses its bulk spherical shape and wets the membrane surface without significantly remodeling the membrane. We note that polyelectrolytes with blocky charge patterns can induce a negative curvature and local lipid demixing on the same membrane, as shown in Figure 5B. Therefore, the membrane composition, as well as the polymer sequence can play a pivotal role in determining the morphology of the system.

To explore larger-scale membrane remodeling processes (Mondal and Cui 2023), we next turned to an implicit solvent model with a higher degree of coarse-graining (Figure 6). The systems contained two types of homopolymers (A and B, with each chain containing 30 monomers) and a single-component liposome. Using the Cooke model (Cooke et al. 2005a; Cooke et al. 2005b), each lipid molecule was modeled with three beads and the polymers as self-repulsive flexible chains with an attractive interpolymeric interaction. By systematically varying the interaction strengths among the polymers and lipids in this simple framework, we were able to recapitulate various membrane transformations observed in different experimental

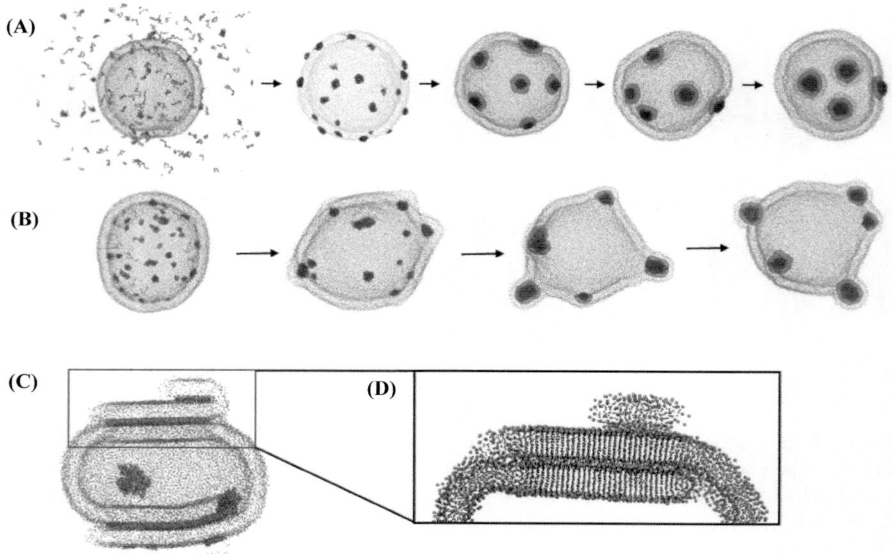

Figure 6. Snapshots of a model system where the interpolymeric attractive interaction is twice that of polymer-lipid interactions illustrate endocytosis and exocytosis processes. Multiple events of (A) endocytosis of coacervate droplets starting from the initial random configuration. The polymers first show scattered adsorption on the outer vesicle surface, followed by partial association to increase their size, and endocytosis after creating a negative curvature. (B) Bud and neck formation of the vesicle by coacervate droplets when the polymers are randomly placed in the interior part of the vesicle. (C) The emergence of multilamellar bilayer structures and local gelation by the act of phase separating polymer mixtures when polymer-lipid-head interactions are stronger than the inter-polymer interactions. In the magnified view, the ordering of the lipid tails is highlighted in the vicinity of the polymers.

studies (Allolio et al. 2018; Lu et al. 2022). Contemporarily, using dissipative particle dynamics (DPD) simulation and continuum mechanics, Lipowsky and others have studied processes (Ghosh et al. 2023; Satarifard and Lipowsky 2023). The interaction of nanoparticles of different shapes with bilayers has been previously studied by using the Cooke lipid model (Vácha et al. 2011).

In the system where the interpolymeric attractive interaction is twice that of polymer-lipid interactions, significant membrane remodeling was observed. As shown in Figure 5A, starting from a randomly distributed polymer configuration outside the liposome, we observed several events of endocytosis (Figure 6A). The sequence of the observed events was as follows. First, the polymer pairs underwent scattered adsorption on the outer layer of the liposome, followed by partial association through surface-mediated diffusion to form larger coacervate droplets. Once the critical size of the droplet was surpassed, the droplets started to generate negative curvature and eventually were engulfed by the bilayer, leading to endocytosis, similar to recent experimental observations (Lu et al. 2022); during our simulation, we observed the endocytosis of three such coacervate droplets.

We also performed a simulation with the same system, but the polymers were randomly distributed inside the liposome (Figure 6B). The polymers formed small clusters that merged into larger coacervates. The coacervates then induced a negative curvature (from the lumen's perspective) and formed buds. However, the buds (or the wrapped configuration) remained stable and did not exhibit spontaneous exocytosis within the timescale explored in our simulations.

The difference between endo/exocytosis was also observed in earlier studies by Weikl and co-workers (Bahrami et al. 2016) on the role of membrane curvature in the engulfment of nanoparticles. Their continuum analyses revealed that endocytosis passes through an unstable intermediate, whereas exocytosis passes through a neck, forming stable intermediate. Such a difference is consistent with our observations that the detachment of the bud (Figure 6B) was slow and not energetically downhill.

In systems where the polymer-lipid interactions are strong (e.g., almost twice the lipid tail-tail interaction), we observed the formation of multilamellar bilayer structures out of a spherical liposome, as shown in Figure 6C. We found that the spherical shape of the coacervate was not retained, and the polymers spread out and folded the bilayer around them to maximize the amount of polymer-lipid contacts. Moreover, we observed an enhanced local gel-like ordering of the lipid tails in the vicinity of the polymer coatings, where the other parts showed liquid-like disordered tails. In the magnified view of a cross-section in the right panel of Figure 6C, we show the enhanced ordering of lipid tails. Since gelation reduced the area-per-lipid, a larger number of lipids could be packed in the vicinity of the polymers, thus providing enthalpic stabilization at the expense of entropy loss. A multi-layer structure was observed in the electron microscopy study of polyarginine/membrane interactions (Allolio et al. 2018), and several recent studies also observed that LLPS of proteins enhanced local lipid ordering (Lee et al. 2022).

Functionalized Nanoparticles and Liposomes

Functionalized nanomaterials have numerous applications in diverse fields including energy, imaging and drug delivery. As a result, there has been an increasing interest in better understanding their environmental and health impacts (Buchman et al. 2019; Murphy et al. 2015). In terms of interaction with biological systems, cell membranes are among the first encounters for nanomaterials. Therefore, a considerable number of experimental and computational studies have been conducted to understand factors that determine the interaction between nanomaterials and cell membranes of varying complexity (Chew et al. 2022; Rossi and Monticelli 2016; Vacha et al. 2011). Common features examined included the charge, hydrophobicity and density of the surface ligands, the shape, size and mechanical properties of the nanomaterials, and the composition of the membrane. In our recent study, taking advantage of progress in nanoparticle surface functionalization (Zhang et al. 2019), we have explored the impact of ligand flexibility by comparing nanoparticles functionalized with molecular ligands and polymeric ligands.

Coarse-grained simulations were conducted to analyze the interaction between nano-diamond particles and liposomes consisting of 9:1 DLPC:DLPG with a radius between 6 and 7 nm. The nano-diamond was functionalized with either alkyl amine ligands of different linker lengths or the polyelectrolyte, poly(allylamine hydrochloride) (PAH). The nano-diamonds displayed a significant number of positive charges due to the functionalization, ranging from +138 to +320 e. Therefore, the

BMW-MARTINI model (Wu et al. 2011), which treats electrostatics more reliably, was used in those simulations.

The key experimental observation (Zhang et al. 2021) was that all studied nano-diamonds featured rather similar ζ potentials of ~ 40 mV, yet they exhibited rather different impacts on the stability of Small Unilamellar Vesicles (SUVs). Nano-diamonds functionalized with PAH polymers induced significant damage to SUVs even at low concentrations, while nano-diamonds functionalized with short amines induced minimal amount of SUV damage across the concentration range studied therein; nanodiamonds functionalized with longer amine ligands induced an intermediate level of SUV damage. Moreover, T_2 relaxation measurements with ^{31}P NMR was used to study the lipid dynamics in the liposome upon interacting with the functionalized nano-diamonds. The results revealed rigid lipid motions in the presence of molecular-amino nano-diamonds, while much less restricted lipid motions upon interacting with PAH-nano-diamonds; these observations were interpreted to suggest that nano-diamonds functionalized with molecular amine ligands associated with the liposome, while PAH-functionalized nano-diamond induced significant rearrangement or repacking of lipids in the SUVs.

The coarse-grained simulations provided further insights into the processes (Zhang et al. 2021). With molecular amine ligands (Figure 7A), the nano-diamonds were observed to bind to the liposome; the binding was stronger with the longer alkyl amine ligands, due to the more favorable packing between these ligands and the hydrophobic lipid tails. In fact, the favorable alkyl linker/lipid tail interactions enabled the nano-diamonds to extract lipids from the liposome, while no lipid extraction was observed for nano-diamonds with short alkyl amine ligands or polar ligands. With PAH-functionalized nano-diamonds, more dramatic perturbations on the liposome structure were observed (Figure 7B), which involved significant shifts in lipid distributions (~ 10 Å) for both the inner and outer lipid layers. It is worth noting that if the PAH chains were made structurally more rigid in the simulations, the nano-diamond merely associated with the liposome with minimal perturbation in the lipid distributions (Figure 7C). Overall, these observations were consistent with the experimental data on liposome damage and lipid dynamics discussed above; they highlighted the role of surface ligand flexibility in determining the interaction between nanoparticles and lipid membranes, while supporting the notion that a higher surface charge density does not necessarily lead to a stronger interaction with lipid membranes.

In another example (Kesner et al. 2023), we studied the interaction between polystyrene nanoparticles and liposomes, which is a model for how nanoplastics impact cell membranes. Nanoplastics are small plastic particles formed through the breakdown of microplastics in the ocean through a variety of physical and chemical processes. This interaction was monitored experimentally with the fluorescence of laurdan embedded in the liposome. The significant blue shift of the laurdan fluorescence was observed when the liposomes were exposed to polystyrene nanoparticles; the "canonical" interpretation (Wang et al. 2008) of the blue shift was that the lipids undergo a fluid-to-gel transition upon interacting with the polystyrene nanoparticles, although a molecular mechanism for how polystyrene could induce such a transition was not available.

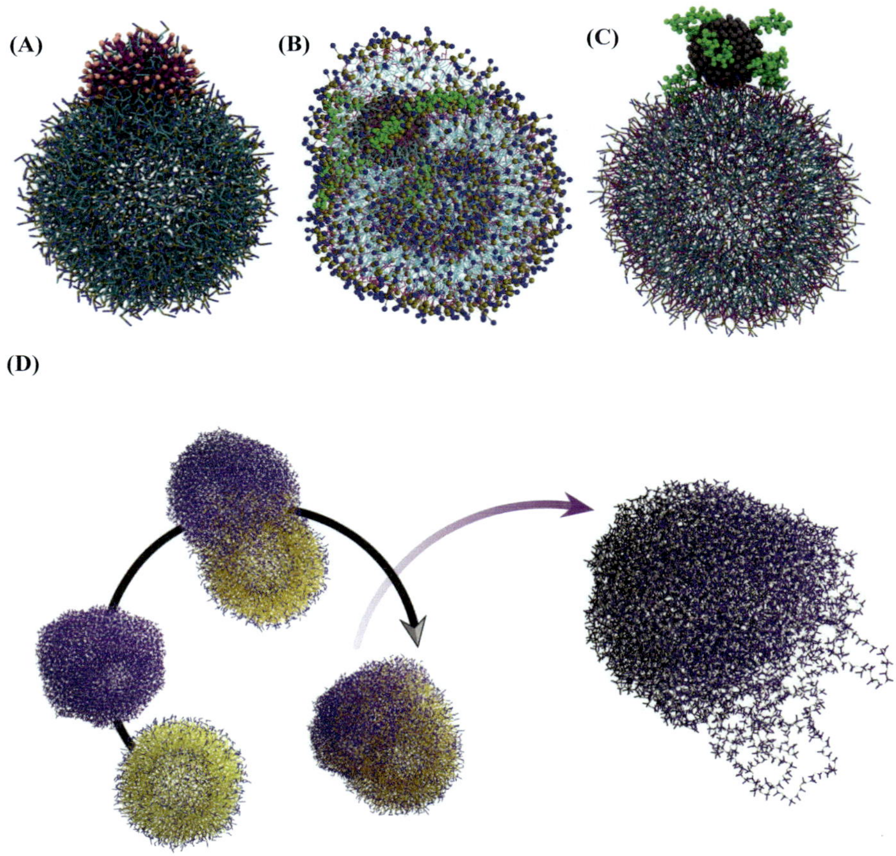

Figure 7. Coarse-grained simulations probe how nanoparticles interact with liposomes. (A-C) For nano-diamonds functionalized with either (A) molecular amine ligands or (C) rigid polymeric ligands, the nanodiamonds are associated tightly with the liposome without significant structural disruptions. By contrast, nano-diamonds functionalized with flexible cationic polymers penetrate deeply into the liposome and induce significant structure distortions. (D) Due to their hydrophobicity and their flexibility, polystyrene penetrates liposomes initially through individual chains. Once these chains contact the hydrophobic membrane core, they dissolve into the liposome. (Black Arrow) Timelapse of this process (Purple Arrow) Representative configuration of polystyrene during liposome penetration. Note the chains that extend into the membrane.

Coarse-grained simulations revealed a rather different picture (Figure 7D). The polystyrene chains were observed to dissolve into the membrane hydrophobic region, leading to a partial merge of the nanoparticle and the liposome. The largely non-polar polystyrene chains remained in the hydrophobic region of the membrane and essentially fully encapsulated the interior of the liposome; this result was consistent with the experimental observation from previous studies, which observed that dye molecules trapped inside the liposome were not released in the presence of the polystyrene nanoparticles. Calculations of the area per lipid from the simulations found no evidence of lipid gelation upon interacting with the polystyrene, in agreement with a recent study that probed lipid bilayer-polystyrene interactions through both experimental calorimetry and atomistic simulations (Bochicchio et al. 2022). Furthermore, a recent all-atom study of dialkyl para poly phenylene ethylene (PPE) "polydots" demonstrated that these particles were likewise able to penetrate lipid membranes and that the degree of penetration could be controlled via functionalization of the polymer with carboxyl groups (Wijesinghe et al. 2023).

This led us to re-examine the origin of the "canonical" interpretation of laurdan fluorescence, which originates from the fact that laurdan fluorescence is heavily dependent on the ability of water molecules hydrogen-bonded to lipids at the glycerol level of the lipid membrane to reorganize around and stabilize the excited state of laurdan. Since gel membranes are more tightly packed than fluid membranes, there is both less population of water within the membrane as well as less room for the remaining water molecules to reorganize, leading to a blue shift. Therefore, the observed blue shifts in laurdan fluorescence in the presence of polystyrene might be due to the change of local hydration level induced by the presence of polystyrene within the membrane rather than a lipid phase transition. To explore this hypothesis, we have recently used atomistic simulations of a POPC membrane (with laurdan) in a bilayer configuration, to which we have added increasing concentrations of 10-mer chains of polystyrene to the hydrophobic region. Importantly, we have found that increasing polystyrene concentration within the membrane leads to dehydration of laurdan through modifications of both the internal conformations of laurdan, as well as through modification of the depth at which laurdan sits within the membrane. Ongoing work is using snapshots from these all-atom configurations in quantum mechanics/molecular mechanics calculations to correlate laurdan hydration environment with their absorption energy. While this hypothesis needs to be further examined through further computational and experimental analysis, the results highlight that care needs to be exercised to interpret laurdan spectral shifts for membrane processes and that molecular simulations are indispensable for stimulating and testing possible mechanistic proposals.

Modulation of Membrane Pore Formation by Peptides

Finally, we briefly discuss another important type of membrane remodeling: pore formation modulated by peptides, which is implicated in many important processes such as virial infection and antimicrobial activities. Since the topic has been discussed rather extensively in the literature (Guha et al. 2019; Schmidt and Wong 2013), we use two of our recent studies to highlight some remaining questions and challenges. Questions of general interest include: (i) what is the minimal number of peptides that lead to instability in the membrane and formation of a pore?(ii) are these membrane pores meta-stable and if so, what is the life-time? (iii) in the presence of multiple peptides that are characteristic of leading to different signs of Gaussian curvature in the membrane, to what degree the effects are additive?

For example, the first two questions arose in our analysis of complexin peptides (Courtney et al. 2022), which are well-known to play a major role in membrane fusion via their interaction with synaptobrevin (Maximov et al. 2009; Sudhof and Rothman 2009). Recent studies suggested the hitherto unrecognized role of complexin's C-terminus in directly remodeling the membrane; this has a major impact on the

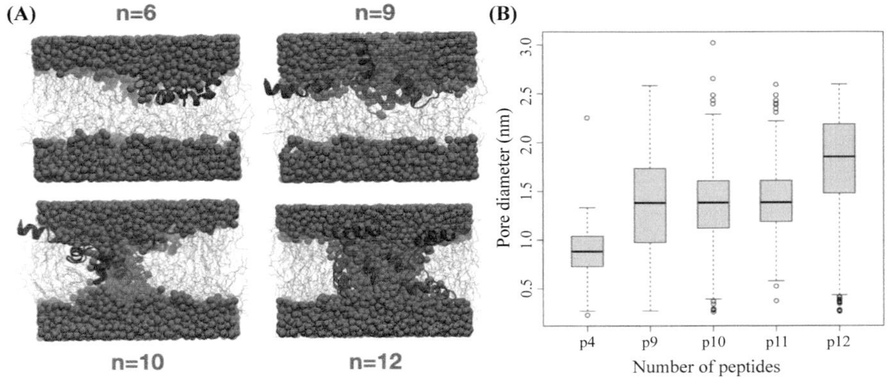

Figure 8. Atomistic simulation of membrane pore formation by complexin C-terminal peptides. (A) Final snapshots from ~ 300–500 ns simulations with different numbers of peptides; the water molecules (as spheres) are used to highlight the pore spontaneously formed with n > 9; (B) Estimate of pore diameter with different numbers of peptides; for n < 9, the pore was obtained by removing a given number of peptides from the final structure in simulation with n = 12.

mechanism of SNARE-catalyzed membrane fusion (Brunger et al. 2018). The degree of membrane remodeling with different stoichiometries of complexin is important in this context, and is difficult to determine experimentally; with atomistic simulations, we were able to assemble multiple copies of complexin C-terminal peptides, which formed and stabilized a membrane pore only with copy numbers larger than 9, as shown in Figure 8A. However, when peptides were systematically removed from the formed pore, the pore remained stable with as few as four copies of peptides during hundreds of nanoseconds of simulations (Figure 8B).

The third question listed above is of particular current interest because if the answer is positive, it might suggest an approach to use designer-peptides to compensate for the negative Gaussian Curvature (NGC) activities of membrane pore-forming peptides in applications such as anti-virial-infection design. Experimental studies (G. C. L. Wong, private communication) indeed observed that a class of "inverse translocation peptides" (ITPs), which generate positive membrane Gaussian curvature, was able to reduce the NGC generated by the histone H4 tail peptide (Silvestre-Roig et al. 2019) and, more recently, by a peptide segment from the SARS-CoV-2 spike protein. The detailed molecular mechanism behind such compensation effect (Dommersnes and Fournier 2002), however, remains poorly understood. In our all-atom molecular dynamics simulations (Figure 9), we observed that 6–9 copies of SARS-CoV-2 spike protein peptides or histone tail peptides were able to stabilize a membrane pore for at least three to five hundred nanoseconds; it was difficult, however, to even qualitatively compare the ability of these peptides to generate and stabilize the pore structure. Moreover, with the addition of 6 copies of ITPs, two out of four independent simulations observed that the pore became destabilized and eventually disappeared after ~ 200–300 ns, supporting the Gaussian curvature compensation effect; in the other two simulations of similar length, however, the pore structure remained.

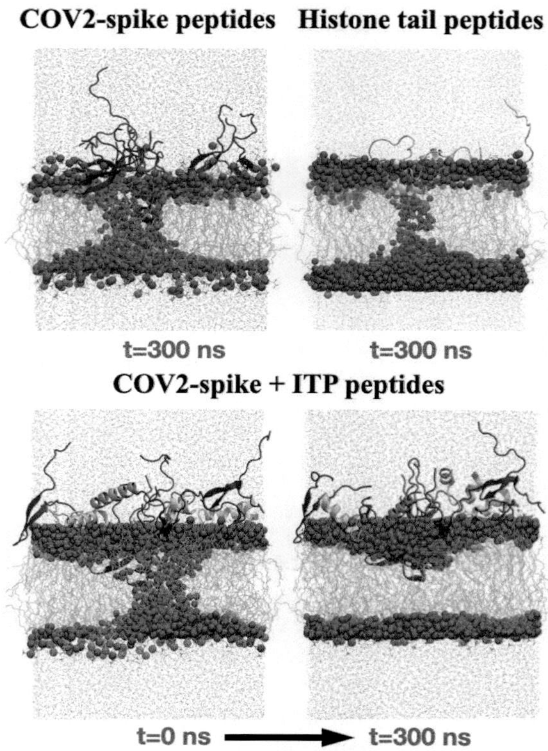

Figure 9. Atomistic simulation of membrane pore stability with different peptides. Top: 6–9 copies of COV2-spike peptides or Histone H4 tail peptides; Bottom: 9 copies of COV2-spike peptides with 6 copies of inverse translocation peptides (ITPs) designed in the Wong lab; the ITPs abolished the pore after 200–300 ns in two out of four independent simulations.

Therefore, while the simulations shown in Figures 8–9 provide hints regarding fascinating phenomena associated with membrane pore modulation, they also highlight the need to develop methodologies that can analyze the underlying process in a quantitative and reproducible manner, such that the physical factors that dictate the modulation effect can be understood and design rules of novel peptide motifs be established.

Concluding Remarks

In this book chapter, we have discussed several recent studies from our group to illustrate how multi-scale computations can be used to help better understand the molecular mechanism of lipid membrane remodeling by proteins, peptides and nanoparticles. The results highlight both common and unique features of various membrane remodeling processes.

First, since the remodeling agent (protein/peptide/nanoparticle) needs to be recruited to the membrane surface, electrostatic interactions are often the key driving force. Since electrostatic interactions are usually treated at a rather approximate level with coarse-grained models, care must be exercised to calibrate such models for the specific system of interest to avoid computational artifacts. Along this line, the spontaneous curvature of charge lipids at the coarse-grained level should also be carefully examined.

Once at the surface, proteins or other (bio)polymers can generate local curvature by inserting structural motifs into the bilayer and perturbing lipid packing. While the importance of insertion depth has been extensively discussed in the literature, our recent analyses of Sar1 highlighted that insertion volume is also an important factor to consider. Moreover, in the case of polyelectrolyte coacervate/membrane interactions, significant membrane bending can be induced without polymer penetration into the bilayer at all; rather, membrane bending is largely driven by the favorable polymer/lipid head group and polymer-counter-ion interactions. The examples of functionalized nanoparticles further illustrate that membrane remodeling can be substantially enhanced by the dynamical feature of surface ligands, in addition to the much discussed surface charge density (Olenick et al. 2018; Troiano et al. 2017) and ligand lipophilicity (Lochbaum et al. 2021).

Another important aspect of membrane remodeling is the co-operative interactions of multiple proteins, which depend critically on their organizations. As illustrated by the discussion of Sar1 and ESCRT-III, while both generate local positive membrane curvature with the insertion of amphipathic helices, the global membrane deformations are very different due to the different organizations of Sar1 and ESCRT-III proteins. The example of SARS-CoV-2 peptides and "inverse translocation peptides" further illustrates that Gaussian curvature can also be additive, leading to the modulation of membrane pore formation.

While it is clear that much can be explored about the molecular mechanism of membrane remodeling with multi-scale computations, many fascinating questions remain and they continue to stimulate the development of novel computational methodologies that integrate models of different resolutions in a more cohesive manner than the studies illustrated in this chapter. For example, several experimental and computational studies have underscored the coupling between the phase behaviors of membranes and proteins (Liu et al. 2023; Rouches et al. 2021; Wang et al. 2023), and how such coupling is manifested in membrane remodeling processes remains to be better elucidated. The discussion of membrane-active peptides also highlighted the need for novel simulation methodologies for quantitatively capturing the thermodynamics and kinetics of membrane pore formation catalyzed by multiple peptides/proteins. Moreover, many membrane remodeling processes *in vivo* are driven far from equilibrium by active processes such as ATP hydrolysis; understanding the non-equilibrium features of these processes (Oh and Cui 2023) demands the development of novel theoretical and computational tools. Finally, given the emerging physical principles of membrane remodeling, it is exciting to employ novel tools such as machine learning algorithms (Lee et al. 2016; Yount et al. 2019) to design peptide/protein/nanoparticles to remodel membranes for important applications including drug delivery and synthetic biology.

Acknowledgments

We appreciate the generous support from the National Science Foundation over the past few years for our developments in the general area of multi-scale simulations of lipid membrane remodeling (grants DMS-1661900, CHE-2001611, CHE-1829555 and CHE-2154804). We also acknowledge all co-workers who have contributed to the work discussed here, especially experimental collaborators. Computational resources from the Extreme Science and Engineering Discovery Environment (XSEDE) (Towns et al. 2014), which is supported by NSF Grant ACI-1548562, are greatly appreciated; part of the computational work was performed on the Shared Computing Cluster, which is administered by Boston University's Research Computing Services (URL: www.bu.edu/tech/support/research/).

References

Allolio, C., A. Magarkar, P. Jurkiewicz et al. 2018. Arginine-rich cell-penetrating peptides induce membrane multilamellarity and subsequently enter via formation of a fusion pore. Proc. Natl. Acad. Sci. U.S.A. 115(47): 11923–11928.

Anthony, A.A., O. Sahin, M.K. Yapici et al. 2022. Systematic measurements of interleaflet friction in supported bilayers. Biophys. J. 121(15): 2981–2993.

Arkhipov, A., Y. Yin and K. Schulten. 2009. Membrane-bending mechanism of amphiphysin n-bar domains. Biophys. J. 97(10): 2727–2735.

Arnarez, C., J.J. Uusitalo, M.F. Masman et al. 2015. Dry martini, a coarse-grained force field for lipid membrane simulations with implicit solvent. J. Chem. Theory Comput. 11: 260–275.

Ayton, G.S. and G.A. Voth. 2010. Multiscale simulation of protein mediated membrane remodeling. Seminars in Cell & Dev. Biol. 21: 357–362.

Bahrami, A.H., R. Lipowsky and T.R. Weikl. 2016. The role of membrane curvature for the wrapping of nanoparticles. Soft Matter 12(2): 581–587.

Bassereau, P., R. Jin, T. Baumgart et al. 2018. The 2018 biomembrane curvature and remodeling roadmap. J. Phys. D: Appl. Phys. 51: 343001.

Bhaskara, R.M., P. Grumati, J. Garcia-Pardo et al. 2019. Curvature induction and membrane remodeling by FAM134B reticulon homology domain assist selective ER-phagy. Nat. Comm. 10: 2370.

Blood, P.D. and G.A. Voth. 2006. Direct observation of bin/amphiphysin/rvs (bar) domain- induced membrane curvature by means of molecular dynamics simulations. Proc. Natl. Acad. Sci. U.S.A. 103(41): 15068–15072.

Boal, D. 2002. Mechanics of the cell. Cambridge, UK: Cambridge University Press.

Bochicchio, D., L. Cantu, M.V. Cadario et al. 2022. Polystyrene perturbs the structure, dynamics, and mechanical properties of dppc membranes: An experimental and computational study. J. Coll. & Interf. Sci. 605: 110–119.

Bremer, A., M. Farag, W.M. Borcherds et al. 2022. Deciphering how naturally occurring sequence features impact the phase behaviours of disordered prion-like domains. Nat. Chem. 14(2): 196–207.

Brunger, A.T., U.B. Choi, Y. Lai et al. 2018. Molecular mechanisms of fast neurotransmitter release. Annu. Rev. Biophys. 47: 469–497.

Buchman, J.T., N.V. Hudson-Smith, K.M. Landy et al. 2019. Understanding nanoparticle toxicity mechanisms to inform redesign strategies to reduce environmental impact. Acc. Chem. Res. 52: 1632–1642.

Campelo, F., H.T. McMahon and M.M. Kozlov. 2008. The hydrophobic insertion mechanism of membrane curvature generation by proteins. Biophys. J. 95: 2325–2339.

Chernomordik, L.V. and M.M. Kozlov. 2008. Mechanics of membrane fusion. Nat. Struct. Mol. Biol. 15: 675–683.

Chew, A.K., J.A. Pedersen and R.C. van Lehn. 2022. Predicting the physicochemical properties and biological activities of monolayer-protected gold nanoparticles using simulation-derived descriptors. ACS Nano 16: 6282–6292.

Chiaruttini, N., L. Redondo-Morata, A. Colom et al. 2015. Relaxation of loaded ESCRT-III spiral springs drives membrane deformation. Cell 163: 866–879.

Chong, P.A. and J.D. Forman-Kay. 2016. Liquid–liquid phase separation in cellular signaling systems. Curr. Opin. Struct. Biol. 41: 180–186.

Cooke, I.R. and M. Deserno. 2005. Solvent-free model for self-assembling fluid bilayer membranes: Stabilization of the fluid phase based on broad attractive tail potentials. J. Chem. Phys. 123: 224710.

Cooke, I.R., K. Kremer and M. Deserno. 2005. Tunable generic model for fluid bilayer membranes. Phys. Rev. E 72(1): 011506.

Corradi, V., B.I. Sejdiu, H. Mesa-Galloso et al. 2019. Emerging diversity in lipid-protein interactions. Chem. Rev. 119: 5775–5848.

Courtney, K.C., L. Wu, T. Mandal et al. 2022. The complexin c-terminal amphipathic helix stabilizes the fusion pore open state by sculpting membranes. Nat. Struct. Mol. Biol. 29: 97–107.

Danielsen, S.P., J. McCarty, J.-E. Shea et al. 2019. Small ion effects on self-coacervation phenomena in block polyampholytes. J. Chem. Phys. 151(3): 034904.

Das, M., U. Dahal, O. Mesele et al. 2019. Molecular dynamics simulation of interaction between functionalized nanoparticles with lipid membranes: Analysis of coarse-grained models. J. Phys. Chem. B 123: 10547–10561.

Das, S., Y.-H. Lin, R.M. Vernon et al. 2020. Comparative roles of charge, π, and hydrophobic interactions in sequence-dependent phase separation of intrinsically disordered proteins. Proc. Natl. Acad. Sci. U.S.A. 117(46): 28795–28805.

Day, K.J., G. Kago, L. Wang et al. 2021. Liquid-like protein interactions catalyse assembly of endocytic vesicles. Nat. Cell Biol. 23: 366–376.

Dignon, G.L., W. Zheng, Y.C. Kim et al. 2018. Sequence determinants of protein phase behavior from a coarse-grained model. PLOS Comput. Biol. 14: e1005941.

Dinic, J., A.B. Marciel and M.V. Tirrell. 2021. Polyampholyte physics: Liquid–liquid phase separation and biological condensates. Curr. Opin. Coll. & Inter. Sci. 54: 101457.

Dommersnes, P.G. and J.B. Fournier. 2002. The many-body problem for anisotropic membrane inclusions and the self-assembly of "saddle" defects into an "egg carton". Biophys. J. 83: 2898–2905.

Ghosh, R., V. Satarifard and R. Lipowsky. 2023. Different pathways for engulfment and endocytosis of liquid droplets by nanovesicles. Nat. Comm. 14(1): 615.

Guha, S., J. Ghimire, E. Wu et al. 2019. Mechanistic landscape of membrane-permeabilizing peptides. Chem. Rev. 119: 6040–6085.

Hamm, M. and M.M. Kozlov. 2000. Elastic energy of tilt and bending of fluid membranes. Eur. Phys. J. E 3: 323–335.

Hanna, M.G., I. Mela, L. Wang et al. 2016. Sar1 gtpase activity is regulated by membrane curvature. J. Biol. Chem. 291(3): 1014–1027.

Hansen, A.G., N. Ramakrishnan, P.B.S. Kumar et al. 2017. Numerical insights into the phase diagram of p-atic membranes with spherical topology. Eur. Phys. J. E. 40: 32.

Hariri, H., N. Bhattacharya, K. Johnson et al. 2014. Insights into the mechanisms of membrane curvature and vesicle scission by the small gtpase sar1 in the early secretory pathway. J. Mol. Biol. 426(22): 3811–3826.

Jarsch, I.K., F. Daste and J.L. Gallop. 2016. Membrane curvature in cell biology: An integration of molecular mechanisms. J. Cell Biol. 214: 375–387.

Kasberg, W., P. Luong, M.G. Hanna et al. 2021. Unconventional COPII-mediated secretory protein trafficking continues in the absence of the Sar1 GTPase. Research Square, https://doi.org/10.21203/rs.3.rs-1168612/v1.

Kesner, L.A., Z.A. Piskulich, Q. Cui et al. 2023. Untangling the interactions between anionic polystyrene nanoparticles and lipid membranes using laurdan fluorescence spectroscopy and molecular simulations. J. Am. Chem. Soc. 145: 7962–7973.

Kim, Y.W. and W. Sung. 2001. Membrane curvature induced by polymer adsorption. Phys. Rev. E 63(4): 041910.

Kozlov, M.M. and J.W. Taraska. 2023. Generation of nanoscopic membrane curvature for membrane trafficking. Nat. Rev. Mol. Cell Biol. 24: 63–78.

Latham, A.P. and B. Zhang. 2022. Unifying coarse-grained force fields for folded and disordered proteins. Curr. Opin. Struct. Biol. 72: 63–70.

Lee, E.Y., B.M. Fulan, G.C.L. Wong et al. 2016. Mapping membrane activity in undiscovered peptide sequence space using machine learning. Proc. Natl. Acad. Sci. U.S.A 113: 13588–13593.

Lee, Y., S. Park, F. Yuan et al. 2022. Lateral compression of lipids drives transbilayer coupling of liquid-like protein condensates. bioRxiv, 2022–12.

Lee, Y., S. Park, F. Yuan et al. 2023. Transmembrane coupling of liquid-like protein condensates. Nature Commun. 14: 8015.

Liu, Z., A. Yethiraj and Q. Cui. 2023. Sensitive and selective polymer condensation at membrane surface driven by positive co-operativity. Proc. Natl. Acad. Sci. U.S.A. 120: e2212516120.

Lochbaum, C.A., A.K. Chew, X. Zhang et al. 2021. Lipophilicity of cationic ligands promotes irreversible adsorption of nanoparticles to lipid bilayers. ACS Nano 15: 6562–6572.

Lu, T., S. Liese, L. Schoenmakers et al. 2022. Endocytosis of coacervates into liposomes. J. Am. Chem. Soc. 144(30): 13451–13455.

Mandal, T., W. Lough, S.E. Spagnolie et al. 2020. Molecular simulation of mechanical properties and membrane activities of the ESCRT-III complexes. Biophys. J. 118: 1333–1343.

Mandal, T., S.E. Spagnolie, A. Audhya et al. 2021. Protein induced membrane curvature in coarse-grained simulations: Binding interface, insertion depth and lipid spontaneous curvature. Biophys. J. 120: 3211–3221.

Marrink, S.J., V. Corradi, P.C.T. Souza et al. 2019. Computational modeling of realistic cell membranes. Chem. Rev. 119: 6184–6226.

Marrink, S.J., L. Monticelli, M.N. Melo et al. 2023. Two decades of martini: Better beads, broader scope. WIREs Comput. Mol. Sci. 13: e1620.

Maximov, A., J. Tang, X.F. Yang et al. 2009. Complexin controls the force transfer from snare complexes to membranes in fusion. Science 323: 516–521.

Mitchell, M.J., M.M. Billingsley and R. Langer. 2021. Engineering precision nanoparticles for drug delivery. Nat. Rev. Drug Discov. 20: 101–124.

Mondal, S. and Q. Cui. 2022. Coacervation of poly-electrolytes in the presence of lipid bilayers: mutual alteration of structure and morphology. Chem. Sci. 13(26): 7933–7946.

Mondal, S. and Q. Cui. 2023. Coacervation induced remodeling of nanovesicles. J. Phys. Chem. Lett. 14: 4532–4540.

Mori, T., N. Miyashita, W. Im et al. 2016. Molecular dynamics simulations of biological membranes and membrane proteins using enhanced conformational sampling algorithms. Biochim. Biophys. Acta - Biomem. 1858: 1635–1651.

Murphy, C.J., A.M. Vartanian, F.M. Geiger et al. 2015. Biological responses to engineered nanomaterials: Needs for the next decade. ACS Central 1: 117–123.

Oh, Y. and Q. Cui. 2023. Non-equilibrium transport of nanoparticles across the lipid membrane. Nanoscale, 15: 12307–12318.

Ohkubo, Y.Z., T.V. Pogorelov, M. Arcario et al. 2012. Accelerating membrane insertion of peripheral proteins with a novel membrane mimetic model. Biophys. J. 102: 2130–2139.

Olenick, L.L., J.M. Troiano, A. Vartanian et al. 2018. Lipid corona formation from nanoparticle interactions with bilayers. Chem. 4: 2709–2723.

Paul, S., A. Audhya and Q. Cui. 2023. Molecular mechanism of gtp binding-and dimerization-induced enhancement of sar1-mediated membrane remodeling. Proc. Natl. Acad. Sci. U.S.A. 120: e2212513120.

Pezeshkian, W., M. Konig, T.A. Wassenaar et al. 2020. Backmapping triangulated surfaces to coarse-grained membrane models. Nat. Comm. 11: 2296.

Pfitzner, A.K., J.M. von Filseck and A. Roux. 2021. Principles of membrane remodeling by dynamic ESCRT-III polymers. Trends Biochem. Sci. 31: 856–868.

Priftis, D. and M. Tirrell. 2012. Phase behaviour and complex coacervation of aqueous polypeptide solutions. Soft Matter 8(36): 9396–9405.

Raote, I., M. Chabanon, N. Walani et al. 2020. A physical mechanism of tango1-mediated bulky cargo export. eLife 9: e59426.

Rideau, E., R. Dimova, P. Schwille et al. 2018. Liposomes and polymersomes: a comparative review towards cell mimicking. Chem. Soc. Rev. 47: 8572–8610.

Rossi, G. and L. Monticelli. 2016. Gold nanoparticles in model biological membranes: A computational perspective. Biochim. Biophys. Acta 1858: 2380–2389.

Rouches, M., S. Veatch and B. Machta. 2021. Surface densities prewet a near-critical membrane. Proc. Natl. Acad. Sci. U.S.A. 118: e2103401118.
Ruff, K.M., T.S. Harmon and R.V. Pappu. 2015. Camelot: A machine learning approach for coarse-grained simulations of aggregation of block-copolymeric protein sequences. J. Chem. Phys. 143: 243123.
Rumyantsev, A.M., N.E. Jackson, B. Yu et al. 2019. Controlling Complex Coacervation via Random Polyelectrolyte Sequences 8(10): 1296–1302.
Satarifard, V. and R. Lipowsky. 2023. Mutual remodeling of interacting nanodroplets and vesicles. Comm. Phys. 6(1): 6.
Saunders, M.G. and G.A. Voth. 2013. Coarse-graining methods for computational biology. Annu. Rev. Biophys. 42: 73–93.
Schmidt, N.W. and G.C.L. Wong. 2013. Antimicrobial peptides and induced membrane curvature: geometry, coordination chemistry, and molecular engineering. Curr. Opin. Solid State Mater. Sci. 17: 151–163.
Shen, Q.T., A.L. Schuh, Y.Q. Zheng et al. 2014. Structural analysis and modeling reveals new mechanisms governing ESCRT-III spiral filament assembly. J. Cell. Biol. 206: 763–777.
Silvestre-Roig, C., Q. Braster, K. Wichapong et al. 2019. Externalized histone h4 orchestrates chronic inflammation by inducing lytic cell death. Nature 569: 236–240.
Simunovic, M., G.A. Voth, A. Callan-Jones et al. 2015. When physics takes over: BAR proteins and membrane curvature. Trends in Cell. Biol. 25: 780–792.
Snead, W.T. and A.S. Gladfelter. 2019. The control centers of biomolecular phase separation: How membrane surfaces, ptms, and active processes regulate condensation. Mol. Cell 76: 295–305.
Sodt, A.J., A.H. Beaven, O.S. Andersen et al. 2017. Gramicidin a channel formation induces local lipid redistribution ii: A 3d continuum elastic model. Biophys. J. 112: 1198–1213.
Sodt, A.J. and R.W. Pastor. 2014. Molecular modeling of lipid membrane curvature induction by a peptide: More than simply shape. Biophys. J. 106: 1958–1969.
Sodt, A.J., R.M. Venable, E. Lyman et al. 2016. Nonadditive compositional curvature energetics of lipid bilayers. Phys. Rev. Lett. 117: 138104.
Souza, P.C., R. Alessandri, J. Barnoud et al. 2021. Martini 3: a general purpose force field for coarse-grained molecular dynamics. Nat. Meth. 18(4): 382–388.
Sudhof, T.C. and J.E. Rothman. 2009. Membrane fusion: Grappling with snare and sm proteins. Science 323: 474–477.
Towns, J., T. Cockerill, M. Dahan et al. 2014. Xsede: Accelerating scientific discovery. Comput Sci. & Engn. 16: 62–74.
Troiano, J.M., A.C. McGeachy, L.L. Olenick et al. 2017. Quantifying the electrostatics of polycation-lipid bilayer interactions. J. Am. Chem. Soc. 139: 5808–5816.
Tsanai, M., P.W.J.M. Frederix, C.F.E. Schroer et al. 2021. Coacervate formation studied by explicit solvent coarse-grain molecular dynamics with the martini model. Chem. Sci. 12: 8521–8530.
Vácha, R., F.J. Martinez-Veracoechea and D. Frenkel. 2011. Receptor-mediated endocytosis of nanoparticles of various shapes. Nano. Lett. 11(12): 5391–5395.
Valsson, O., P. Tiwary and M. Parrinello. 2016. Enhancing important fluctuations: Rare events and metadynamics from a conceptual viewpoint. Annu. Rev. Phys. Chem. 67: 159–184.
Vietri, M., M. Radulovic and H. Stenmmark. 2020. The many functions of escrts. Nat. Rev. Mol. Cell Biol. 21: 25–42.
Wang, B., L.F. Zhang, S.C. Bae et al. 2008. Nanoparticle-induced surface reconstruction of phospholipid membranes. Proc. Natl. Acad. Sci. U.S.A. 105: 18171–18175.
Wang, H.-Y., S.H. Chan, S. Dey et al. 2023. Coupling of protein condensates to ordered lipid domains determines functional membrane organization. Sci. Adv., 9: eadf6205.
Weikl, T.R. 2018. Membrane-mediated cooperativity of proteins. Annu. Rev. Phys. Chem. 69: 521–539.
Wijesinghe, S., C. Junghans, D. Perahia et al. 2023. Polydots, soft nanoparticles, at membrane interfaces. RSC Advances 13(28): 19227–19234.
Wu, Z., Q. Cui and A. Yethiraj. 2011. A New Coarse-grained Force Field for Membrane- peptide Simulations. J. Chem. Theory Comput. 7: 3793–3802.
Wu, Z. and K. Schulten. (2014, September). Synaptotagmin's role in neurotransmitter release likely involves ca^{2+} -induced conformational transition. Biophys. J. 107(5): 1156–1166.

Yoo, J. and Q. Cui. 2013. Three-dimensional stress field around a membrane protein: atomistic and coarse-grained simulation analysis of gramicidin A. Biophys. J. 104: 117–127.

Yoo, J., M.B. Jackson and Q. Cui. 2013. A comparison of coarse-grained and continuum models for membrane bending in lipid bilayer fusion pores. Biophys. J. 104: 841–852.

Yount, N.Y., D.C. Weaver, E.Y. Lee et al. 2019. Unifying structural signature of eukaryotic alpha-helical host defense peptides. Proc. Natl. Acad. Sci. U.S.A. 116: 6944–6953.

Yu, H. and K. Schulten. 2013. Membrane sculpting by f-bar domains studied by molecular dynamics simulations. PLoS Comput. Biol. 9(1): e1002892.

Yuan, F., H. Alimohamadi, B. Bakka et al. 2021. Membrane bending by protein phase separation. Proc. Natl. Acad. Sci. USA, 118: e2017435118.

Yuan, F., C.T. Lee, A. Sangani et al. 2023. The ins and outs of membrane bending by intrinsically disordered proteins. Sci. Adv. 9: eadg3485.

Zhang, Y.Q., U. Dahal, Z.V. Feng et al. 2021. Influence of hydrophobic length and flexibility of surface ligands on phospholipid membrane disruption by cationic nanoparticles. Langmuir 37: 7600–7610.

Zhang, Y.Q., A.A. Tamijani, M.E. Taylor et al. 2019. Molecular surface functionalization of carbon materials via radical-induced grafting of terminal alkenes. J. Am. Chem. Soc. 141: 8277–8288.

Zhao, Y.G. and H. Zhang. 2020. Phase separation in membrane biology: the interplay between membrane-bound organelles and membraneless condensates. Dev. Cell 55: 30–44.

Zheng, W., G.L. Dignon, N. Jovic et al. 2020. Molecular details of protein condensates probed by microsecond long atomistic simulations. J. Phys. Chem. B 124(51): 11671–11679.

Zhu, C., C.T. Lee and P. Rangamani. 2022. Mem3DG: Modeling membrane mechanochemical dynamics in 3D using discrete differential geometry. Biophys. Rep. 2: 100062.

Zhu, L., J. Wu, Y. Liu et al. 2016. Gating mechanism of mechanosensitive channel of large conductance: A coupled continuum mechanical-continuum solvation approach. Biomechan. & Modeling in Mechanobiol. 1–20.

7

Bicontinuous Inverted Cubic Phases as Tools for Studying the Mechanism of Membrane-Remodeling Proteins

David P. Siegel

Introduction

Membrane trafficking is an essential process by which patches of protein-containing lipid membranes are exchanged between organelles in a controlled manner. The process involves proteins that generate membrane curvature, driving budding and fusion of cell membrane compartments (recently reviewed in Kozlov and Taraska 2023). This involves membrane curvature-dependent binding of proteins to membranes and induction of curvature in the membranes by various processes. The processes include the insertion of hydrophobic and amphipathic protein moieties into the lipid membranes ("wedging"), the scaffolding of the membranes with mechanically stiff protein networks that impose a local curvature, and the adsorption of proteins that impose a local electrostatic or entropy-driven bending force. Many of the proteins involved in these processes bind preferentially to membranes with particular curvatures (e.g., vesicles of smaller diameters as opposed to larger ones).

The processes of vesicle fusion and budding (fission) require the transient formation of continuous bilayer-lined pores with a spool-like shape. These pores have curvatures generally, unlike those of phospholipid bilayers at equilibrium. Deforming the bilayers into such shapes can require the input of energy. The amount of energy necessary is a function of the mean and Gaussian curvatures of the bilayers in the final and initial states (Helfrich 1973, 1990). The energy of fusion/fission pores (which I will refer to in this work henceforth as pores) is a complicated function of the curvature elastic properties of the membranes, constraints on the pore size, and the membrane tension. Here, my purpose is not to explicate fusion/fission pore

7113 Gregory Creek Lane, West Chester, OH 45069, USA. Email: davesiegel20@fuse.net

energetics as a function of these parameters, which others have done elsewhere (e.g., Su et al. 2023). I merely show how peptides could affect the pore curvature energy and demonstrate the unique curvature properties of the pores once they come into existence. Both factors may affect the curvature-dependent binding of trafficking proteins.

Bicontinuous inverted cubic (Q_{II}) phases are composed of continuous, curved lipid bilayers, which form the walls of interpenetrating labyrinths of water channels (Seddon 1990; Tenchov and Koynova 2017). Q_{II} phases exhibit curvatures similar to fusion/fission pores and have similar curvature energies per lipid molecule (Siegel 2006, 2008, 2010). This is consistent with the observations that pores are the elements that assemble into the Q_{II} phase during lamellar-to-Q_{II} phase transitions (Siegel et al. 1989; Squires et al. 2002) and that fusion rates increase in proximity to the lamellar-to-Q_{II} phase transition in model membrane systems (reviewed in Siegel 2005). It has been often proposed that fusion-catalyzing proteins *in vivo* partially stabilize the curvature found in pores (recently reviewed in Siegel 2022). There is some evidence that viral fusion peptides act partly in this manner (Tenchov et al. 2013). The effects of a peptide on pore curvature energy can be estimated by measuring the effects of the peptide on the temperature and composition ranges of Q_{II} phase stability and on the Q_{II} phase lattice constant (Siegel 2022). That is the first reason to study peptides in Q_{II} phases in the context of membrane trafficking.

As discussed in detail in Siegel (2022), previous authors tried to relate the activity of antimicrobial peptides to the ability of the peptides to stabilize Q_{II} phases in certain host lipid systems. Much of that work suffered from fundamental confusion about the contribution of Gaussian curvature to the curvature energy of Q_{II} phases, as opposed to the curvature energy of the relevant lipid-protein intermediates in cell permeation. Moreover, the contribution of the host lipid curvature elastic properties to the systems was often neglected: I showed that the host lipid systems most often used in that work adopt a Q_{II} phase at nearby temperatures in the absence of peptides.

The second reason to study the Q_{II} phase in the context of membrane trafficking is that the highly-curved surfaces of pores may have unique protein-binding properties. As I will show, the monolayers of pores have qualitatively and quantitatively different curvatures and curvature gradients than the model systems, which are widely used to study curvature-dependent protein binding. The conventional systems are membrane vesicles and the tubules that are pulled out of them (Has and Das 2021). In curvature-dependent binding, proteins interact mostly with the facing monolayer of the bilayer and only penetrate more deeply into the membrane for proteins that insert transmembrane domains or translocate across the membrane (Kozlov and Taraska 2023). The curvature of the monolayer facing the solution of proteins is thus most important for determining their propensity to bind.

It is, therefore, possible that the protein-binding properties of pore surfaces are quite different than those of model membranes. Proteins may have evolved a special affinity for such surfaces, which could be important in the dynamics of such pores in membrane trafficking. To put it crudely, it is known that curvature-dependent binding of trafficking proteins is involved in pore evolution, but we have yet to study the binding of those proteins to membranes that have pore-like curvature. It should also be easier to study peptide-lipid interactions unique to pores in Q_{II} phases rather than in conventional systems. Pores are transient and make up a small fraction of the total membrane area. In contrast, all the lipid-water interfaces in a Q_{II} phase have pore-like curvature (although there are gradients in curvature within the unit cell of Q_{II} phases).

Theoretical Methods

Peptides Can Affect the Curvature Energies of Pores and Q_{II} Phases

The curvature energy G_c of a patch of phospholipid monolayer or bilayer can be expressed in terms of the Helfrich model (Helfrich 1973);

$$G_C = k\left(\frac{1}{R_1}+\frac{1}{R_2}-\frac{1}{R_0}\right)^2+\bar{\kappa}\left(\frac{1}{R_1 R_2}\right). \tag{1}$$

Equations pf the same form as Equation 1 are used to model the curvature energy of either monolayers or bilayers. Obviously, different values of the elastic constants and curvatures are applied in each case. k is a bending elastic constant, $\bar{\kappa}$ is Gaussian curvature elastic modulus, and R_0 is the spontaneous curvature of the given surface. R_1 and R_2 are the principal radii of curvature of the surface. R_0 is the spontaneous curvature of the interface. The constants k, $\bar{\kappa}$, and R_0 are composition- and temperature-dependent. The radii of curvature are evaluated with reference to the midplane of the bilayer in the bilayer case or at the neutral surface of the monolayer in the monolayer case. The neutral plane is defined as the plane at which the monolayer does not change area as the monolayer is bent. Tilt elasticity is assumed to not contribute to the curvature energy. The elastic constants of mono- and bilayer systems are measured experimentally, by physical measurements in bilayer systems or, in the case of lipid monolayers, by x-ray diffraction experiments on lipid or peptide-lipid systems in one or more lipid phases. In particular, the monolayer elastic constants k_m, R_0, and $\bar{\kappa}$ are obtained by measuring the unit cell constants of lamellar, inverted hexagonal (H_{II}), and Q_{II} phases as a function of temperature and osmotic stress (Siegel 2010; Rand et al. 1990; Kozlov 2007; Siegel and Kozlov 2004).

The curvature energy of pores is closely related to that of Q_{II} phases. The lamellar-to-Q_{II} phase transition occurs through the formation of intermediates, which are essentially pores (Siegel et al. 1989; Squires et al. 2002). When a high enough number accumulates in a stack of flat lamellar phase bilayers, they re-arrange into a Q_{II} phase lattice. Lipid and lipid-peptide mixtures that form equilibrium Q_{II} phases under certain conditions form pores easily under the same conditions.

Thus, the effects of peptides and proteins on Q_{II} phase stability can be used to estimate the effects on pore stability (Siegel 2022). If a peptide binds to a membrane surface, it must affect the curvature energy of the membrane. A membrane-inserted peptide must affect at least one and probably all of the curvature elastic constants. This is because the adsorption/insertion of the peptide must affect the lateral stress profile in the membrane, which is the distribution of lateral force as a function of height along a normal to the bilayer or monolayer. The first and second moments of this stress profile determine the values of the curvature elastic constants (Seddon 1990). Hence, peptide binding must change the values of some of these constants and, thus, the curvature energy (Equation 1). Some preliminary data links peptide effects on Q_{II} phase stability to peptide fusogenic activity (Tenchov et al. 2013). Many membrane-active peptides have been shown to change bending elastic constants (e.g., by low concentrations of viral fusion peptides; Tristram-Nagle and Nagle 2007; Tristram-Nagle et al. 2010). The Gaussian component of the curvature energy is particularly relevant to membrane fusion and fission (Siegel 2008).

Curvature Properties of Pores

Since many proteins exhibit curvature-dependent binding, it is important to determine the curvatures on the outer surfaces of pores. Here, I am not concerned with the mechanism of pore formation: it is important only that they exist and participate in trafficking processes. They can come about via the action of proteins remote from the actual site of pore formation, which introduce curvature stresses that favor pore formation (Kozlov and Taraska 2023). I model the pores as lipid bilayer structures containing integral or peripheral proteins with uniform curvature elastic coefficients but do not have a rigid protein scaffold.

Bilayers of symmetric composition have a bilayer spontaneous curvature, C_s, which is = 0. The mid-planes of such pores will conform to catenoid minimal surfaces because that minimizes the bilayer curvature energy of the pore (Figure 1). The catenoid surface is cylindrically symmetric about the axis of the (dashed line) and has a minimum "throat" diameter, *a*, which ranges between several nm to tens of nm *in vivo*. I assume no work against area elasticity is done in creating or destroying pores.

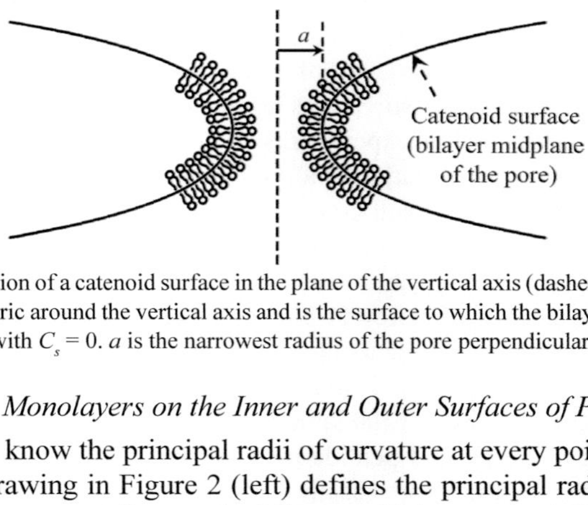

Figure 1. Cross-section of a catenoid surface in the plane of the vertical axis (dashed line). The catenoid is cylindrically symmetric around the vertical axis and is the surface to which the bilayer midplane conforms in a pore with $C_s = 0$. *a* is the narrowest radius of the pore perpendicular to the axis.

Curvature of the Monolayers on the Inner and Outer Surfaces of Pores

It is necessary to know the principal radii of curvature at every point on the surfaces of a pore. The drawing in Figure 2 (left) defines the principal radii of curvature of the constant curvature surface at the bilayer midplane. z is defined as the height along the pore vertical axis above a plane equidistant between the two original flat bilayers, and x is defined as the distance in the plane perpendicular to the pore axis. For a catenoidal pore ($C_s = 0$), the principal radii of curvature, R_p and R_m, are equal in magnitude and opposite in sign at any point.

However, the radii of curvature of the monolayers on each side of the midplane differ: the neutral surface of each monolayer lies on a surface that is vertically displaced from the midplane by a constant distance δ (Figure 2, right). This constant monolayer thickness property of minimal surfaces has been verified experimentally by an x-ray diffraction study of Q_{II} phases, where the bilayers lie on surfaces with zero mean curvature (Oka et al. 2018).

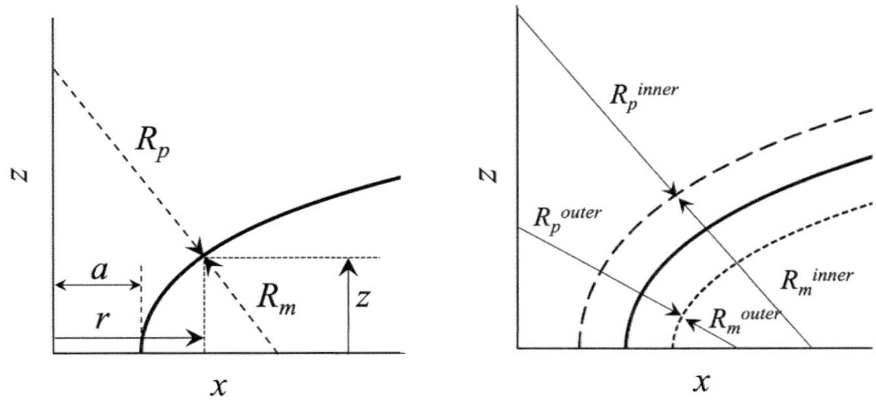

Figure 2. (Left) Definitions of the principal radii of curvature of the catenoid surface of a pore, which corresponds to the bilayer midplane, and (Right) of the inner and outer monolayers of a pore. δ is the distance along a local monolayer normal between the constant curvature surface at the bilayer midplane and the neutral plane of each monolayer. These geometric definitions of the curvatures are used for pores with all of C_s values discussed in the present work.

Because the value of d is comparable in size to the principal radii of curvature near the neck ($r \approx a$), the mean curvature of the two monolayers and the constant curvature surface can be quite different. This difference in curvatures is most severe for pores with small values of a (i.e., early-stage fusion pores and late-stage fission pores).

Monolayers of Catenoidal Pores ($C_s = 0$)

The curvatures of the inner and outer monolayers of the catenoid can be evaluated in terms of the curvature of the bilayer midplane of the catenoid, which is a catenoid surface. For any bilayer surface, the principal curvatures C_m and C_p are the inverses of each of the principal radii of curvature R_m and R_p, evaluated at the bilayer midplane. The mean curvature is the sum of two principal curvatures, and the Gaussian curvature is the product of the two:

$$C_m = -\frac{1}{R_m}, C_p = \frac{1}{R_p}, C_m + C_p = C_s, K = C_m C_p. \tag{2}$$

The midplane of the bilayer comprising the pore lies on a catenoid surface, for which $C_p + C_m = 0$ and $C_p = -C_m$. It is straightforward to express the principal curvatures at any point on this surface ($C_{m\text{-}cat}$ and $C_{p\text{-}cat}$) in terms of r and z. Recalling that for a point on a catenoid surface, $r = a \cosh\left(\frac{z}{a}\right)$, the principal curvatures are:

$$C_{m-cat} = -\frac{1}{\left[a\cosh^2\left(\frac{z}{a}\right)\right]}, C_{p-cat} = \frac{1}{\left[a\cosh^2\left(\frac{z}{a}\right)\right]}. \tag{3}$$

It is possible to express the curvatures of the monolayers on either side of the catenoid surface in terms of the curvatures of the catenoid surface. The monolayer directly surrounding the water channel along the cylindrical axis of the pore is designated the inner monolayer, and the outer monolayer has the greatest radius from the pore axis along the *x*-axis. By simple geometry (see Figure 2, right), the corresponding curvatures of the outer and inner monolayers on each side of the catenoid are:

$$C^{outer}_{m-cat} = -\frac{1}{\left[a\cosh^2\left(\frac{z}{a}\right)-\delta\right]},\ C^{outer}_{p-cat} = -\frac{1}{\left[a\cosh^2\left(\frac{z}{a}\right)+\delta\right]},\ K_{outer-cat} = C^{outer}_{m-cat}C^{outer}_{p-cat}. \tag{4}$$

and

$$C^{inner}_{m-cat} = -\frac{1}{\left[a\cosh^2\left(\frac{z}{a}\right)+\delta\right]},\ C^{inner}_{p-cat} = -\frac{1}{\left[a\cosh^2\left(\frac{z}{a}\right)-\delta\right]},\ K_{inner} = C^{inner}_{m-cat}C^{inner}_{p-cat} \tag{5}$$

Note that the monolayer neutral surfaces are not congruent with catenoid surfaces: at every point, the surfaces on which the neutral planes lie are displaced from the catenoid surface at the bilayer midplane by a constant distance $= \pm\,\delta$ along a normal to the catenoid surface. The sum of the principal curvatures of each such monolayer surface is non-zero. Note also that *z* is evaluated on the catenoid surface in Equations 4 and 5 and not on the individual monolayers.

Pores expand or contract during the membrane trafficking process. Fusion pores form at small values of *a* from small stalk-like intermediate structures and expand radially and along the *z*-axis, while fission pores start as constrictions around the waist of membrane buds, which successively contract. For a constant bilayer thickness of 4 nm, values of $a = 3$ nm thus correspond to early-stage fusion pores or late-stage fission pores, with aqueous pores with a minimal diameter of ca. 2 nm. Pores with $a =$ 10 nm correspond to late-stage fusion pores or early-stage fission pores, with aqueous pores with minimal radii of about 6 nm. The outer diameter of the narrowest part of the pore along the vertical axis (the "neck" diameter) and the minimum diameter of the water channel through pores with different values of *a* are given in Table 1.

Table 1. Dimensions of pores with different values of *a*.

	***a* = 3 nm** (Early-stage fusion pores, late-stage fission pores)	***a* = 5 nm** (Middle-stage fusion and fission pores)	***a* = 10 nm** Late-stage fusion pores, early-stage fission pores)
Outer diameter of the pore "neck"	10 nm	14 nm	24 nm
Minimum diameter of the water channel	2 nm	6 nm	16 nm

Monolayers of Pores With $C_s \neq 0$

Bilayers that are asymmetric in composition can have a non-zero value of C_s (Hossein and Deserno 2020). Many biomembranes are asymmetric. There is a family of constant mean-curvature surfaces (CMCS), which have a specific non-zero value of mean curvature, C_s, everywhere on the surface ($C_s = C_m + C_p$). Thus, to minimize the curvature energy of the resulting pores, I assume that bilayers with non-zero C_s will form pores in which the bilayer midplane lies on a CMCS with the same mean curvature. C_s can be either negative or positive. These surfaces are qualitatively similar to catenoids in the shape within their pore-like segments. When C_s is > 0, the CMCS is referred to as unduloids, and the pores correspond to the "necks" of such surfaces.

Cross-sections through pores based on CMCS are shown in Figure 3. The curves depict the height along z of the constant curvature surface (the bilayer midplane) as a function of r; for $a = 3$, 5, and 10 nm, respectively. Curves with $C_S = 0$ are catenoid curves, shown for comparison. Curves are also displayed for CMCS with $C_s = \pm\, 0.01$ and $\pm\, 0.03$ nm^{-1}. This range of bilayer spontaneous curvature covers a wide range of possible composition asymmetry. For comparison, a bilayer with one leaflet composed of pure DOPC and the other of pure POPC would have $|C_s| = 0.017$ nm^{-1}, with the sign depending on which side the POPC is on. A bilayer composed of pure DOPE in one leaflet and pure DOPC in the other would have $|C_s| = 0.1$ nm^{-1} (Hossein and Deserno 2020).

The geometry and principal curvatures of the monolayers on CMCS are calculated using the method by Markin and Albanesi (2002). Values of $z(r)$ for a surface with a given value of C_s are obtained by numerical integration of the following expression, with small increments for r values near a:

$$\frac{dz}{dr} = \left\{ \frac{1}{\left[\frac{rC_s}{2} + a\left(1 - \frac{C_s a}{2}\right)(1 \backslash r) \right]^2} - 1 \right\}^{-1/2} \tag{6}$$

Let C_p and C_m be the curvatures of the bilayer midplane on the surface as a function of r, using the same geometric definitions as in Figure 2. C_s is the mean curvature of the CMCS. Then:

$$C_p = \frac{C_s}{2} + (a/r^2)[1 - aC_s/2]; \qquad C_m = C_s - C_p. \tag{7}$$

By the definitions in Figure 2, the principal curvatures for the inner and outer monolayers, $C_m^{\,inner}$, $C_p^{\,inner}$, $C_m^{\,outer}$, and $C_p^{\,outer}$ are given by:

$$C_m^{\,inner} = \frac{C_m}{\delta C_m - 1}, \qquad C_p^{\,inner} = \frac{C_p}{\delta C_p - 1}, \tag{8}$$

$$C_m^{\,outer} = \frac{C_m}{\delta C_m + 1}, \qquad C_p^{\,outer} = \frac{C_p}{\delta C_p + 1} \tag{9}$$

Again, the mean curvature of a surface at a given point is the sum of the two principal curvatures, and the Gaussian curvature is the product of the two inverses. The profiles of the pores with $C_s < 0$ become continuous with the surrounding planar bilayers at a value of r (r_{max}) given by:

$$r_{max} = \sqrt{a^2 - \frac{2a}{C_s}} \tag{10}$$

For pores with $C_s \geq 0$, the value of z grows continuously with increasing r, so there is no smooth transition to the surrounding bilayers, which are assumed

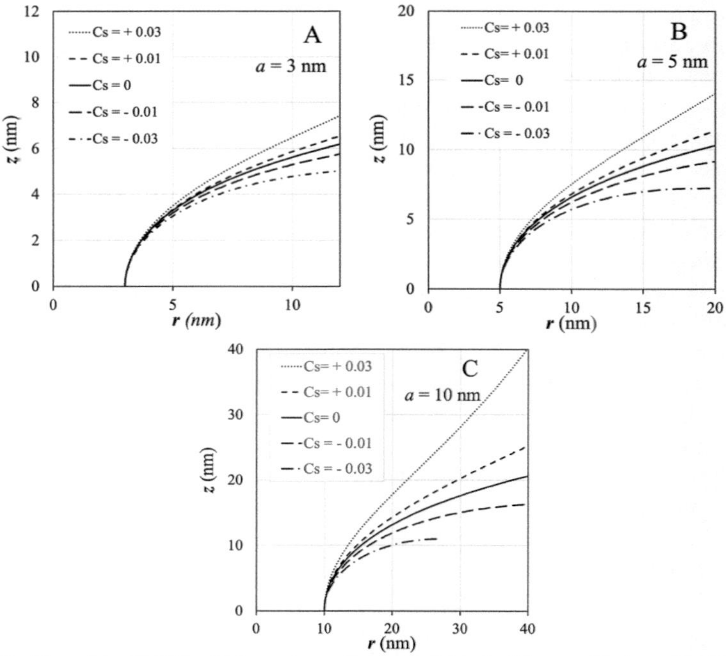

Figure 3. Profiles of the CMCS, which are the surfaces on which the bilayer mid-planes lie, for systems with non-zero values of C_s. (**A**) $a = 3$ nm. (**B**) $a = 5$ nm. (**C**) $a = 10$ nm. The catenoid surface ($C_s = 0$) is shown for comparison.

to be planar, at large values of r. The modest curvature of the membranes of the fusing compartments *in vivo* may match with the non-planar membranes at the pore periphery, eliminating the need for surfaces with intermediate values of curvature. If each of the compartments is a segment of a sphere with radius = R_m, there is always some value of r_{max} for which they become smoothly continuous with the pore at that value of r for which $C_p = 1/R_m$. Moreover, tension exerted on the membrane during pore evolution affects its shape (Su et al. 2023), an effect that is not considered here. Our primary interest is the high-curvature region of the pore at $r \approx a$.

Curvature Properties of Model Systems Presently Used for Protein-Binding Experiments

It is useful to compare the range of K predicted for CMCS-based pores with the local values of K found in vesicles and tubules used in protein-binding experiments. Let R_v and R_t be the radii of bilayer vesicles and tubules used in protein-binding experiments, respectively, and h equals the bilayer thickness (typically = 4 nm). For tubules, the principal curvature in the plane of the length of the tube is essentially zero. The inner surfaces of tubules can be used for protein-binding experiments if the tubules are pulled out from cells or protein-filled GUVs. Then, the values of mean curvature and Gaussian curvature on the inner and outer surfaces are approximately:

$$\text{Vesicles: } C^{inside}_{vesicle} = -\frac{2}{(R_v - h)}; \quad C^{outside}_{vesicle} = +\frac{2}{(R_v + h)}; \qquad (11)$$

$$K^{inside}_{vesicle} = \frac{2}{(R_v - h)^2}; \quad K^{outside}_{vesicle} = + \frac{1}{(R_v + h)^2} \tag{12}$$

$$\text{Tubules:} \quad C^{inside}_{tubule} = - \frac{1}{(R_t - h)}; \quad C^{outside}_{tubule} = + \frac{1}{(R_t + h)}. \tag{13}$$

$$K_{tubule} = 0 \text{ (inside and outside).} \tag{14}$$

Generally, in binding experiments, R_{tublue} is ≥ 15 nm and $R_{vesicle}$ is typically ≥ 25 nm; generally, only the outside surfaces of the vesicles and tubules are available for binding (Has and Das 2021). However, in some cases, the lumens of tubules might be accessible to proteins within cells and GIUVs. With these values, the outside surfaces of vesicles offer mean curvature in the range ≈ 0 to 0.08 nm^{-1} and $K \approx 0$ to 0.002 nm^{-2}. Tubules pulled from cells of GUVs offer $C <$ ca. 0.1 on their outer surfaces and $C > -0.17$ nm^{-1} in their lumens. In both cases, the K of the tubule monolayers is 0 because the principal radius of curvature along the tub axis is essentially infinite over most of the tubule. Except for regions near the site where the tubule is pulled from a cell of GUV, the gradients in mean curvature and Gaussian curvature are small in these systems.

Equilibrium Curvature and Dimensions of Q_{II} Phases

In Q_{II} phases, the monolayers have an area-average negative mean curvature and a negative Gaussian curvature. The equilibrium values of mean and Gaussian curvature values in phospholipid Q_{II} phases are close to the values predicted for CMCS-based pores.

Q_{II} phases of three different geometries are observed in phospholipid systems. Each is based on a different infinite periodic minimal surface. Their space groups Im3m, Pn3m, and Ia3d; the three phases are usually referred to as P, D, and Q phases, respectively. More than one is frequently observed simultaneously; P and D most commonly. The values of K vary across the unit cell of all three phases. The area-average Gaussian curvature of a monolayer within a unit cell of Q_{II} phase, $<K_m>$, and the area-average mean curvature $<J_m>$, of the monolayers in Q_{II} phases are given by (Siegel 2006, 2010):

$$\langle K_m \rangle \simeq \frac{S_1}{S_0 c^2} \tag{15}$$

and

$$\langle J_m \rangle \simeq 2\delta \langle K_m \rangle \tag{16}$$

(The average curvatures of the two monolayers of each bilayer in Q_{II} phases are equal.) S_1 and S_0 are geometric constants that are defined by the geometry of each of the three phases, and c is the unit cell constant of the particular Q_{II} phase. c is the Euler characteristic of each phase. Values of the relevant constants for each of the commonly-observed Q_{II} phases are given in Table 2.

The average radius of the water channels in the Q_{II} phase, R_w, is given by:

$$R_w = \left[\sqrt{\frac{S_0}{|S_1|}}(c) - l \right] \tag{17}$$

Table 2. Values of S_0, S_1, and S_2 for commonly-observed Q_{II} phases.

	D (Pn3m)	P (Im3m)	G (Ia3d)
χ	–2	–4	–8
S_0	1.91889	2.34510	3.09144
S_1	-4π	-8π	-16π
S_2	100.294	328.270	996.071

The value of K varies within the unit cell of the Q_{II} phases, ranging between 0 and a value more negative than given by the area-average value in Equation 15. The area-averaged standard deviation in local values of K within the unit cell, SD_K, is given by:

$$SD_K \simeq \left(\sqrt{\frac{S_2 S_0}{|S_1^2|}} - 1 \right) \left[\frac{S_1}{S_0 c^2} \right] \tag{18}$$

A similar estimate for the gradient in mean curvature is derived from Equation 16 by multiplying SD_K by 2δ. A rough estimate of the gradient in Gaussian curvature on the monolayer surface is therefore:

$$\frac{dJ_m}{d_x} \approx \frac{2\delta(\langle K \rangle + SD_K)}{R_w} \tag{19}$$

where x is the distance along the monolayer neutral plane. Finally, the total lipid-water interface area in a unit cell of Q_{II} phase is approximately $2S_0c^2$.

Results and Discussion

Studying Q_{II} Phase Stability Can Show How Polypeptides Affect Membrane Curvature and Curvature Energy

Numerous studies have claimed to relate curvature-related effects of peptides on Q_{II} phase formation to peptide activity in model systems and *in vivo*. Many of these are reviewed elsewhere (Siegel 2022), especially those related to the activity of antimicrobial peptides (AMP), membrane transiting peptides, and viral fusion peptides. Interpretation of the results has often been confused by inadequate assessment of the phase behavior of the host lipid itself and by confusion about the role that curvature elastic energy could play in a given process (Siegel 2022). Nevertheless, theory and data support the possibility that fusion-catalyzing peptides act, at least in part, by reducing the curvature energy of Q_{II} phases and pores.

Some experiments showed that peptides associated with fusion do stabilize Q_{II} phases and thus pores. In both, fairly hydrophobic peptides were incorporated into the lipids by conventional methods (co-solvation and drying), as opposed to equilibrating with a pre-formed, pure-lipid Q_{II} phase. Tenchov et al. (2013) made lipid-fusion peptide mixtures using viral fusion peptides from influenza hemagglutinin and used temperature cycling to form the Q_{II} phase. Using SAXS, they showed that peptides corresponding to hemagglutinins with fusion activity stabilized Q_{II} phases, while an inactive mutant had almost no effect. This would be consistent with the peptides reducing the curvature energy of fusion pores. Ideally, this result should be tested in more biomembrane-mimetic systems and more thoroughly quantified (i.e., what mole fraction of bound peptide is required to achieve a given reduction in fusion pore curvature energy?). A large influence on pore energy would suggest that curvature energy reduction is a principal part of the mechanism. If so, this would be a new basis for developing fusion peptide structure/function relationships. The extent of the reduction in pore energy can be estimated by the change in lamellar/Q_{II} phase transition temperature achieved by the peptide (Siegel and Kozlov 2004). ^{31}P-NMR can be used as a relatively fast way of estimating the size of this effect. Samples containing bits of Q_{II} phases or even a few percent of the membrane area present as pores give characteristic isotropic resonances (Tenchov and Koynova 2017).

Another study (Siegel et al. 2006) used WALP peptides (Weiss et al. 2003) as models of the membrane-spanning regions of integral membrane proteins. The host lipid was N-mono-methylated dioleoylphosphatidylethanolamine or DOPE-Me. The aim was to determine if the transmembrane domains of proteins surrounding fusion sites play a role in catalyzing membrane fusion. Transmembrane domains can stress the membrane if their hydrophobic domains are thinner or thicker than the equilibrium lipid bilayer thickness. WALP proteins effectively lowered the lamellar-Q_{II} phase transition temperature in a membrane thickness mismatch-dependent manner. The reduction was as much as 25 K at a concentration of only 0.5 mol % for the shortest and longest peptides. In contrast, the peptides did not affect the lattice dimensions of the H_{II} phase (which normally coexists in pure-lipid samples of DOPE-Me in the relevant temperature range, Cherezov et al. 2003). The longest peptide, W31, even suppressed H_{II} phase formation. When taken together, these results suggest that the small membrane thickness gradients set up by the WALP peptide/bilayer thickness mismatch destabilized the lamellar phase and stabilized the Q_{II} phase. The bilayers in a protein-free Q_{II} phase have a nearly constant thickness (Oka et al. 2018). The effect on lamellar-to-Q_{II} phase transition temperature corresponds to a large effect on fusion pore energy. Using data for pure DOPE-Me (Siegel 2008), the effect of 0.5 mol % of WALP19 corresponds to a ca. 27 k_BT reduction in fusion pore energy. This implies that the mere presence of length-mismatched TMDs at the periphery of the fusion or fission site can reduce the energy required to create a pore.

These examples show that the Q_{II} phase can illuminate the surprising effects of peptides on the curvature energy of peptide-lipid intermediates in fusion and other membrane processes.

Comparison of the Curvature Properties of Pores, Q_{II} Phases, and Conventional Model Systems

Local Mean Curvature on a CMCS-Based Pore Surface

Figure 4 contains plots of the local mean curvature, evaluated at the constant curvature surface, at different values of z on pores with different values of C_s, for (Figure 4A) $a = 3$, (Figure 4B) $a = 5$, and (Figure 4C) $a = 10$ nm.

The disparity between the mean curvature of the midplane and the mean curvature of the monolayers at the neck of the pore is predicted to be greatest for pores with small minimum diameters (e.g., $a = 3$ nm, Figure 4A). In Figure 4A, the minimal local value of the mean curvature of the outer monolayer is less than -0.4 nm^{-1}. This value is even smaller than the mean curvature of phospholipid H_{II} phases, which is typically about -0.4 to -0.33 nm^{-1}. This may seem surprising at first since, by definition, for the catenoid, the curvature of the bilayer midplane is zero. However, we must recall that one of the two principal radii of curvature is always negative, and this radius is quite small at the neck. For the outer monolayers, with $a = 3$, $d = 1.3$ nm, and $C_s = 0$, at the neck of the pore $C_m^{outer} = -1/(3 - d) = -1/1.7\ nm^{-1} = -0.59\ nm^{-1}$, $C_p^{outer} = +\ 0.23\ nm^{-1}$ (Equation 8). For all values of a and C_s, the most severe curvature occurs in a narrow band of z near the neck of the pore. At larger r, the mean curvature approaches C_s of the constant curvature surface. Also, for larger values of a (Figures 4B and 4C), the values of mean curvature near the neck also approach zero. For a given value of a, the mean curvature is smallest for negative values of C_s and increases as C_s increases. However, note that the local mean curvature at the neck is still considerably more negative than the value of C_s for the CMCS in each case.

The local mean curvature of the inner monolayers depends in a qualitatively similar manner on *a*, C_s, and *z* (data not shown). On the inner monolayer near the neck, one principal radius of curvature of the CMCS is negative and small compared to the corresponding principal radius of the CMCS near the neck of the pore. Hence,

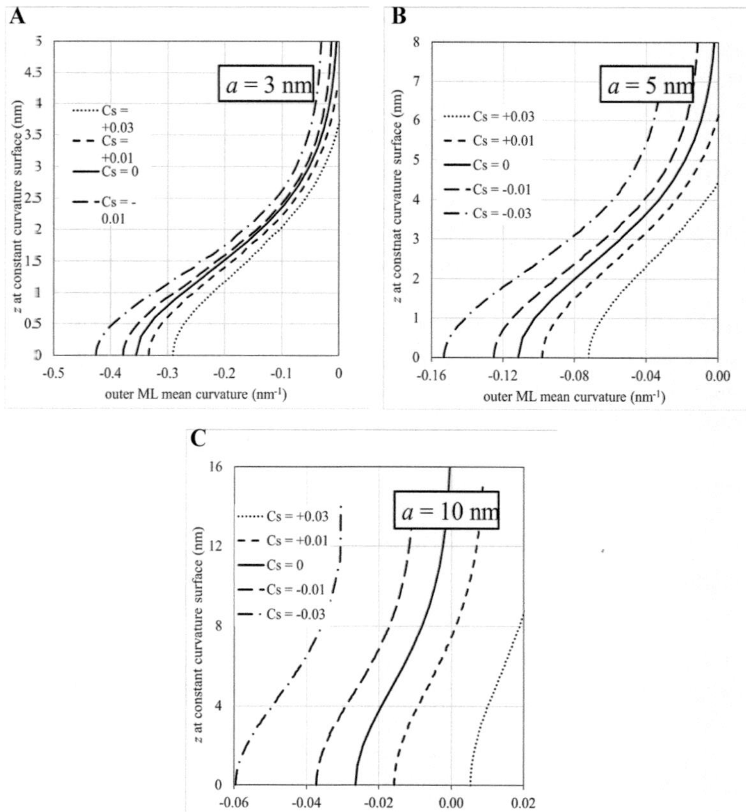

Figure 4. Plots of the mean curvature of the outer monolayers of CMCS-based pores as function of height *z*, for *d* = 1.3nm and three different values of the minimal pore diameter, a. Values of C_s are given in units of nm^{-1}. **A:** *a* = 3 nm. **B:** *a* = 5 nm. **C:** *a* = 10 nm.

the local mean curvature of the inner monolayer is negative and comparable to that of the inner monolayer for the indicated range of C_s.

There is a large gradient in the mean curvature near the neck of the pores along the z direction. Such gradients might influence peptide binding if they are on the length scale of a typical peptide. Consider the curvature gradient over a length comparable to the size of a typical alpha helix. The gradient in mean curvature over a distance of 1 nm along the z-axis is 0.083, 0.013, and 0.0005 nm^{-2} for $C_s = 0$ and $a = 3$, 5, and 10 m, respectively. An alpha helix is ca. 1 nm in width and 6.7 residues per nm length of a stable alpha helix. Depending on the orientation of the bound peptide's axis with respect to the pore axis, a distance along the z-axis of 1 nm can correspond to a longer distance along the curved surface of the pore.

For the $a = 3$ nm case, the size of the curvature change over the length of the peptide can be visualized as follows: this gradient is equivalent to the first reside of a helix experiencing a flat membrane locally (zero curvature), while a residue 14 or 15 residues further along the helix would experience a local curvature of 0.16 nm^{-1}, consistent with the surface of a spherical vesicle with a near-minimal radius of only 12 nm (as in sonicated unilamellar vesicles). It is also possible that sequences have evolved to bind preferentially to surfaces with large curvature gradients, which would be relevant to early-stage fusion pores and late-stage fission pores.

Local Gaussian Curvature on a CMCS-Based Pore

The total Gaussian curvature of a pore is determined only by the topology of the system. The formation of a pore lowers the Gaussian curvature of the entire system (the fused vesicles compared to the two initial vesicles) by an amount which, integrated over the total surface area, equals –4p. However, the local value of the $\bar{\kappa}$ can vary over the surface depending on the local composition. Therefore, the local value of $\bar{\kappa}$, and the Gaussian curvature of a given patch of membrane determines the Gaussian curvature elastic energy necessary to deform that patch into its final shape (Equation 1). The local value of K of a patch of monolayer surface may also influence the binding of peptides to that patch. For example, several different membrane-bound proteins sort according to the Gaussian curvature of the host membranes, preferring vesicles ($K > 0$) to cylindrical tubules ($K = 0$), which have the same value of mean curvature as the vesicles (Larsen et al. 2020). Hence, it is important to determine how the value of K changes on the pore surface.

Figure 5 contains plots of the local Gaussian curvature, evaluated at the constant curvature surface, at different values of z on pores with different values of C_s, for (Figure 5A) $a = 3$, (Figure 5B) $a = 5$, and (Figure 5C) $a = 10$ nm. The most negative values of K occur near the necks of the pores, as in the case of the mean curvature. About 90% of the area-integrated Gaussian curvature of a catenoidal pore, for example, is found on the surface within values of z between 0 and ± 1.47a. The local Gaussian curvature is < 0 everywhere on the monolayer surface and is the smallest (most negative) for smaller values of a and the most negative values of C_s. The values of K and the distribution of K on the inner monolayer of the pore are similar (data not shown).

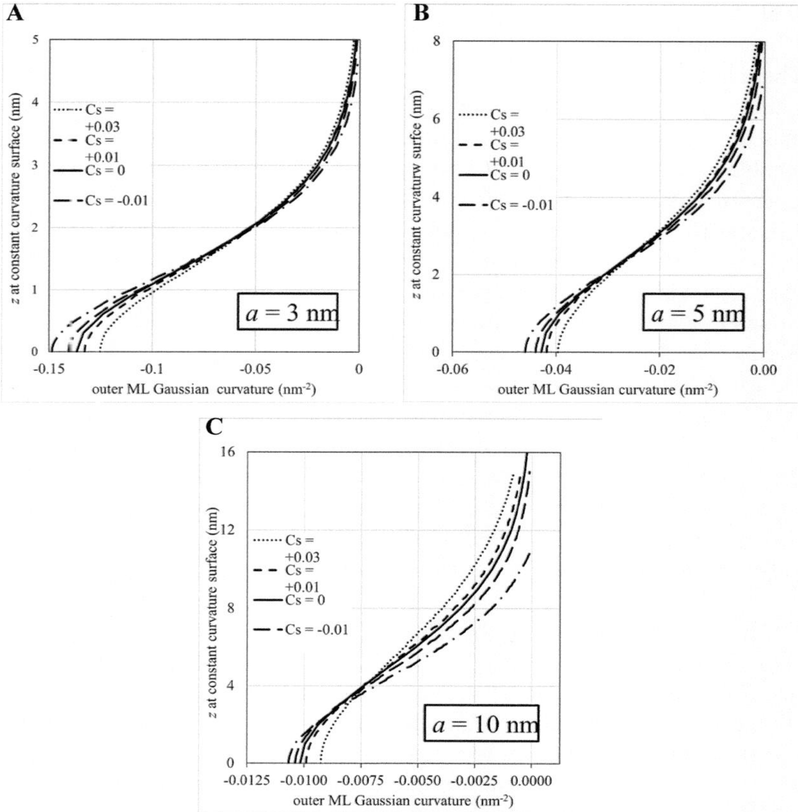

Figure 5. Plots of the local Gaussian curvature of the outer monolayers of CMCS-based pores as a function of height, z, for d = 1.3 nm and different values of a. **A**: a = 3 nm. **B**: a = 5nm. **C**: a = 10 nm.

Table 3. Values of curvature and curvature gradient predicted for fusion/fission pores, compared to value in the Q_{II} phase and model systems.

	"Constricted" (a = 3 nm)	**"Middle stage" (a = 5 nm)**	**"Expanded" (a = 10 nm)**	**Q_{II} phase** (Pn3m, c = 11–28 nm)	**Liposomes & tubules** (R_{ves} > 25 nm, R_{tub} > 15 nm)
K (nm^{-2})	> –0.13	> –0.04	–0.01	–0.054 to –0.0084	0 to 0.002
C (nm^{-1})	> –0.36	> –0.11	-0.025	–0.14 to –0.022	0 to 0.08 (in tubule lumens, > –0.17)
Gradient in *C* (for the pores, over Δz = 1 nm, measured at the neck; nm^{-2})	+ 0.08	+ 0.009	+ 0.0005	+ 0.019 to + 0.029	essentially none

Comparing the Curvatures of Pores With Q_{II} Phases and Conventional Model Systems

The range of mean curvature and Gaussian curvature and the gradient in mean curvature, predicted for pores, is compared with the same quantities observed in Q_{II} phases and in vesicles and tubules in Table 3. Each comparison is done for three values of a, corresponding to constricted pores (a = 3 nm), middle-stage pores (a = 5 nm), and expanded pores (a = 10 nm). The values displayed for the Q_{II} phase are the range of values found in Pn3m phases over the range of frequently observed unit cell sizes (11–28 nm).

The values in Table 3 make four things clear. First, pores have significant negative Gaussian curvature and share this property with Q_{II} phases, while the model systems do not. The model systems have zero or positive Gaussian curvature. Second, both negative mean curvature and negative Gaussian curvature are simultaneously present in both pores and Q_{II} phases. Third, the constricted pores express even more negative curvature than the lumens of tubules. Finally, pores and Q_{II} phases exhibit significant gradients in curvature, which are essentially absent in the model systems.

It is possible that proteins involved in membrane trafficking have evolved to bind preferentially, or differently, to such surfaces compared to those found in the model systems. The effects of such differences in curvature are predicted to be largest for early-stage fusion pores or late-stage fission pores.

Using Phospholipid Q_{II} Phases to Study Curvature-Dependent Membrane Binding

It is clearly desirable to use phospholipid Q_{II} phases as substrates in protein-binding studies. To the author's knowledge, this has not yet been attempted in phospholipid systems. One reason may be because it is not straightforward: Q_{II} phases form in a small subset of phospholipid systems; the transitions between lamellar and Q_{II} phases (during the preparation of the system) can be slow and hysteretic and frequently more than one Q_{II} phase co-exist. In addition, it can be difficult to equilibrate the Q_{II} phase with proteins while maintaining the proteins in their native state using conventional sample preparation methods (i.e., preparing lipid/peptide phases de novo by co-solvation and freeze-thawing). Finally, diffraction patterns obtained from Q_{II} phases are often weak, making SAXS experiments time-consuming.

Why are Phospholipid Q_{II} Phases a Better Model System Than Monoglycerides

Q_{II} Phases? Monoglycerides are well-studied (Czeslick et al. 1995; Briggs et al. 1996; Qui and Caffrey 2002) and easily form Q_{II} phases near room temperature. However, monoglyceride systems are poor models for biomembranes. Obviously, trafficking proteins have evolved to affect the dynamics of phospholipid-based membranes. There are also important physical chemical differences between monoglyceride and phospholipid Q_{II} phases. In monoglycerides, there are substantial attractive van der Waals forces between the closely-spaced bilayers of the lamellar phase (Pezron et al. 1990), which are a major barrier to Q_{II} phase formation (Siegel and Tenchov 2008) in bulk lipid phases. This can introduce substantial additional hysteresis in forming the Q_{II} phase. It also means that the effects of added proteins on inter-membrane forces would have to be considered. Finally, the inter-bilayer water layers in monoglyceride lamellar phases are much thinner than in phospholipids (Pezron et al. 1990), and the water channels in the Q_{II} phase are also comparatively small (Czeslick et al. 1995; Briggs et al. 1996). As we will see below, this limits the size of the peptides that can diffuse into the Q_{II} phase. Moreover, since the water contents of the lamellar and Q_{II} phases are different, water must be extruded or imbibed during these transitions. The small spacings slow the rate of water transport, adding some additional hysteresis. When studying the influence of a given peptide on pore stability when using monoglyceride systems (e.g., Surmeier et al. 2022), it is hard to deconvolute these kinetic effects of a peptide from the effects on curvature properties. The larger inter-bilayer spacings and Q_{II} phase water channel diameters in phospholipid systems greatly reduce the adhesion energy and water transport effects.

One advantage of monoglycerides and their derivatives is that they are more chemically stable (Ischenko et al. 2017) than phospholipids (Evistigneeva 1993). Phospholipids can hydrolyze slowly in water. Hydrolysis produces lysolipids, fatty acids, and glycerides, which change the electrostatic charge density and curvature elastic constants of the system. Ideally, studies using phospholipids should be conducted within hours or days to minimize this effect.

An Assay for Curvature-Dependent Protein Binding Using Q_{II} Phase as a Membrane Substrate

I propose an experiment using columns of ca. 0.5 cm in size of pre-formed phospholipid Q_{II} phase in narrow capillary tubes. It should be possible to equilibrate these lipid samples with polypeptides by diffusion from an overlying aqueous phase. The extent of polypeptide incorporation into the Q_{II} phase lipid samples could be determined using native UV-VIS absorbance or fluorescently labeled polypeptides. The amount of polypeptide binding would be normalized to the amount of lipid present in the capillary and compared to the extent of binding in a conventional liposomal dispersion of nearly identical lipid composition. (The Q_{II} phase-forming lipid can be stabilized in the lamellar phase for this purpose by, for example, adding a few mole % of phosphatidylcholine or a smaller amount of lipid coupled to a high molecular-weight PEG. In comparing the binding, the membrane-sidedness of liposomal systems also has to be taken into account.) Fully-swollen Q_{II} phases have a low volume fraction of water phase (generally less than 80 wt. % water) (Siegel 2010). The water contains a negligible amount of solubilized peptide compared to the amount bound to the lipid if the protein-binding constant is significant. The time scale of the experiment is limited mostly by the lipid-polypeptide equilibration time, t. To protect the sample from extensive hydrolysis, we should limit t to several days.

Many capillaries could be made at once. A concentrated stock solution of the lipid in a volatile solvent like chloroform/methanol could be injected into each via a Hamilton syringe. The capillaries would be dried under a vacuum and then individually rehydrated by syringe and subjected to appropriate lipid equilibration protocols in parallel, including freeze-thaw and temperature cycling. Concentrated polypeptide stock solutions could be introduced in a thin layer over the Q_{II} phase columns in the capillaries similarly. Then, the capillaries should be flame-sealed or sealed with a cement or a wax plug.

The capillaries should be monitored via SAXS before and after protein addition to determine the Q_{II} phase composition of the sample, requiring access to an intense X-ray source with an area detector. The three commonly observed Q_{II} phases (Pn3m, Im3m, and Ia3d) are expected to all have the same value of $<K_m>$ and area/lipid at the water-lipid interface in the presence of excess water (Siegel 2008, 2010). Note that $<K_m>$ has the same value in the three commonly observed Q_{II} phases (Pn3m, Im3m, and Ia3d) when the phases have different values of c (Equation 15, Table 2). At equilibrium in excess water, the different phases should all have the same value of $<K_m>$, and thus their unit cell constants are inter-related (e.g., Siegel 2008, 2010): $c_{Im3m} = 1.28c_{Pn3m}$ and $c_{Ia3d} = 1.58c_{Pn3m}$. The value c is determined by the curvature elastic constants of the peptide-lipid composition (Siegel 2006, 2010, 2022). If the value of $<K_m>$ decreases upon exposure to a peptide (e.g., if c of the Q_{II} phases decreases), then the peptide energetically stabilizes the Q_{II} phase and also stabilizes pores.

What Determines the Equilibration Time?

Proteins with a surprisingly large hydrodynamic radius should be able to diffuse into short columns of phospholipid Q_{II} phase since R_w is in the range of 2.5–9 nm (DOPE-Me Q_{II} phases, using data from (Cherezov et al. 2003) in Equation 17). The value of t is determined by how fast the test polypeptide can diffuse into the entire length column of the Q_{II} phase. Water-soluble molecules of substantial size can enter and diffuse within samples of monoglyceride Q_{II} phases over samples as large as 6 mm in length (Meikle et al. 2017; Ghambari et al. 2017). This transport process has been analyzed theoretically (Assenza and Mezzenga 2018). Molecules having hydrodynamic radii less than R_w should be able to equilibrate with all the lipid/water interface within a column of pre-formed Q_{II} phase and diffuse into it with an effective diffusion coefficient $D_{eff} \approx D_{water}/4$, where D_{water} is the coefficient in bulk water. Proteins with molecular weights of tens and even hundreds of kDa have Stokes radii in the size range of R_w (Erickson 2009).

What Range of Polypeptide Molecular Weights Is Usable?

A 1 mm-diameter column of Q_{II} phase about 5 mm tall is a sufficient volume of lipid phase, which can be reliably examined by SAXS. It is also large enough for quantitative measurement of the fluorescence intensity of incorporated proteins (for example, in an adapted SDS gel scanner). The random-walk diffusion time t of a protein down the axis of a column of the Q_{II} phase of height H is $t \sim H^2/2D_{eff}$. For $H = 5$ mm, proteins with $D_{eff} > 3.5 \times 10^{-7}$ cm^2/s should be able to equilibrate with the entire Q_{II} phase sample within less than about 100 hours (Polypeptide binding is driven by a favorable free energy change, which would accelerate transport. Therefore, equilibration probably occurs in a shorter time). Via the Stokes-Einstein equation, a polypeptide with $D_{eff} = 3.5 \times 10^{-7}$ cm^2/s has a Stokes radius of about 1.5 nm and a molecular weight of roughly 5 kDa. Thus, the proposed method would work best for proteins with molecular weights of ca. 5 kDa or less. Of course, it is possible to diffuse small-molecule cofactors like GTP into the Q_{II} phase, along with or after equilibration with the protein in question, without substantially affecting the value of t for the polypeptide.

The equilibration time is also affected by the concentration of polypeptide in the aqueous stock solution. The concentration should be as high as possible, and/or one or two changes to the overlying solution should be made during equilibration. The apparent concentration of lipid in a volume of the Q_{II} phase will be > 0.2 mM. If the target peptide/lipid ratio (mol/mol) = 1/100, then 2 micromoles of protein/cc must be delivered. If the volume of the peptide stock solution is approximately twice that of the column of the Q_{II} phase, this means that the concentration of polypeptide in the stock solution must ≈ 10 mM.

Which Phospholipid Systems Are Likely to Have the Required Phase Behavior?

Ideally, the lipid composition should mimic biomembranes involved in trafficking and be almost completely in the Q_{II} phase at near-physiological pH, ionic strength, and temperature. Moreover, there must be enough water so that Q_{II} phases can exist in excess water conditions. Phospholipid Q_{II} phases are generally made by dispersing lipids in the lamellar phase and then heating the system to form the Q_{II} phase; sometimes by temperature cycling the system across the lamellar/inverted hexagonal (H_{II}) phase boundary (Tenchov and Koynova 2017; Cherezov et al. 2003). Q_{II} phase formation can be quite hysteretic in such systems. The influence of polypeptides may evolve over hours during or after equilibration. It is likely that temperature cycling and perhaps non-physiological concentrations of certain salts in the buffer may be required (Tenchov and Koynova 2017).

Candidate Lipid Systems

Based on previous observations, I propose that at least three other systems with a biomembrane-mimetic composition are probably capable of forming nearly pure Q_{II} phase samples. Some fine-tuning of the compositions may be necessary. Detection of isotropic ^{31}P-NMR resonances has proven to be a good experimentally convenient indicator of sponge and/or Q_{II} phase formation in phospholipid systems for initial studies. SAXS studies are required to confirm Q_{II} phase formation and measure *c*.

DOPE/DOPC/Cholesterol/Anionic Lipid: Tilcock et al. (1982) used ^{31}P-NMR to show that a sponge phase or some Q_{II} phase coexisted with a lamellar phase in equimolar DOPE/DOPC/Cholesterol at 40°C. This mole fraction of phosphatidylethanolamine may need to be adjusted, and a small mole fraction of anionic lipid may be necessary to both stabilize the Q_{II} phase and reflect the presence of PS, PI, PIP_n, or PA in biomembranes.

POPE/POPG/Cholesterol: A 3/1 mixture of POPE and POPG (1-palmitoyl-2-oleoyl-phosphatidlyethanolamen and phosphatidylglycerol, respectively) was recently shown to form sponge or Q_{II} phases via ^{31}P-NMR at approximately 60°C in excess water (Siegel 2022). This system is of interest because it is often used as a mimic of bacterial membrane composition. The observations suggest that adding some more PE, or with a small amount of divalent cation, or a small mole fraction of cholesterol, will stabilize the Q_{II} phase at lower temperatures.

Either of the Two Above Compositions With Shorter, Unsaturated Acyl Chains. Q_{II} phase forms more quickly in di-(C16:1-D9)-PE than in di-(C18:1-D9)-PE upon temperature cycling at around room temperature (Tenchov and Koynova 2017).

It has been observed that decreasing the chain length tends to stabilize the Q_{II} phase in many inverted phase-forming lipid compositions (Tenchov and Koynova 2012).

Conclusion

The outer surfaces of pores have unique combinations of curvatures that are not presented by the conventional model systems used to study curvature-dependent protein binding. It is known that curvature-dependent binding of trafficking proteins is involved in pore evolution, but the binding of proteins to membranes with pore-like curvature has not been studied. Known or presently unknown trafficking/remodeling proteins may have evolved to bind preferentially to such surfaces. Successful implementation of an assay using the Q_{II} phase as the substrate would cover an important blind spot in our knowledge of trafficking and membrane remodeling. It could produce further insights into the structure-function relationships of proteins involved in those processes. Moreover, some of the effects of peptides on pore energies could be studied by investigating the effects of the peptides on Q_{II} phase stability.

References

Anderson, D.M., S.M. Gruner and S. Leibler. 1988. Geometrical aspects of the frustration in the cubic phases of lyotropic liquid crystals. Proc. Nat'l. Acad. Sci. 85: 5364–5368.

Assenza, S. and R. Mezzenga. 2018. Curvature and bottlenecks control molecular transport in inverse bicontinuous cubic phases. J. Chem. Phys. 148: 054902.

Briggs, J., H. Chung and M.C. Caffrey. 1996. The temperature-composition phase diagram and mesophase structure characterization of the monoolein/water system, J. Phys. II France. 6: 723–751.

Cherezov, V., D.P. Siegel, W. Shaw et al. 2003. The kinetics of non-lamellar phase formation in DOPE-Me: relevance to biomembrane fusion. J. Membr. Biol. 195: 165–182.

Czeslick, C., R. Winter, G. Rapp et al. 1995. Temperature- and pressure-dependent phase behavior of monoacylglycerides monoolein and monoelaidin. Biophys. J. 68: 1423–2429.

Erickson, H.P. 2009. Size and shape of protein molecules at the nanometer level determined by sedimentation, gel filtration and electron microscopy. Biol. Proc. Online. 11: 32–51.

Evstigneeva, R. 1993. Chemical Stability. pp. 323–334. *In*: Cevc, G. [ed.]. Phospholipids Handbook. CRC Press, Boca Raton, Florida, USA.

Ghambari, R., S. Assenza, A. Saha et al. 2017. Diffusion of polymers through periodic networks of lipid-based nanochannels. Langmuir. 33: 3491–3498.

Has, C. and S.L. Das. 2021. Recent developments in membrane curvature sensing and induction by proteins. BBA Gen. Subj. 1865: 29971.

Helfrich, W.Z. 1973. Elastic properties of lipid bilayers: theory and possible experiments. Naturforsch. C: Biosci. 28: 693–703.

Helfrich, W. and H. Rennschuh. 1990. Landau theory of the lamellar- to-cubic phase transition. J. Phys. 51: C7189–C7195.

Hossein, A. and M. Deserno. 2020. Spontaneous curvature, differential stress, and bending modulus of asymmetric lipid membranes. Biophys. J. 118: 624–642.

Ishchenko, A., L. Peng, E. Zinovev et al. 2017. Chemically stable lipids for membrane protein crystallization. Crystal Growth and Design. 17: 3502–3511.

Kozlov, M.M. 2007. Determination of lipid spontaneous curvature from x-ray examination of inverted hexagonal phases. Methods Molec. Biol. 400: 355–366.

Kozlov, M.M. and J.W. Taraska. 2023. Generation of nanoscopic membrane curvature for membrane trafficking. Nature Reviews of Molecular Cell Biology 24: 63–78.

Larsen, J.B., K.R. Rosholm, C. Kennard et al. 2020. How membrane geometry regulates protein sorting independently of mean curvature. ACS Cent. Sci. 6: 1159–1168.

Markin, V.S. and J.P. Albanesi. 2002. Membrane fusion: stalk model revisited. Biophsy. J. 82: 693–712.

Meikle, T.G., Y. Shenggen, Z. Alexandrou et al. 2017. Predicting the release profile of small molecules from within the ordered nanostructured lipidic bicontinuous cubic phase using translational diffusion coefficients determined by PFG-NMR. Nanoscale. 9: 2471–2478.

Oka,T., N. Ohta, N. and S. Hyde. 2018. Polar-nonpolar interfaces of inverse bicontinuous cubic phases in phytantriol/water system are parallel to triply periodic minimal surfaces. Langmuir. 34: 15462–15469.

Pezron, I., E. Pezron, B.A. Berganståhl et al. 1990. Repulsive pressure between monoglyceride bilayers in the lamellar and gel states. J. Phys. Chem. 94: 8255–8261.

Qui, H. and M.C. Caffrey. 2002. The phase diagram of the monoolein/water system: metastability and equilibrium aspects. Biomaterials 21: 223–234.

Rand, R.P., N.P. Fuller, S.M. Gruner et al. 1990. Membrane curvature, lipid segregation, and structural transitions for phospholipids under dual-solvent stress. Biochemistry 29: 76–87.

Seddon, J.M. 1990. Structure of the inverted hexagonal (HII) phase, and non-lamellar phase transitions of lipids. Biochim. Biophys. Acta. 1031: 1–69.

Siegel, D.P., J.L. Burns, M.H. Chestnut et al. 1989. Intermediates in membrane fusion and bilayer/nonbilayer phase transitions imaged by time-resolved cryo-transmission electron microscopy. Biophys. J. 56: 161–169.

Siegel, D.P. and M.M. Kozlov. 2004. The Gaussian curvature elastic modulus of N-mono-methylated dioleoylphosphatidylethanolamine: relevance to membrane fusion and lipid phase behavior. Biophys. J. 87: 366–374.

Siegel, D.P. 2005. The relationship between bicontinuous inverted cubic phases and membrane fusion. pp. 59–98. *In*: Lynch, M.L. and P.T. Spicer [eds.]. Bicontinuous Liquid Crystals (Surfactant Science Series; vol. 127). Taylor and Francis: Boca Raton, Florida, USA.

Siegel, D.P., V. Cherezov, D.V. Greathouse et al. 2006a. Transmembrane peptides stabilize inverted cubic phase in a biphasic length-dependent manner: implications for protein-induced membrane fusion. Biophys. J. 90: 200–211.

Siegel, D.P. 2006b. Determining the ratio of the Gaussian curvature and bending elastic moduli of phospholipids from QII phase unit cell dimensions. Biophys. J. 91: 608–616.

Siegel, D.P. and B.G. Tenchov. 2008a. Influence of lamellar phase binding energy on the relative stability of lamellar and inverted cubic phases. Biophys. J. 94: 3987–3995.

Siegel, D.P. 2008b. The gaussian curvature elastic energy of intermediates in membrane fusion. Biophys. J. 95: 5200–5215.

Siegel, D.P. 2010. Fourth-order curvature energy model for the stability of bicontinuous inverted cubic phases in amphiphile-water systems. Langmuir. 26: 8673–8683.

Siegel, D.P. 2022. Bicontinuous inverted cubic phase stabilization as an index of antimicrobial and membrane fusion peptide activity. BBA Biomembranes. 1864: 183815.

Squires, A.M., R.H. Templer, J.M. Seddon et al. 2002. Kinetics and mechanism of the lamellar to gyroid inverse bicontinuous cubic phase transition. Langmuir. 18: 7384–7392.

Su, R., S. Wang, Z. McDargh et al. 2023. The membrane fusion ore families determine the pathway to pore dilation. Biophys. J. (in press).

Surmeier, G., S. Dogna-Surmeier, M. Paulus et al. 2022. The interaction of viral fusion peptides with lipid membranes. Biophys. J. 121: 3811–3825.

Tenchov, B. and R. Koynova, R. 2012. Cubic phases in membrane lipids. Eur. Biophys. J. 10: 841–850.

Tenchov, B.G., R.C. MacDonald and B.R. Lentz. 2013. Fusion peptides promote formation of bilayer cubic phases in lipid dispersions. An x-ray diffraction study. Biophys. J. 104: 1029–1037.

Tenchov, B. and R. Koynova. 2017. Cubic phases in phosphatidylethanolamine dispersions: formation, stability and phase transitions. Chem. Phys. Lipids 208: 65–74.

Tilcock, C.P.S., M.B. Bally, S.B. Farren et al. 1982. Influence of cholesterol on the structural preferences of dioleoylphosphatidylethanolamine-dioleoylphosphatidylcholine systems: a Phosphorous-31 and Deuterium nuclear magnetic resonance study. Biochemistry 21: 4596–4601.

Tristram-Nagle, S. and J.F. Nagle. 2007. HIV-1 fusion peptide decreases bending energy and promotes curved fusion intermediates. Biophys. J. 93: 2048–2055.

Tristram-Nagle, S., R. Chan, E. Koojiman et al. 2010. HIV fusion peptide penetrates, disorders, and softens T-cell membrane mimics. J. Mol. Biol. 402: 139–153.

Weiss, T.M., P.C.A. van der Wel, J.A. Killian et al. 2003. Hydrophobic mismatch between helices and lipid bilayers. Biophys. J. 84: 379–385.

8

Recognition of Membrane Curvature by the Amphipathic Lipid Packing Sensor (ALPS) Motif

*Antoine Reynaud,** *Romain Gautier* and *Bruno Antonny*

Introduction

The discovery that some proteins can sense lipid membrane curvature dates from the 2000s with, on the one hand, the discovery of this sensitivity in ArfGAP1 (Bigay et al. 2003), which led to the discovery of the amphipathic lipid packing sensor (ALPS) motif (Bigay et al. 2005; Drin et al. 2007), and, on the other hand, the discovery of the banana-like structure of the BAR domain (proteins from the Bin/Amphyphysin/RVS family) and its preferential interaction with curved membranes (Peter et al. 2004). In this chapter, we review the ALPS motif, a subfamily of amphipathic helices (AH) with a polar face rich in three small polar residues (glycine, serine, and threonine). The first two sections present the starting observations on ArfGAP1 and the mechanism of membrane curvature sensitivity of its ALPS motif. The following sections list all proteins in which the presence of an ALPS motif has been experimentally tested, notably as regard to membrane curvature sensitivity. The final section compares ALPS motifs with other AH and also highlights differences within the ALPS family itself.

Université Côte d'Azur and CNRS, Institut de Pharmacologie Moléculaire et Cellulaire, 660 route des lucioles, 06560 Valbonne, France.
* Corresponding author: antonny@ipmc.cnrs.fr

ArfGAP1 and Its Yeast Orthologue Provide First Examples of Curvature Sensitivity

ArfGAP1 (UniProt accession number: Q8N6T3), a GTPase-activating protein (GAP) of 406 amino acids, catalyses GTP hydrolysis in Arf1 (ADP ribosylation factor 1), a small G protein involved in the formation of COPI vesicles (Cukierman et al. 1995). In cells, ArfGAP1 has a perinuclear localisation, corresponding to the Golgi apparatus, but also presents a large cytosolic fraction.

The GDP/GTP cycle of Arf1 controls the assembly/disassembly cycle of the COPI coat (Taylor et al. 2022). Arf1 is activated by the GEF Guanosine exchange factor GBF1 located at the cis-Golgi. In the GTP-bound form, Arf1 is strongly anchored to the membrane of the Golgi apparatus and enables the recruitment of coatomer, a 7-subunit complex. This complex self-assembles into the COPI coat, which induces the budding of transport vesicles from the membrane. Once the vesicle is formed, the coat remains stable until GTP hydrolysis occurs. The hydrolysis of GTP in Arf1 by ArfGAP1 leads to the detachment of Arf1 and, therefore, of the COPI coat.

The catalytic activity of ArfGAP1 is carried by its N-terminal region, which is structured around a zinc finger fold (Figure 1). This domain defines the ArfGAP family (Kahn et al. 2008). The rest of ArfGAP1 is intrinsically disordered and has no homology with other proteins. In 1997, it was observed that ArfGAP1 was sensitive to the arrangement of lipids, and in 2003, it was also sensitive to membrane curvature (Antonny et al. 1997; Bigay et al. 2003). The measurements followed the Arf1 GTPase reaction, catalysed by ArfGAP1, on liposomes of defined composition and size. The GTPase activity increased ≈ 50-fold when the liposome radius decreased from 150 nm to 35 nm.

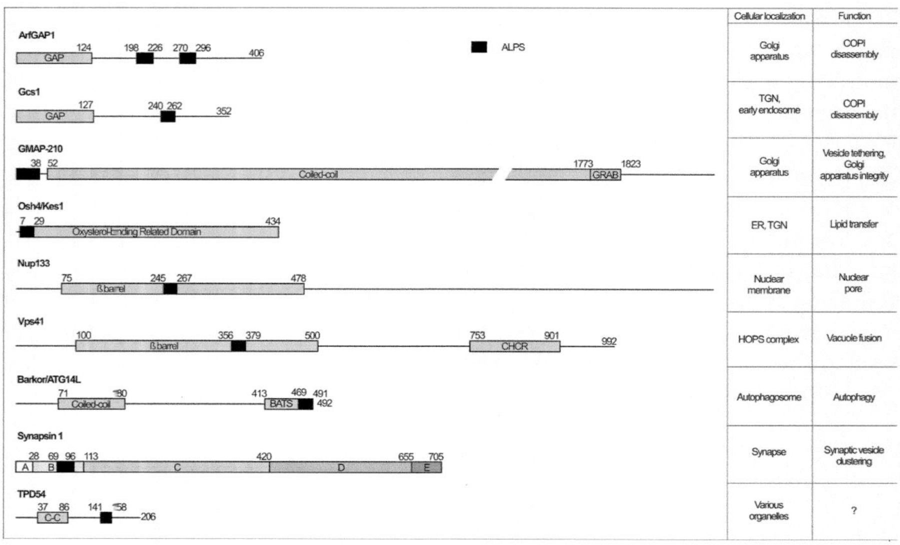

Figure 1. ALPS motif-containing Proteins that have been experimentally tested. Left: position of the ALPS motif in the primary sequence. Right: Summary of the localisation and function of the protein.

The yeast protein Gcs1p (UniProt P35197) exhibits sequence conservation with ArfGAP1 not only for the N-terminal catalytic domain as other members of the ArfGAP family but also for a central region (240–262) (Figure 1). This homology was key for the discovery of the ALPS motif. Indeed, Gcs1 is extremely sensitive to membrane curvature, whereas another protein of the ArfGAP family is not (Bigay et al. 2003; 2005). In addition, the binding of Gcs1 to liposomes is very sensitive to lipid unsaturation and curvature (Antonny et al. 1997). On this basis, J Bigay postulated that the conserved central region between ArfGAP1 and Gcs1p contains a sensor of membrane curvature.

By using various constructs, it was shown that the central region (198–226) of ArfGAP1 is necessary and sufficient for membrane curvature recognition (Bigay et al. 2005). This sequence has an amphipathic character with a hydrophobic moment [(<μH>) equal to 0.422] and a hydrophobicity <H> of 0.380. Flotation techniques with reconstituted liposomes, as well as fluorescence measurements, demonstrated the membrane curvature-dependent binding properties of the sequence, whereas limited proteolysis and circular dichroism techniques indicated that a structural change accompanies liposome binding: a complete shift from a random coil to a helical conformation (Bigay et al. 2005; Drin et al. 2007). Binding affinity depends on membrane curvature and lipid unsaturation but is poorly sensitive to lipid charge (Bigay et al. 2005; Quartino et al. 2018). Later, using membrane tubes of defined radius revealed a membrane curvature threshold (Ambroggio et al. 2010). At curvature below 1/35 nm^{-1}, no binding of ArfGAP1 is observed. Above this value, binding increases linearly with membrane curvature (1/radius). All these features led to the name ArfGAP1 Lipid Packing Sensor (ALPS). Later, the discovery of sequences with similar properties prompted us to modify the meaning of the acronym into Amphipathic Lipid Packing Sensor (Drin et al. 2007).

The sensitivity of ArfGAP1 to membrane curvature suggested a feedback loop based on membrane geometry: coat assembly induces membrane curvature; in turn, membrane curvature promotes coat disassembly (Figure 2B). However, the relevance of ArfGAP1 sensitivity to membrane curvature in cells has been poorly studied. In contrast, more information has been collected on Gcs1p (Xu et al. 2013). In yeast, Gcs1 decorates puncta structures corresponding to TGN and early endosome (EE). Gcs1 becomes soluble when harbouring a mutation that abrogates the amphipathic character of its ALPS motifs. Surprisingly, the membrane localisation of Gcs1 requires the activity of Drs2, a P4-ATPase that catalyses the movement of PS from the lumenal leaflet to the cytoplasmic one at the TGN and EE level. Although the ALPS motif is poor in cationic residues, a region upstream of the ALPS motif and including a critical lysine residue (K210) is important for Gcs1 localisation to highly curved membrane structures *in vivo* as well as for its binding to curved and negatively charged liposomes (Xu et al. 2013). Thus, Gcs1 combines electrostatic- and curvature-sensing mechanisms to recognise TGN and EE structures. Functionally, the Drs2- and ALPS-dependent recruitment of Gcs1 is important for EE-to-TGN cargo transport (Xu et al. 2013).

ArfGAP1 contains another sequence of about 20 amino acids (270–296) that has similar physicochemical features as the first ALPS motif. This sequence is also sufficient to bind lipid membranes in a curvature and lipid unsaturation-dependent manner (Mesmin et al. 2007). Cellular and biochemical experiments indicate that the two ALPS work in tandem and that they both contribute to ArfGAP1 sensitivity to membrane curvature and Golgi localisation. Interestingly, splice variants of ArfGAP1 exist that affect the ALPS1-ALPS2 tandem region, which might reflect an adaptation to the lipid composition of specific tissues (Levi et al. 2008). In the case of Gcs1, another amphipathic motif downstream of the first ALPS motif has been found, although its physical chemistry is quite distant from canonical ALPS motifs (Zendeh-boodi et al. 2013).

Since the initial discovery of ArfGAP1 and its sensitivity to membrane curvature (Bigay et al. 2003; Cukierman et al. 1995), other functional and structural studies have suggested that ArfGAP1 has a large spectrum of activities and that its role is not limited to COPI vesicle regulation. In fact, other GAPs for Arf1 exist that are not sensitive to curvature and make direct protein-protein interaction with the COPI coat, which conditions their GAP activity (Taylor et al. 2022; Weimer et al. 2008). Recent advance in the structure of the COPI coat by cryo-EM suggests different environments for Arf1-GTP molecules in the coat, which might require different Arf GAPs for proper regulation (Arakel et al. 2019; Taylor et al. 2022).

Molecular Basis for the Membrane Curvature Sensitivity of the ALPS Motif of ArfGAP1

The ALPS motif of ArfGAP1 differs from classical AH by the abundance of small polar residues (Gly, Ser, and Thr) and the paucity of positively charged residues. Extensive site-directed mutagenesis in the polar face indicated that this feature is key to membrane curvature sensing. Notably, mutants of ArfGAP1 that contain positively charged residues at the border between the polar and non-polar faces, a feature recurrent in many membrane-interacting AH, are much less sensitive to membrane curvature; they tend to bind equally well to small and large liposomes (Drin et al. 2007). These properties suggested a model for the membrane curvature sensitivity of ArfGAP1: the paucity of positive charge eliminates the possibility of strong interaction with the lipid polar heads, which are frequently negative and, as such, exacerbates the role of hydrophobic insertions, which should be facilitated by membrane curvature.

Since then, molecular dynamics (MD) simulations have provided atomic descriptions of how lipid packing defects allow the coupling between membrane insertion and folding of the ALPS motif (Cui et al. 2011; Gautier et al. 2018; González-Rubio et al. 2011; Karmakar and Klauda 2022; Vamparys et al. 2013; Vanni et al. 2014; 2013) (Figure 2A). Notably, such studies illustrate the additive effect of positive curvature and monounsaturated lipids on forming packing defects, accommodating the large hydrophobic residues from the AH. In addition, single liposome measurements show that membrane curvature mostly increases the number of binding sites suitable for hydrophobic insertion (Hatzakis et al. 2009).

HeliQuest, a Bioinformatics Tool to Discover Other ALPS Motifs

The observation that the curvature sensitivity of ArfGAP1 depends on the peculiar amino-acid composition of its ALPS motif motivated the design of a bioinformatics

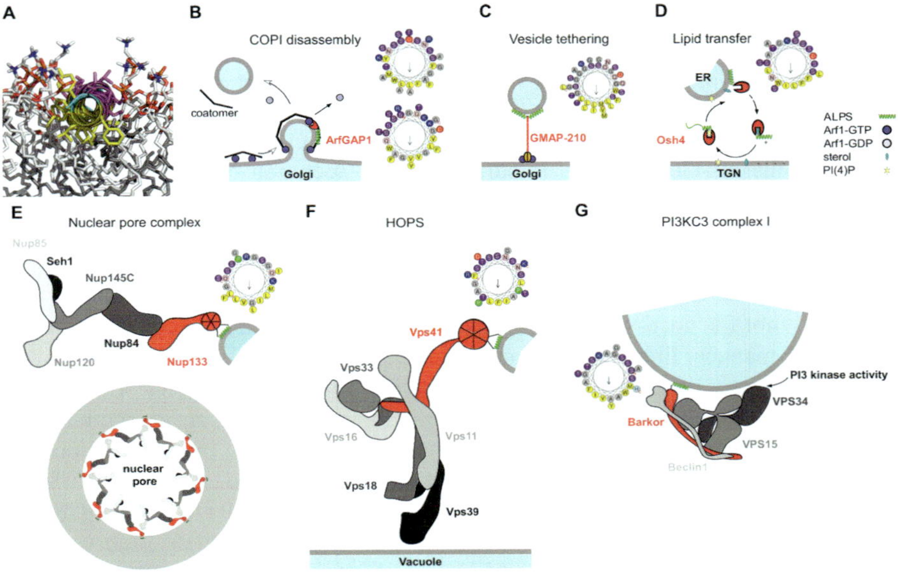

Figure 2. Structural and functional aspects of the ALPS motif. A. Front view of the ALPS1 motif of ArfGAP1 in a lipid membrane with packing defects as assessed by the atom molecular dynamics simulations. **B-G**. Schematic view of the various cellular contexts in which ALPS motifs control membrane-associated mechanisms. The ALPS motif is involved in the tethering of small vesicles (e.g., GMAP-210, Vps41), in the coupling between membrane curvature sensing and catalytic activity (e.g., ArfGAP1, Osh4, PI3KC3) or in the attachment of a complex at the nuclear pore membrane (Nup133 in the Nup84 complex). In each case, a *HeliQuest* view of the ALPS motif is shown.

tool to look for similar motifs in other proteins (Drin et al. 2007; Gautier et al. 2008). In contrast to classical homology algorithms (e.g., BLAST), this tool named *HeliQuest* (https://heliquest.ipmc.cnrs.fr/) does not consider the exact sequence but its putative amphipathic properties. R Gautier and G Drin built and used *HeliQuest* to identify in the full human and yeast proteomes 18 aa sequences that fulfil the following criteria: (i) hydrophobicity between 0.28 and 0.6; (ii) hydrophobic moment > 0.35; (iii) not more than four charged residues with a sum of the charges between –1 and 2; (iv) a sum of serine, threonine and glycine greater than 5. After adding geometric constraints and manually sorting sequences to focus on cytosolic proteins, we obtained about 200 potential ALPS motifs in the yeast and human proteomes (Drin et al. 2007; Gautier et al. 2008). From this list, a few proteins looked interesting candidates for being sensitive to curvature and were subjected to experimental tests. They are summarised in the following sections as well as in Figure 1 for the position of the ALPS motif in the sequence and Figure 2 for the functional role of the ALPS motif. However, the majority of proteins present in the list have not been tested yet.

The Golgin GMAP-210

Golgi microtubule-associated protein 210 (UniProt Q15643) is a long dimeric protein of the Golgin family, which is anchored to the Golgi apparatus by its C-terminal-210

GRAB domain (GRIP-related Arf binding) via Arf1-GTP. GMAP-210 forms a long dimeric molecular string of ≈ 200 nm (1,700 aa) owing to long coiled-coil regions. The N-terminus of GMAP-210 (1–38) was detected by the *HeliQuest* screening and is the longest ALPS sequence (Drin et al. 2007; Gautier et al. 2008) (Figure 1). GMAP-210 is involved in vesicular transport between the endoplasmic reticulum (ER) and the cis-Golgi and is important for the secretion of some extracellular matrix molecules, in particular, Perlecans (Roboti et al. 2015; Smits et al. 2010). The presence of an ALPS motif at the end of a golgin offered a straightforward mechanism for capturing transport vesicles.

Biochemical studies using various GMAP-210 constructs demonstrated that the ALPS motif is necessary and sufficient to bind lipid membranes in a curvature-dependent manner. First, a construct encompassing the N-terminus ALPS motif followed by part of the coiled-coil region binds to small liposomes with a lipid composition mimicking that of cis-Golgi membranes (Drin et al. 2007). Circular dichroism measurements showed that binding is accompanied by the structuration of the ALPS sequence into a helix (Drin et al. 2007). Second, a mini GMAP-210 coiled-coil construct encompassing the two ends of GMAP-210 (ALPS and GRAB) is capable of promoting the aggregation of two liposome populations: highly curved Golgi-like liposomes through the ALPS motif and liposomes containing Arf1-GTP, through the GRAB domain (Drin et al. 2008). This tethering, which can be monitored by dynamic light scattering and fluorescence microscopy, is reversible if the liposomes on which Arf1-GTP is bound are themselves highly curved. Indeed, the Arf1-GTP-GRAB interaction is disrupted by the addition of ArfGAP1, which is curvature-sensitive (Drin et al. 2008). Thus, these reconstitution experiments suggest two ALPS-based regulations: one positive and direct via the N-terminal ALPS motif of GMAP-210; one negative and indirect via the action of ArfGAP1. Thus, GMAP-210 could attach small transport vesicles to flat cisternae and contribute to the stable but dynamics architecture of the Golgi apparatus (Figure 2C).

Several studies have addressed the involvement of the ALPS motif of GMAP-210 as well as other regions, in the localisation and function of this Golgin. Both the ALPS motif and the GRAB domain enable Golgi targeting, although the two regions do not display the same localisation when expressed separately and do not show the same sensitivity to the drug BFA, which prevents Arf1 activation (Cardenas et al. 2009). In addition, the ability of GMAP-210 to tether vesicles at the Golgi depends on the ALPS motif (Sato et al. 2015). Strikingly, artificially targeting GMAP-210 N-terminus constructs to mitochondria promotes the attachment of small vesicles, which are homogenous in size, as assessed by electron microscopy (Wong and Munro 2014). The mechanism of vesicle tethering at the Golgi by the ALPS motif of GMAP-210 also explains an old surprising observation: when small liposomes are added to perforated cells, they concentrate in the Golgi area (Kobayashi and Pagano 1988); this activity depends on the presence of GMAP-210 (Magdeleine et al. 2016).

Whether recognition of membrane curvature by the ALPS motif of GMAP-210 fully explains the selection of transport vesicles is not clear. One experimental strategy to test this hypothesis is to reverse the ALPS sequence, such as keeping its amphipathic character while disrupting possible specific protein-protein interactions. In cells, the localisation of a construct carrying such an inverted sequence overlaps almost exactly with that of a construct carrying the wild-type ALPS motif, suggesting that membrane curvature recognition is the main mechanism of vesicle selection (Magdeleine et al. 2016; Pranke et al. 2011). However, a recent study suggests that GMAP-210 selects transport vesicles by the ALPS motif and a specific mechanism involving a highly conserved N-terminus "SWL" sequence (Wong et al. 2017).

GMAP-210 participates in the structural integrity of the Golgi cisternae. Silencing GMAP-210 causes the scattering of the Golgi apparatus with a reduction in the number of intact cisternae, an increase in the number of vesicles, and a decrease in the length of the cisternae (Sato et al. 2015). This phenomenon has been observed both by optical and electron microscopy (Magdeleine et al. 2016; Sato et al. 2015) Rescue expression of GMAP-210 enables the reformation of the Golgi apparatus in terms of both the number and size of the cisternae and vesicles. In contrast, the expression of truncated chimeras for the different domains of GMAP-210 shows that the ALPS motif is necessary but not sufficient for the proper organisation of the Golgi apparatus. The length of the coiled-coil region, as well as the presence of Rab-binding domains, are also important (Sato et al. 2015).

As aforementioned, ALPS motifs are rich in Ser, Thr, and Gly residues and poor in charged residues. Analysis of the GMAP-210 ALPS sequence revealed an additional feature. Each helical turn contains a single bulky hydrophobic residue. This contrasts with the AH of Arf1, which contains two hydrophobic residues per turn and is marginally sensitive to membrane curvature (Ambroggio et al. 2010). Shortening and condensing the ALPS motif of GMAP-210 by pairing its hydrophobic residues makes it much more promiscuous for various organelles in the cell.

The ALPS of GMAP-210 is currently the most studied motif of this family and illustrates the extent of possible techniques to validate the functionality of an ALPS motif.

An ALPS Motif Defines the Lid of the Lipid Transfer Protein Osh4/Kes1

Osh4 (UniProt P35844) or Kes1 is a lipid transfer protein from the oxysterol-binding protein-related protein family in yeast, which can exchange ergosterol from the reticulum against the phosphoinositide PI(4)P at the TGN (de Saint-Jean et al. 2011; Im et al. 2005; Lipp et al. 2020; Moser von Filseck et al. 2015). Osh4 contributes to the enrichment of sterol along the late secretory pathway, and its activity also impacts vesicular transport reactions (Fang et al. 1996; Lipp et al. 2020).

The *HeliQuest* screening detected an ALPS motif in the N-terminus of Osh4, which corresponds to the lid covering the lipid binding site in the structure (Figure 2D). Initial experiments performed with the apo form showed that Osh4 binds preferentially to small liposomes through its ALPS motif (Drin et al. 2007). Subsequent studies revisited this interaction by considering the various lipid-bound forms of Osh4 during the lipid exchange cycle (de Saint-Jean et al. 2011; Moser von Filseck et al. 2015). When bound to sterol, Osh4 uses the ALPS motif as a lid, which is thus poorly available for membrane interaction. Instead, the Osh4-sterol complex interacts with negatively charged membranes, leading to the opening of the lid/ALPS and ergosterol release into the acceptor TGN membrane. Osh4 captures PI(4)P, which is then transported back and released preferentially into neutral and weakly ordered membranes (e.g., ER). At this stage, the ALPS motif is more available for membrane interaction: kinetics measurements of PI(4)P release suggests that this step is favoured by unsaturated and curved membranes. Furthermore, the presence of ergosterol increases lipid exchange, and PI4P is ultimately hydrolysed in the ER membrane by Sac1 (Moser von Filseck et al. 2015). Overall, the membrane interaction of Osh4 seems finely tuned to be neither too weak nor too strong, such as to allow the protein to bind lipid membranes but also to detach from them, hence ensuring fast lipid transfer reactions (Lipp et al. 2019). The ALPS motif seems important for anchoring Osh4 at the ER membranes during exchange of PI4P by sterol; conversely, electrostatic interactions predominate at the TGN membranes during exchange of sterol by PI4P (de Saint-Jean et al. 2011; Moser von Filseck et al. 2015; Lipp et al. 2019).

All-atom MD simulations have been performed to get plausible views of the engagement of the ALPS motif of Osh4 and other protein residues in membrane interactions. They confirm the preferential insertion of the ALPS motif into loosely packed membranes but also suggest various possible conformations (Karmakar and Klauda 2022; Monje-Galvan and Klauda 2018).

An ALPS Motif to Anchor the Nuclear Pore at the Nuclear Membrane

Nup133 (UniProt Q8WUM0) is a component of the nuclear pore heptameric subcomplex Nup84. This complex displays a characteristic Y shape, with Nup133 forming the base of the structure (Figure 2E). The complex localises through Nup133 to the equatorial region of the pore, that is, at the junction of the inner and outer nuclear membranes where membrane curvature is the highest. The first part of Nup133 (75–478) forms a β barrel structure, which has been solved by X-ray crystallography except for an unstructured loop. This loop of 23 amino acids (245–267), which is isolated from the rest of the structure by two prolines at position 246 and 268, was detected as an ALPS motif by *HeliQuest* (Figures 1 and 2E) (Drin et al. 2007).

The Nup133 ALPS peptide adopts a helical structure upon binding to small liposomes as measured by circular dichroism (Drin et al. 2007). Flotations experiments using liposomes of defined size showed increasing binding of Nup133 with membrane curvature, whereas deletion of the ALPS motif or introducing a negative charge in the hydrophobic side prevents Nup133 from binding to small liposomes. A complementary approach consisting of incubating liposomes of defined size with Ni-NTA magnetic beads coated with His tagged Nup133 confirmed a striking preference of Nup133 for small liposomes and a key role of the ALPS motif (Doucet et al. 2010).

In vivo experiments demonstrated the importance of the ALPS motif for the localisation of Nup133 and, more generally, for nuclear pore formation (Doucet et al. 2010). Structural analysis and modelling suggest that the ALPS motif of Nup133 is partially conserved in yeast and is well-positioned to anchor the Nup84 complex on the nuclear pore membrane (Kim et al. 2014; Nordeen et al. 2020). The yeast yNup133 (252-270) sequence is also an unstructured loop exiting the N-terminal-β-barrel region (Nordeen et al. 2020). This loop does not fulfil all criteria for ALPS because of a smaller hydrophobic moment compared to human ALPS. Notwithstanding, it can bind liposomes made of yeast polar lipids in a curvature-dependent manner. In addition, yeast Nup133 not only senses membrane curvature but also promotes liposome deformation, suggesting that the corresponding AH induces curvature rather than senses it. The presence of pairs of large hydrophobic residues in several helical turns might make this AH strong enough to deform membranes.

Further studies have been performed to address the importance of the position of the ALPS motif in the structure (Doucet et al. 2015). When the ALPS motif of Nup133 is introduced as a loop in GFP (Green Fluorescent Protein), it leads to a perinuclear and cytoplasmic localisation. In contrast, adding Nup133 ALPS motif at the C-terminus of GFP leads to a soluble construct. When present at the N-terminal position of a coiled-coil dimer (133-ACC-GFP), that is, with a geometry resembling that of GMAP-210, an almost exclusive perinuclear concentration was observed. These results suggest that the ALPS motif is not the only determinant of membrane selectivity. Other parameters, such as the degree of freedom in the structure, also seem decisive.

The nuclear pore might be a site of concentration of ALPS motifs. Indeed, among the 30 nucleoporins, the *HeliQuest* screening predicted ALPS motifs in three human nucleoporins (Nup133, Nup214, and Nup50) and in five nucleoporins in yeast (Nup188, Nup170, Nup1, Nup85, and Nup120) (Drin et al. 2007). Of note, Nup120, which is another subunit of the Nup84 complex, displays a loop at the periphery of a β-barrel structure akin to Nup133 (Kim et al. 2014).

An ALPS Motif in Vps41 Regulates the Attachment of Organelles to the Vacuole in Yeast

Yeast Vps41 (UniProt P38959) is a subunit of the HOPS complex (HOmotypic vacuole fusion Protein Sorting), which is responsible for the fusion of various organelles, including late endosomes, autophagosomes, and AP-3 vesicles with the vacuole (Shvarev et al. 2022).

The core of the HOPS complex is composed of Vps11, Vps16, Vps18, and Vps33. One functional end is represented by Vps39, which can specifically bind the vacuole by interacting with membrane Ypt7p, the yeast form of the GTPase Rab7 (Figure 2F). The other end is composed of Vps41. This subunit is responsible for selecting the structures that fuse with the vacuole. Like Vps39, Vps41 has a Ypt7p-binding domain, explaining the ability of the HOPS complex to assemble and then fuse vacuoles. In addition, Vps41 can bind endosomes, thus promoting their fusion with the vacuole. The selectivity of Vps41 can be modulated by phosphorylation by casein kinase Yck3 (S367, S368, S371, and S372); the phosphorylation of these sites leads to the recruitment to the vacuole of transport vesicles by the AP-3 pathway, a non-endocytic transport pathway (Cabrera et al. 2010).

The N-terminal half of Vps41 forms a β-barrel, whereas the rest of the sequence forms an alpha solenoid. Three loops emerge from the β-barrel, one corresponding to an ALPS motif (aa 356–379) detected by *HeliQuest* (Figures 1 and 2F). As in

Nup133, two prolines mark out the boundaries of the ALPS loop. Strikingly, all phosphorylation sites by Yck3 are found in the ALPS sequence.

Flotation experiments of the whole protein or purified fragments validate the ability of Vps41 to bind liposomes in a curvature-dependent fashion. In addition, the L366D mutation in the hydrophobic face of the ALPS motif or deletion of the ALPS motif disrupts Vps41 binding to small membranes (Cabrera et al. 2010). Measurements by Electron paramagnetic resonance (EPR) of the distance between two amino acids that are seven residues apart in the ALPS sequence (366 and 373) give a value of 1.1 nm, which fits well with two helical turns.

The membrane tethering capacity of the HOPS complex and the relative role of the ALPS motif and Ypt7 binding sites have been assessed by elegant *in vitro* experiments (Ho and Stroupe 2016). The experiment involves mixing two liposome populations of defined features (high vs. low curvature and presence or absence of Ypt7), each labelled with a different fluorophore (green or red). The liposomes are too small to be visible in the microscope, but their clustering can be observed and quantified by determining Pearson's coefficient of the green and red fluorescence signals. These experiments indicate that the purified HOPS complex can tether small naked liposomes to larger ones covered by Ypt7 and that this activity is lost when the ALPS motif of Vps41 is phosphorylated by Yck3. Moreover, removing the Vps39 end of the HOPS complex responsible for binding to Ypt7p also leads to loss of tethering. These experiments suggest that the HOPS complex can tether highly curved membranes through its Vps41 ALPS motif to Ypt7-GTP-containing membranes through its Vps39 subunit (Figure 2F). However, other combinations are possible, and the binding of Ypt7 by the Vps41 subunit is functionally important (Lürick et al. 2017. The recent elucidation of the structure of the HOPS complex should help better decipher the different membrane tethering modes of the HOPS complex and the contribution of the ALPS motif (Shvarev et al. 2022).

An ALPS Motif in the Autophagy Protein Barkor/ATG14L

Autophagy is the process by which cells engulf cytoplasmic compounds, including organelles in a double-membrane organelle called the phagophore, which, after closure, forms the autophagosome, which is then directed towards lysosomes to be degraded. This pathway, essential to cells, involves more than 30 players as well as different membrane sources, some of which are highly curved. In addition, the double membrane of the phagophore undergoes several dynamic changes. Therefore, the biophysical properties of the membranes, including curvature, vary greatly during autophagy. Several curvature-sensing mechanisms have been reported, one involving an ALPS motif (Brier et al. 2019; Fan et al. 2011; Ma et al. 2017; Ohashi et al. 2020).

The class III phosphatidylinositol 3-kinase (PI3KC3) is a complex composed of Vps34, p150/Vps15, and Beclin 1/Atg6. The complex has a Y shape with one catalytic arm made of Vps34 and Vps15 and one adaptor arm made of Vps15 and Beclin 1. The recruitment of the PI3KC3 complex to autophagosome membranes is directed by Barkor/ATG14L (Uniprot Q6ZNE5), a protein of 492 amino acids, which tightly interacts with Beclin 1 to form the type I PI3KC3 complex (Figure 2G).

To determine which part of Barkor is required for localisation to autophagosomes, Fan et al. expressed truncated chimeras with a fluorescent tag and followed their intracellular localisation in U2OS cells (Fan et al. 2011). Colocalization with LC3, a classical autophagy marker, was only obtained when the C-terminal 80 amino acids of Barkor were present. Conversely, deletion of this region, named BATS for Barkor Autophagosome Targeting Sequence, led to a reduction in the lipidation of LC3, indicating that it is necessary for proper autophagy.

A phylogenetic alignment shows that the BATS sequence is highly conserved in vertebrates and contains an AH. This helix is an ALPS motif (Figure 1), although it was not detected in the initial ALPS screening because Barkor/ATG14L was not present in the curated databases at that time (Drin et al. 2007; Fan et al. 2011; Gautier et al. 2008). Mutations that reduce the hydrophobic character of the helix reduce Barkor/ATG14L localisation to autophagosomes, suggesting that Barkor/ATG14L localises to autophagosomes through its ALPS motif.

Various reconstitution experiments demonstrated the key role of the ALPS motif of Barkor/Atg14L in directing the PI3KC3 complex to highly curved and/or unsaturated lipid membranes (Figure 2G). These include sedimentation or flotation experiments with liposomes of defined size and composition, reconstitution of the PI3K reaction and LC3 lipidation, hydrogen-deuterium exchange mass-spectrometry and structural analysis (Brier et al. 2019; Fan et al. 2011; Ma et al. 2017; Ohashi et al. 2020). Interestingly, the PI3KC3 complex is involved in other cellular contexts. Complex I, which includes Barkor/ATG14L, is restricted to autophagy, whereas complex II, which does not contain Barkor/ATG14L but UVRAG in the adaptor arm, recognises other physiochemical features of the membranes and is involved in other subcellular functions (Ohashi et al. 2020).

As aforementioned, autophagy involves numerous membrane sources, membrane remodelling reactions, and membrane-protein interactions. If the role of the ALPS motif of Barkor/ATG14L is well established, other autophagy reactions involve membrane-interacting AH that are not ALPS motifs (Horenkamp et al. 2015; Jensen et al. 2022; Nath et al. 2014). In the future, it will be interesting to compare these motifs and determine if their physicochemical features fit with the different membrane interfaces at play during autophagy.

Synapsin1 Contains an ALPS Motif to Cluster Synaptic Vesicles

Synaptic vesicles (SV) have a small diameter (≈ 40 nm) and form dense clusters at the synapse. These clusters act as SV pools to enable neurotransmitter release by exocytosis during repetitive axon stimulation. Synapsin-1 (UniProt P17600) is one of the most abundant proteins found in these clusters. The *HeliQuest* screening detected a conserved ALPS motif in the N-terminal region (B domain) of various Synapsins (Drin et al. 2007; Gautier et al. 2008) (Figure 1). This prediction motivated a molecular and functional study of the ALPS motif of Synapsin–1 (Krabben et al. 2011).

First, *in vitro* experiments using liposomes immobilised on a glass plate and incubated with GFP-tagged mutated or wild-type synapsin 1 established that Synapsin–1 can sense membrane curvature owing to the ALPS motif. Complementary flotation experiments were carried out and confirmed the role of the ALPS motif in binding to SVs or small artificial liposomes (Krabben et al. 2011). Second, *in vitro* experiments were conducted to reconstitute the role of Synapsin 1 in synaptic vesicle clustering. The addition of synapsin 1 to small liposomes led to the formation of liposome clusters visible by fluorescence microscopy. In the absence of its ALPS motif, synapsin 1 significantly lost its clustering efficiency (Krabben et al. 2011). Finally, functional experiments were conducted in hippocampal neurons expressing a GFP fusion of wild-type or ΔALPS synapsin–1. Cycles of axonal dispersion and re-clustering of the constructs were elicited by neuronal stimulation. These real-time imaging experiments demonstrated that the ALPS motif of synuclein-1 contributes to the ability of the protein to control the formation and disassembly of SV clusters (Krabben et al. 2011).

O-GlcNAcylation (O-GlcNAc) is a post-translational modification that corresponds to adding an N-Acetylglucosamine group to a serine or threonine. It has been suggested that O-GlcNAc regulates presynaptic plasticity. Interestingly, mass spectrometry identified numerous O-GlcNAc sites in Synapsin, including one in the ALPS sequence (T87) (Skorobogatko et al. 2014). To assess the impact of this post-translational modification on the activity of synapsin 1, Skorobogatko et al. compared the axonal distribution of wild-type synapsin 1 with that of a T87A mutant, which should prevent O-GlcNAc of the ALPS motif. The T87A mutation increased the number of SV clusters and their size, suggesting that O-GlcNAc negatively regulates the ability of the ALPS motif of synapsin to recognise SVs (Skorobogatko et al. 2014). Because the ALPS motif of synapsin-1 is known to undergo a coil to α-helix transition upon binding to highly curved membranes (Krabben et al. 2011), it will be interesting to perform circular dichroism experiments with the O-GlcNAc form of the ALPS peptide to directly assess the impact of this modification on ALPS structural flexibility and membrane interaction.

TPD54 Defines a New Class of Small Vesicles and Contains a Functional ALPS Motif

Tumour Protein D54 (UniProt O43399) belongs to a family of four members (TPD52, TPD53, TPD54, and TPD55) of enigmatic function, although they are found overexpressed in several cancers (Boutros et al. 2004). In 2019, Larocque et al. reported a striking cellular observation (Larocque et al. 2019). Confounded by the disparate cellular localisation of endogenous TPD54 (cytoplasm, perinuclear, and dotted structures), they artificially targeted the protein to mitochondria, which are not connected to the other organelles by vesicular trafficking. This tethering caused a spectacular phenotype: the mitochondria clustered through a dense matrix formed by small vesicles (35 nm) about half the size of classical transport vesicles (e.g., COPI and COPII). Further analysis showed that these vesicles exist in the cytoplasm and contain numerous markers of authentic transport vesicles (Larocque et al. 2019).

TPD54, which mostly consists of intrinsically disordered regions, contains a predicted ALPS motif (Drin et al. 2007) (Figure 1). Recently, we investigated whether this ALPS motif contributes to the recognition of intracellular nanovesicles by TPD54 (Reynaud et al. 2022). Both *in vitro* and cellular experiments indicate that the ALPS is key for membrane curvature recognition. However, the ALPS motif is not the only amphipathic motif of the protein. Three other AH are also present and form, with the ALPS motif, an extended region of about 100 amino acids, upstream of a small coiled-coil involved in protein dimerisation. Upon binding to small liposomes, the full amphipathic region becomes protected from proteolysis, suggesting that structuration of this region extends beyond the sole ALPS motif (Reynaud et al. 2022). Furthermore, cellular experiments show that TPD54 constructs localise to different perinuclear regions than the Golgin GMAP-210 (Reynaud et al. 2022). What allows GMAP-210 and TPD54 to recognise different types of vesicles remains to be explored.

Are all ALPS Motifs the Same, and What Makes Them Different From Other AHs?

In the bioinformatics sense, the ALPS family is large, with several hundred members, since the *HeliQuest* algorithm was set in a way to identify motifs with physicochemical features close to, but not necessarily identical to, the motif found in ArfGAP1 (Drin et al. 2007; Gautier et al. 2008). However, only a few sequences have been experimentally tested, as reviewed here. In all cases, the recognition of membrane curvature appears as a property provided by the ALPS motif and seems functionally important. In the future, other candidates will be interesting to test. One example is the ALPS motif of PI4KIIIβ, an enzyme that produces the phosphoinositide PI4P at the Golgi apparatus. Its C-terminal ALPS motif contributes to intracellular localisation and sensitivity to lipid unsaturation (Mesmin et al. 2017).

The membrane contexts in which ALPS motifs act are very diverse (nuclear envelope, ER, Golgi, and endosomes). Because ALPS motifs are not defined by their exact sequence but by trends in their amphipathic properties, it is possible that some intrinsic features that remain to be assessed (e.g., the number of hydrophobic residues per helical turn or the exact number of cationic residues) make each of them adapted to different membrane environments (Magdeleine et al. 2016; Zhang et al. 2022). In addition, their position regarding other structural features seems important (e.g., loop in β-barrel vs. free ends flanking coiled-coil regions) (Doucet et al. 2015).

Despite these differences, ALPS motifs form a particular family of AH, which strongly differs from others involved in organelle targeting. Careful comparison between AH families might highlight differences between the membrane interfaces on which they act (Drin and Antonny 2010; Giménez-Andrés et al. 2018, 2021; Reynaud et al. 2022; van Hilten et al. 2023). An informative example is provided by the comparison between the ALPS motif of GMAP-210 and α-synuclein, a cytosolic protein well known for its role in Parkinson's disease and which contains a long AH capable of binding synaptic vesicles. When expressed in yeast, human α-synuclein and human GMAP-210 N-terminal regions bind to different small vesicles; the former endocytic vesicles, the latter Golgi vesicles (Pranke et al. 2011). *In vitro*, α-synuclein preferentially binds liposomes that combine three traits: (i) high curvature, (ii) high content in negatively charged lipids, and (iii) high level of polyunsaturated phospholipids (Jensen et al. 2011; Manna and Murarka 2021; Pinot et al. 2014; Pranke et al. 2011). In contrast, the ALPS of GMAP-210 recognises vesicles that combine high curvature and lipid monounsaturation (Drin et al. 2007; Magdeleine et al. 2016; Pranke et al. 2011). These differences might explain the different localisation that the two proteins have in cells and also illustrate some key differences in the geometry of the packing defects induced by polyunsaturated lipids vs. monounsaturated ones in curved membranes (Manna and Murarka 2021; Manni et al. 2018; Moss et al. 2023; Pinot et al. 2014; Tiberti et al. 2020; Vamparys et al. 2013). Another instructive comparison is provided by the small G protein Arf vs. some of its effectors or regulators containing ALPS motifs, i.e., GMAP-210 and ArfGAP1. As aforementioned, the AH of Arf1 is short but has a high density of hydrophobic residues per helical turn, which makes Arf1 quite insensitive to membrane curvature and lipid unsaturation. However, this parameter was ignored in the initial definition of the ALPS motif. In the future, it will be interesting to refine the analysis, which might help better understand the continuum of membrane curvature sensing and induction between AH.

Recently, an original approach was used to compare AH in a well-defined physiological context (Zhang et al. 2022). Endophilin is a BAR-domain-containing protein that participates in the recycling of synaptic vesicles. The BAR domain is flanked by a short and cationic N-terminal AH. In endophilin—null mutant worms, neuronal expression of full-length mammalian endophilin restores the locomotion rates and synaptic currents at neuromuscular junctions. In contrast, a truncated endophilin form that lacks the AH completely fails to restore the wild-type phenotype. On this basis, the authors made numerous endophilin chimera, where they replaced the endogenous N-terminus with various AH, including ALPS motifs. Surprisingly, they found that the ALPS motif of Nup133 but not the ALPS motif of Osh4 makes the endophilin chimera functionally active. However, the authors notice that the ALPS of Nup133 has a net positive charge (+2), whereas the ALPS motif of Osh4 is neutral. Strikingly, introducing two positive charges in the ALPS motif of Osh4 completely restores the wild-type phenotype both at the ultrastructural level (the number of synaptic vesicles) and at the electrophysiological level.

Succeeding in finding the physiological functionality of a protein by artificially replacing its membrane-binding sequence with a sequence similar only in terms of physical chemistry strongly encourages the discovery of membrane determinants conducive to common binding selectivity.

One key challenge in the field of ALPS motifs will be to get a precise view of their interaction with curved membranes, which remains mostly inferred from MD simulations so far. The recent success of the combination of cryo-EM and brominated lipids to determine how hydrophobic residues insert in lipid membranes (Moss et al. 2023) suggests that this challenge might soon be overcome.

Acknowledgements

We thank Joëlle Bigay, Guillaume Drin, Jean-François Casella, Sylviane Robineau, Pierre Gounon, Thomas Boehmer, and Thomas U Schwartz for their outstanding contribution to the initial research that led to the discovery of the ALPS motif family.

References

Ambroggio, E.E., B. Sorre, P. Bassereau et al. 2010. ArfGAP1 generates an Arf1 gradient on continuous lipid membranes displaying flat and curved regions. The EMBO Journal 29: 292–303.

Antonny, B., I. Huber, S. Paris et al. 1997. Activation of ADP-ribosylation factor 1 GTPase-activating protein by phosphatidylcholine-derived diacylglycerols. Journal of Biological Chemistry 272: 30848–30851.

Arakel, E.C., M. Huranova, A.F. Estrada et al. 2019. Dissection of GTPase-activating proteins reveals functional asymmetry in the COPI coat of budding yeast. Journal of Cell Science 132.

Bigay, J., J.-F. Casella, G. Drin et al. 2005. ArfGAP1 responds to membrane curvature through the folding of a lipid packing sensor motif. The EMBO Journal 24: 2244–2253.

Bigay, J., P. Gounon, S. Robineau et al. 2003. Lipid packing sensed by ArfGAP1 couples COPI coat disassembly to membrane bilayer curvature. Nature 426: 563–566.

Boutros, R., S. Fanayan, M. Shehata et al. 2004. The tumor protein D52 family: many pieces, many puzzles. Biochem. Biophys. Res. Commun. 325: 1115–1121.

Brier, L.W., L. Ge, G. Stjepanovic et al. 2019. Regulation of LC3 lipidation by the autophagy-specific class III phosphatidylinositol-3 kinase complex. Mol. Biol. Cell 30: 1098–1107.

Cabrera, M., L. Langemeyer, M. Mari et al. 2010. Phosphorylation of a membrane curvature-sensing motif switches function of the HOPS subunit Vps41 in membrane tethering. The Journal of Cell Biology 191: 845–859.

Cardenas, J., S. Rivero, B. Goud et al. 2009. Golgi localisation of GMAP210 requires two distinct cis-membrane binding mechanisms. BMC Biol. 7: 56.

Cui, H., E. Lyman and G.A. Voth. 2011. Mechanism of membrane curvature sensing by amphipathic helix containing proteins. Biophys. J. 100: 1271–1279.

Cukierman, E., I. Huber, M. Rotman et al. 1995. The ARF1 GTPase-activating protein: zinc finger motif and Golgi complex localization. Science 270: 1999–2002.

de Saint-Jean, M., V. Delfosse, D. Douguet et al. 2011. Osh4p exchanges sterols for phosphatidylinositol 4-phosphate between lipid bilayers. The Journal of Cell Biology 195: 965–978.

Doucet, C.M., N. Esmery, M. de Saint-Jean et al. 2015. Membrane Curvature Sensing by Amphipathic Helices Is Modulated by the Surrounding Protein Backbone. PLoS ONE 10: e0137965.

Doucet, C.M., J.A. Talamas and M.W. Hetzer. 2010. Cell cycle-dependent differences in nuclear pore complex assembly in metazoa. Cell 141: 1030–1041.

Drin, G. and B. Antonny. 2010. Amphipathic helices and membrane curvature. FEBS Letters 584: 1840–1847.

Drin, G., J.-F. Casella, R. Gautier et al. 2007. A general amphipathic alpha-helical motif for sensing membrane curvature. Nat. Struct. Mol. Biol. 14: 138–146.

Drin, G., V. Morello, J.-F. Casella et al. 2008. Asymmetric tethering of flat and curved lipid membranes by a golgin. Science 320: 670–673.

Fan, W., A. Nassiri and Q. Zhong. 2011. Autophagosome targeting and membrane curvature sensing by Barkor/Atg14(L). Proc. Natl. Acad. Sci. U.S.A. 108: 7769–7774.

Fang, M., B.G. Kearns, A. Gedvilaite et al. 1996. Kes1p shares homology with human oxysterol binding protein and participates in a novel regulatory pathway for yeast Golgi-derived transport vesicle biogenesis. The EMBO Journal 15: 6447–6459.

Gautier, R., A. Bacle, M.L. Tiberti et al. 2018. PackMem: a versatile tool to compute and visualize interfacial packing defects in lipid bilayers. Biophys. J. 115: 436–444.

Gautier, R., D. Douguet, B. Antonny et al. 2008. HELIQUEST: a web server to screen sequences with specific alpha-helical properties. Bioinformatics 24: 2101–2102.

Giménez-Andrés, M., A. Čopič and B. Antonny. 2018. The many faces of amphipathic helices. Biomolecules 8: 45.

Giménez-Andrés, M., T. Emeršič, S. Antoine-Bally et al. 2021. Exceptional stability of a perilipin on lipid droplets depends on its polar residues, suggesting multimeric assembly. eLife 10.

González-Rubio, P., R. Gautier, C. Etchebest et al. 2011. Amphipathic-Lipid-Packing-Sensor interactions with lipids assessed by atomistic molecular dynamics. Biochim. Biophys. Acta 1808: 2119–2127.

Hatzakis, N.S., V.K. Bhatia, J. Larsen et al. 2009. How curved membranes recruit amphipathic helices and protein anchoring motifs. Nature Chemical Biology 5: 835–841.

Ho, R. and C. Stroupe. 2016. The HOPS/Class C Vps complex tethers high-curvature membranes via a direct protein-membrane interaction. Traffic 17: 1078–1090.

Horenkamp, F.A., K.J. Kauffman, L.J. Kohler et al. 2015. The legionella anti-autophagy effector RavZ targets the autophagosome via PI3P- and curvature-sensing motifs. Developmental Cell 34: 569–576.

Im, Y.J., S. Raychaudhuri, W.A. Prinz2 et al. 2005. Structural mechanism for sterol sensing and transport by OSBP-related proteins. Nature 437: 154–158.

Jensen, L.E., S. Rao, M. Schuschnig et al. 2022. Membrane curvature sensing and stabilization by the autophagic LC3 lipidation machinery. Sci. Adv. 8: eadd1436.

Jensen, M.B., V.K. Bhatia, C.C. Jao et al. 2011. Membrane curvature sensing by amphipathic helices: a single liposome study using α-synuclein and annexin B12. J. Biol. Chem. 286: 42603–42614.

Kahn, R.A., E. Bruford, H. Inoue et al. 2008. Consensus nomenclature for the human ArfGAP domain-containing proteins. The Journal of Cell Biology 182: 1039–1044.

Karmakar, S. and J.B. Klauda. 2022. Modeling the membrane binding mechanism of a lipid transport protein Osh4 to single membranes. Biophys. J. 121: 1560–1575.

Kim, S.J., J. Fernandez-Martinez, P. Sampathkumar et al. 2014. Integrative structure-function mapping of the nucleoporin Nup133 suggests a conserved mechanism for membrane anchoring of the nuclear pore complex. Mol. Cell Proteomics 13: 2911–2926.

Kobayashi, T., R.E. and R.E. Pagano. 1988. ATP-dependent fusion of liposomes with the Golgi apparatus of perforated cells. Cell 55: 797–805.

Krabben, L., A. Fassio, V.K. Bhatia et al. 2011. Synapsin I senses membrane curvature by an amphipathic lipid packing sensor motif. J. Neurosci. 31: 18149–18154.

Larocque, G., P.J. La-Borde, N.I. Clarke. 2019. Tumor protein D54 defines a new class of intracellular transport vesicles. The Journal of Cell Biology.

Levi, S., M. Rawet and L. Kliouchnikov. 2008. Topology of amphipathic motifs mediating Golgi localization in ArfGAP1 and its splice isoforms. Journal of Biological Chemistry 283: 8564–8572.

Lipp, N.-F., R. Gautier, M. Magdeleine et al. 2019. An electrostatic switching mechanism to control the lipid transfer activity of Osh6p. Nature Communications 10: 3926–17.

Lipp, N.-F., S. Ikhlef, J. Milanini et al. 2020. Lipid exchangers: cellular functions and mechanistic links with phosphoinositide metabolism. Front. Cell Dev. Biol. 8: 663.

Lürick, A., J. Gao, A. Kuhlee et al. 2017. Multivalent Rab interactions determine tether-mediated membrane fusion. Mol. Biol. Cell 28: 322–332.

Ma, M., J.-J. Liu, Y. Li et al. 2017. Cryo-EM structure and biochemical analysis reveal the basis of the functional difference between human PI3KC3-C1 and -C2. Cell Res. 27: 989–1001.

Magdeleine, M., R. Gautier, P. Gounon et al. 2016. A filter at the entrance of the Golgi that selects vesicles according to size and bulk lipid composition. eLife 5: 292.

Manna, M. and R.K. Murarka. 2021. Polyunsaturated fatty acid modulates membrane-bound monomeric α-Synuclein by modulating membrane microenvironment through preferential interactions. ACS Chem. Neurosci. 12: 675–688.

Manni, M.M., M.L. Tiberti, S. Pagnotta et al. 2018. Acyl chain asymmetry and polyunsaturation of brain phospholipids facilitate membrane vesiculation without leakage. eLife 7: 15114.

Mesmin, B., J. Bigay, J. Polidori et al. 2017. Sterol transfer, PI4P consumption, and control of membrane lipid order by endogenous OSBP. The EMBO Journal 36: 3156–3174.

Mesmin, B., G. Drin, S. Levi et al. 2007. Two Lipid-Packing Sensor Motifs Contribute to the Sensitivity of ArfGAP1 to Membrane Curvature †. Biochemistry 46: 1779–1790.

Monje-Galvan, V. and J.B. Klauda. 2018. Preferred binding mechanism of Osh4's amphipathic lipid-packing sensor motif, insights from molecular dynamics. J. Phys. Chem. B 122: 9713–9723.

Moser von Filseck, J., S. Vanni, B. Mesmin et al. 2015. A phosphatidylinositol-4-phosphate powered exchange mechanism to create a lipid gradient between membranes. Nature Communications 6: 6671.

Moss, F.R., J. Lincoff, M. Tucker et al. 2023. Brominated lipid probes expose structural asymmetries in constricted membranes. Nat. Struct. Mol. Biol. 30: 167–175.

Nath, S., J. Dancourt, V. Shteyn et al. 2014. Lipidation of the LC3/GABARAP family of autophagy proteins relies on a membrane-curvature-sensing domain in Atg3. Nat. Cell. Biol. 16: 415–424.

Nordeen, S.A., D.L. Turman and T.U. Schwartz. 2020. Yeast Nup84-Nup133 complex structure details flexibility and reveals conservation of the membrane anchoring ALPS motif. Nature Communications 11: 6060–12.

Ohashi, Y., S. Tremel, G.R. Masson et al. 2020. Membrane characteristics tune activities of endosomal and autophagic human VPS34 complexes. eLife 9.

Peter, B.J., H.M. Kent, I.G. Mills et al. 2004. BAR domains as sensors of membrane curvature: the amphiphysin BAR structure. Science 303: 495–499.

Pinot, M., S. Vanni, S. Pagnotta et al. 2014. Lipid cell biology. Polyunsaturated phospholipids facilitate membrane deformation and fission by endocytic proteins. Science 345: 693–697.

Pranke, I.M., V. Morello, J. Bigay et al. 2011. -Synuclein and ALPS motifs are membrane curvature sensors whose contrasting chemistry mediates selective vesicle binding. The Journal of Cell Biology 194: 89–103.

Quartino, P.Y., G.D. Fidelio, J.-B. Manneville et al. 2018. Detecting phospholipase activity with the amphipathic lipid packing sensor motif of ArfGAP1. Biochem. Biophys. Res. Commun. 505: 290–294.

Reynaud, A., M. Magdeleine, A. Patel et al. 2022. Tumor protein D54 binds intracellular nanovesicles via an extended amphipathic region. J. Biol. Chem. 298: 102136.

Roboti, P., K. Sato and M. Lowe. 2015. The golgin GMAP-210 is required for efficient membrane trafficking in the early secretory pathway. Journal of Cell Science 128: 1595–1606.

Sato, K., P. Roboti, A.A. Mironov et al. 2015. Coupling of vesicle tethering and Rab binding is required for in vivo functionality of the golgin GMAP-210. Mol. Biol. Cell 26: 537–553.

Shvarev, D., J. Schoppe, C. König et al. 2022. Structure of the HOPS tethering complex, a lysosomal membrane fusion machinery. eLife 11.

Skorobogatko, Y., A. Landicho, R.J. Chalkley et al. 2014. O-linked β-N-acetylglucosamine (O-GlcNAc) site thr-87 regulates synapsin I localization to synapses and size of the reserve pool of synaptic vesicles. J. Biol. Chem. 289: 3602–3612.

Smits, P., A.D. Bolton, V. Funari et al. 2010. Lethal skeletal dysplasia in mice and humans lacking the golgin GMAP-210. N. Engl. J. Med. 362: 206–216.

Taylor, R.J., G. Tagiltsev and J.A.G. Briggs. 2022. The structure of COPI vesicles and regulation of vesicle turnover. FEBS Letters.

Tiberti, M.L., B. Antonny and R. Gautier. 2020. The transbilayer distribution of polyunsaturated phospholipids determines their facilitating effect on membrane deformation. Soft Matter 16: 1722–1730.

Vamparys, L., R. Gautier, S. Vanni et al. 2013. Conical lipids in flat bilayers induce packing defects similar to that induced by positive curvature. Biophysj. 104: 585–593.

van Hilten, N., J. Methorst, N. Verwei et al. 2023. Physics-based generative model of curvature sensing peptides; distinguishing sensors from binders. Sci Adv. 9: eade8839.

Vanni, S., H. Hirose and H. Barelli . 2014. A sub-nanometre view of how membrane curvature and composition modulate lipid packing and protein recruitment. Nature Communications 5: 4916.

Vanni, S., L. Vamparys, R. Gautier et al. 2013. Amphipathic lipid packing sensor motifs: probing bilayer defects with hydrophobic residues. Biophysj. 104: 575–584.

Weimer, C., R. Beck, P. Eckert et al. 2008. Differential roles of ArfGAP1, ArfGAP2, and ArfGAP3 in COPI trafficking. The Journal of Cell Biology 183: 725–735.

Wong, M. and S. Munro. 2014. Membrane trafficking. The specificity of vesicle traffic to the Golgi is encoded in the golgin coiled-coil proteins. Science 346: 1256898–1256898.

Wong, M., A.K. Gillingham and S. Munro. 2017. The golgin coiled-coil proteins capture different types of transport carriers via distinct N-terminal motifs. BMC Biol. 15: 3.

Xu, P., R.D. Baldridge, R.J. Chi et al. 2013. Phosphatidylserine flipping enhances membrane curvature and negative charge required for vesicular transport. The Journal of Cell Biology 202: 875–886.

Zendeh-boodi, Z., T. Yamamoto, H. Sakane et al. 2013. Identification of a second amphipathic lipid-packing sensor-like motif that contributes to Gcs1p function in the early endosome-to-TGN pathway. J. Biochem. 153: 573–587.

Zhang, L., Y. Wang, Y. Dong et al. 2022. The endophilin curvature-sensitive motif requires electrostatic guidance to recycle synaptic vesicles *in vivo*. Developmental Cell 57: 750–766.e755.

9

The Bending Rigidity of the Red Blood Cell Cytoplasmic Membrane

Sebastian Himbert[1,2,*] and *Maikel C. Rheinstädter*[1,2,†]

Red Blood Cells and Their Shell

Red blood cells (RBCs; erythrocytes) are the most abundant cell type in the human body (Ukidve et al. 2020) and an integral part of blood. Blood circulates through the vascular system and supplies cells with oxygen and nutrients. It can be separated into a cellular and non-cellular fraction through centrifugation. The non-cellular fraction is the plasma, while the cellular fraction is comprised of white blood cells, platelets and RBCs (Müller and Frings 2009).

RBCs typically appear under the microscope as round disks with a diameter of ≈ 7 μm and a central indent (Figure 1A). Cells with this shape are commonly referred to as discocytes (Popescu et al. 2006). Oxygen diffuses across the cell's outer shell and binds reversibly to hemoglobin. The cells' interior lacks internal structures, such as a nucleus and mitochondria, in favor of more volume for this oxygen-binding protein. Hemoglobin is also responsible for the cell's red color (Müller and Frings 2009). Cellular functions, such as mobility, division and vesicle trafficking, are intrinsically related to a cell's ability to comply with deformation (Kirchhausen 2012; McMahon and Boucrot 2015; McMahon and Gallop 2005). In the case of RBCs that have no internal structure, this ability depends upon their two-dimensional shell, which consists of a spectrin network tethered to a cytoplasmic membrane (Liu et al. 1987) (RBC_{cm}). This composite structure can be seen in electron microscopy images (as shown in Figure 1B). The cytoskeleton forms a triangular filament network parallel to the RBC_{cm}. The distance between tethers is ≈ 80 nm (Liu et al. 1987).

[1] Department of Physics and Astronomy, McMaster University, Hamilton, ON L8S 4M1, Canada.
[2] Origins Institute, McMaster University, Hamilton, ON L8S 4M1, Canada.
* Corresponding author: himberts@mcmaster.ca
† Email: rheinstadter@mcmaster.ca

A long-standing popular model of the RBC_{cm} is the fluid mosaic model (Singer and Nicolson 1972), which describes this structure as a two-dimensional fluid-like lipid bilayer with embedded proteins. More than 50 of these membrane proteins have been characterized for the RBC_{cm} (Mohandas and Gallagher 2008). The lipid bilayer is a $\approx$ 5 nm (Himbert et al. 2017) thick membrane formed by two layers (leaflets) of lipid molecules (Figure 2A).

Membrane lipids are amphiphilic, i.e., they consist of a hydrophilic and a hydrophobic part (Van Meer et al. 2008). The molecules orient themselves such that the hydrophobic parts of both leaflets face toward each other while the hydrophilic parts are exposed to the aqueous environment. There are many different types of lipids in a typical mammalian membrane mostly represented by glycerophospholipids (PL), sphingomyelin (SM) and cholesterol.

PLs are built around a glycerol moiety. Two of the carbon atoms are esterified to two fatty acids chains (tails), with the third carbon atom bound to a polar headgroup (see Figure 2B). Common headgroups include Phosphatidylcholine (PC), Phosphatidylethanolamine (PE), Phosphatidylserine (PS), Phosphatidylglycerol (PG), Phosphatidic Acid (PA) and Phosphatidylinositol (PI) (Van Meer et al. 2008). The fatty acid tails can vary in length, i.e., the number of carbon atoms per tail, and in the degree of saturation, i.e., the number of double bonds between the carbon atoms in the tail. SM is built around sphingosine with an attached fatty acid chain and a PC headgroup (Figure 2C). Unlike PL and SM, cholesterol consists of a rigid structure formed by hydrocarbon rings (Figure 2D). It is highly abundant in cell membranes, with typical molar ratios between 20 mol% and 50 mol% (Falck et al. 2004).

The lipid composition (lipidomics) of the RBC_{cm} has been determined by mass spectroscopy (Reisz et al. 2019). The abundance of PL and SM is shown in Figure 2E. PC and PE glycerophospholipids are the most abundant species in the membrane, followed by SM. PS, PG, PA; PI lipids account for $\approx$ 20 % of the membrane. Importantly, these lipids are asymmetrically distributed between the two leaflets (Lorent et al. 2020; Mohandas and Gallagher 2008). PC and SM lipids are predominantly found in the outer leaflet of the membrane, while the majority of

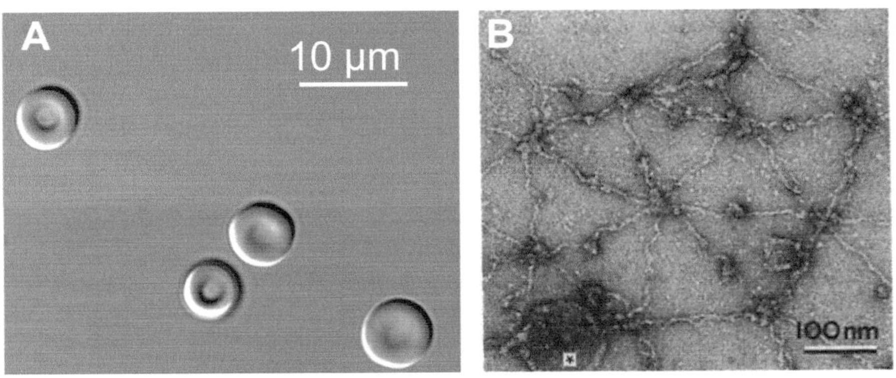

Figure 1. **(A)** RBCs under a DIC-Microscope: The cells are characterized by a discocyte shape with a thicker outer rim and an indent in their center. **(B)** Electron microscopy image of the RBC's outer shell which is comprised of a cytoskeleton tethered to a cytoplasmic membrane (adapted from (Liu et al. 1987)).

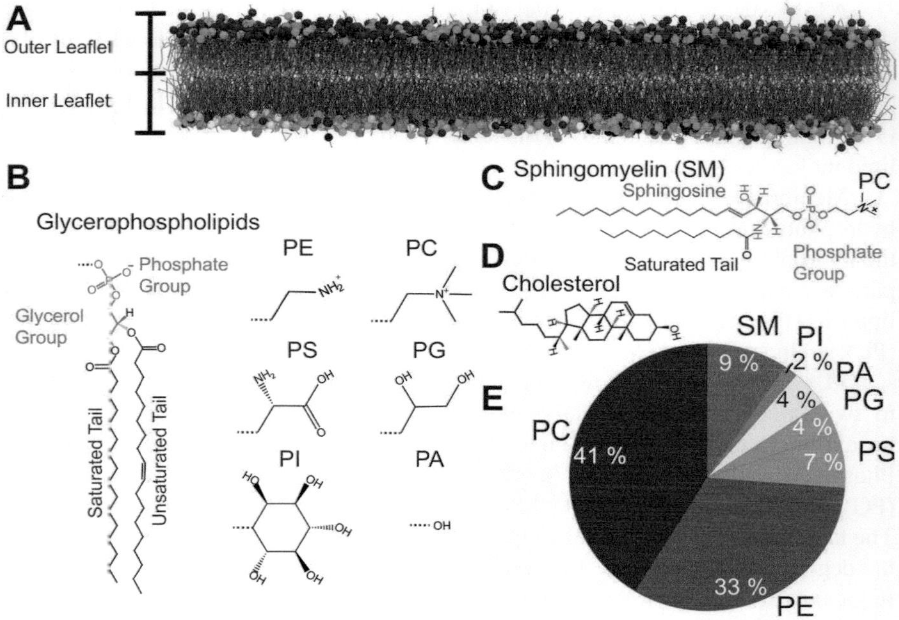

Figure 2. **(A)** A lipid membrane is formed by two layers (leaflets) of lipids molecules. **(B)** Glycerophospholipids and sphingomyelin consist of a hydrophilic head group and two hydrophobic tails. **(B-E)** Chemical structures of glycerophospholipids, sphingomyelin and cholesterol. Common head groups are Phosphatidylcholine (PC), Phosphatidylethanolamine (PE), Phosphatidylserine (PS), Phosphatidylglycerol (PG), Phosphatidic Acid (PA) and Phosphatidylinositol (PI). **(F)** Lipid distribution of the RBC cytoplasmic membrane.

PE and PI lipids as well as all PS and PG lipids are located on the inner leaflet (Mohandas and Gallagher 2008). The cholesterol content of RBC has been reported to be ≈ 50 mol% (Almizraq et al. 2013; Dodge and Phillips 1967).

Experiments in model lipid bilayers that contain saturated, unsaturated PLs and cholesterol (known as a raft-forming mixture) (Kučerka et al. 2010: Nickels et al. 2019: Rheinstädter and Mouritsen 2013) have shown that membranes containing cholesterol form a patchy structure. The cholesterol molecule is preferably located in areas with saturated lipid tails where it straightens the lipid tails and leads to a reduced area per lipid (Pan et al. 2008a). These cholesterol-rich patches have been named "rafts" and are manifestations of the liquid ordered (l_o) lipid phase. The surrounding, less ordered lipid patches are liquid disordered (l_d) domains (Simons and Gerl 2010). Heterogeneities have also been reported in the RBC_{cm} (Ciana et al. 2014; Mikhalyov and Samsonov 2011). The nanoscopic out-of-plane (Figure 3A) and in-plane (Figure 3 B) structure of the RBC_{cm} was determined by high-resolution X-ray diffraction and nanometer-sized (diameter < 3 nm) l_o and l_d lipid domains, and protein domains, were reported (Himbert et al. 2017). Schematics are shown in Figures 3 C and D. In those scattering experiments, the scattering signal was measured over several hours and the observed domain structure was a time-average structure. Molecular Dynamics (MD) simulations, however, suggest that these domains are dynamic and the result of ongoing mixing and de-mixing processes in the membrane (Himbert

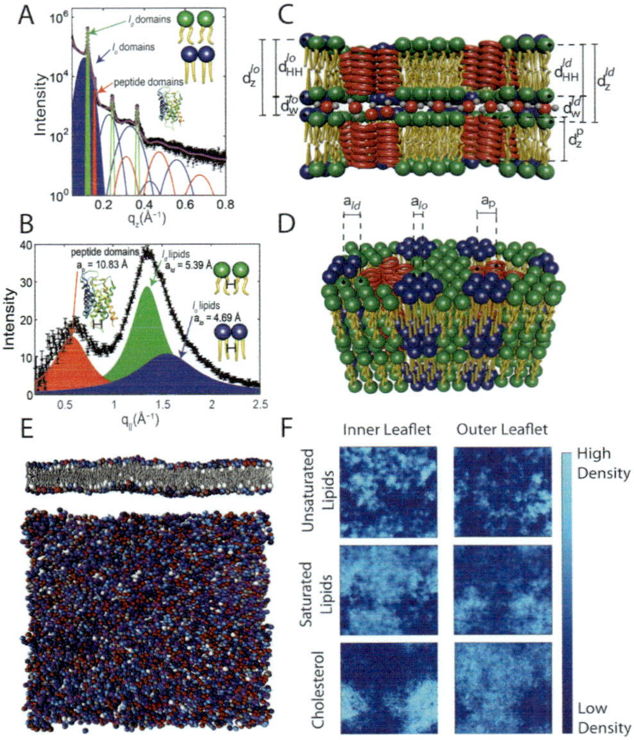

Figure 3. (A, B) XRD experiments reveal the RBC membrane as a patchy structure consisting of liquid-ordered (l_o) and liquid-disordered (l_d) lipid and protein domains (Himbert et al. 2017). Structural parameters, such as the lamellar spacing, d_z, the head group to head group thickness of the membranes, d_{HH}, the thickness of the water layer, d_w, and in-plane distances between lipid tails and integral peptides, a, and domain sizes, ξ, can be determined. **(C, D)**. **(E)** Snapshot of an MD simulation of RBC_{cm} after $t = 5$ μs. **(F)** In-plane density maps for cholesterol, saturated and unsaturated lipids calculated from MD simulations indicating the formation of liquid-ordered and liquid-disordered lipid domains (Himbert et al. 2020).

et al. 2020, Ingólfsson et al. 2014). The membrane composition in those MD simulations matches the lipidomics of the RBC_{cm} and the models consist of hundreds of different lipid species; snapshots are shown in Figures 3E and F. The MD simulations give access to time-resolved domain structures. Density maps that isolate cholesterol, saturated and unsaturated lipids present the membrane as a patchy structure, as shown in Figure 3F.

Bending of Membranes

Many biological processes, such as mobility, cell division and vesicle trafficking, are intrinsically related to a cell membrane's ability to undergo deformation (Kirchhausen 2012; McMahon and Boucrot 2015; McMahon and Gallop 2005). In this section, we will discuss the mathematical foundations to describe membrane bending and compression and introduce the bending modulus κ and the area compression modulus K_A. We will first focus our discussion on model lipid bilayers before elaborating on the deformation of the RBC_{cm}. In-plane refers to the coordinates parallel to the membrane surface ($x - y$-plane), and the out-of-plane direction is defined as an axis perpendicular to the membrane surface (z-axis). A suite of techniques has been used to study cell elasticity. Mechanical properties on cellular length scales were measured by micropipette aspiration (Evans 1983), while atomic force microscopy (AFM) (Radmacher 2002) can probe elastic behavior on the nanoscale. Cell stiffness is also studied indirectly by spectral analysis of flickering of cells under a microscope (Brochard and Lennon 1975; Strey et al. 1995; Zilker et al. 1992), as well as optical interferometric techniques (Park et al. 2010a; Popescu et al. 2006).

Deforming a Membrane

A simple picture to model the deformation of lipid membranes is depicted in Figure 4. Lipid molecules can be thought of as being connected by springs that undergoe compression and stretching when the membrane is deformed (Phillips et al. 2012). The two types of deformation are in-plane compression and in-plane expansion, as well as bending. Intuitively, in-plane compression can be understood as the compression of springs between the lipid molecules (or expansion of the springs in the case of expansion). This change comes with an energy cost and can be expressed by (Phillips et al. 2012):

$$F_{Compress} = \frac{K_A}{2}\frac{\Delta a^2}{2}, \tag{1}$$

where Δa refers to the change of the membrane area relative to the uncompressed area a_0. K_A is a material property known as the area compression modulus (Phillips et al. 2012). $F_{compress}$ has the unit of energy and K_A is in units of *force/length*. K_A can be understood as a measure of the force needed per dimension to compress a lipid bilayer by a certain length.

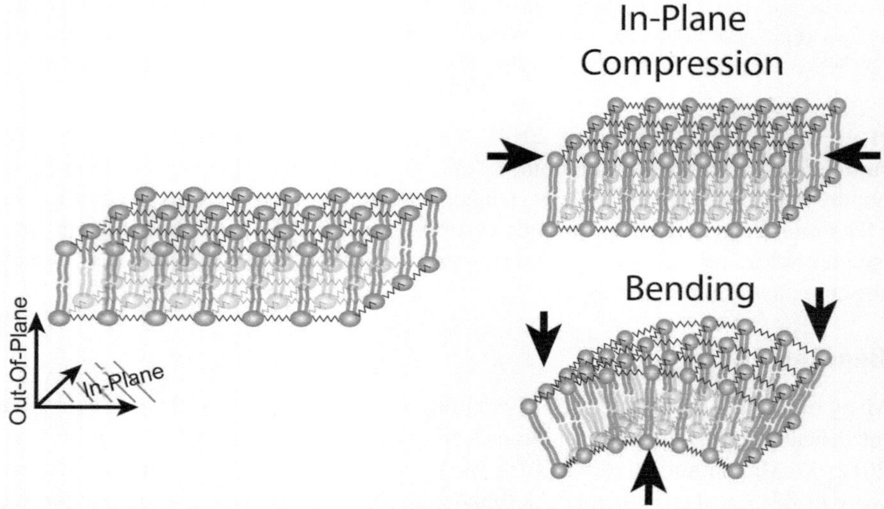

Figure 4. A lipid membrane can be pictured as a lattice of lipid molecules connected by springs. The two types of deformations, in-plane compression or in-pane expansion and bending, can be understood as the stretching and compression of these springs.

The in-plane compression/expansion deforms both monolayers evenly. This changes when a bilayer is bent from its flat state. By returning to the rudimentary model of springs between lipids, bending requires an uneven stretching and compression of springs along the out-of-plane coordinate. This is illustrated in Figure 4, where springs in the lower leaflet are compressed while those in the upper leaflet are stretched. The energy cost resulting from bilayer bending can described by (Phillips et al. 2012):

$$F_{Bend} = \frac{\kappa}{2}\int_A da(\nabla^2 u(x,y))^2, \tag{2}$$

where $u(x, y)$ are local spatial deviations of the bilayer center in the out-of-plane direction, κ is the membrane's bending modulus, and A is the membrane area. Equation (2) is known as the Helfrich-Canham-Evans functional (Phillips et al. 2012). κ is an additional material property that measures the amount of energy needed to bend a membrane.

The area compression modulus of synthetically assembled membranes ranged between 230 and 290 mN/m and was found to vary little with respect to tail saturation and tail length (Phillips et al. 2012; Rawicz et al. 2000b). In contrast, the bending modulus κ has been reported to vary significantly with both lipid characteristics. Rawicz et al. (2000b) showed that single-component lipid bilayers become stiffer with increasing tail length in saturated and mono-unsaturated tails, from $\approx$ 12 k_BT for diC 13:0 PC to $\approx$ 30 k_BT for diC 22:1 PC. No significant difference was observed for lipid molecules with a single mono-unsaturated tail (SOPC) as compared to molecules with two unsaturated tails (DOPC) (Rawicz et al. 2000b).

Both K_A and κ are measures of the membrane's deformability and several experimental methods have been developed to measure these quantities. A particularly popular technique is micropipette aspiration (MA), which can be applied to whole cells and artificially formed liposomes. A small bulge is formed by sucking a section of the membrane into a micropipette with an opening of a few micrometers. This deformation is then visually inspected under a microscope and both material properties can be determined from a shape analysis of the formed bulge (Dimova 2014; Evans and Rawicz 1990; Evans et al. 1976; Evans 1983; Evans et al. 2008). This approach is taken further by the formation of membrane nanotubes (NT). A nanometer-sized section (diameter $\approx$ 100 nm) of the membrane is pulled out of a liposome with an optical tweezer, and κ can be determined from the applied pulling force (Campillo et al. 2013; Dimova 2014, Li et al. 2011). The membrane's elastic properties can also be measured indirectly from spectral analysis of flickering of cells or liposomes under a microscope (Brochard and Lennon 1975; Strey et al. 1995; Zilker et al. 1992), as well as optical interferometric techniques (Park et al. 2010a; Popescu et al. 2006).

In addition, techniques such as X-ray diffuse scattering (XDS) (Kučerka et al. 2006; Tristram-Nagle et al. 2010) and NSE spectrometry (Kelley et al. 2020b; Nagao et al. 2017; Rheinstädter et al. 2006) have become standard methods to determine membrane bending moduli from thermal membrane fluctuations. However, they have been limited thus far to synthetic lipid bilayers (Bouvrais 2012; Nagle et al. 2015, Pabst et al. 2010) as they either require a large volume (a typical NSE sample consists of 20 ml with a membrane mass concentration of approximately 20 mg/ml) or well-organized solid supported membrane stacks (for XDS). Values can vary when comparing different experimental methods. For instance, the bending modulus for DMPC has been reported over a wide range between $\kappa = 7$ to 30 k_BT (Table 2 in (Rappolt and Pabst 2008)).

Deformation of Red Blood Cells

Table 1 shows values for the bending modulus κ of RBCs that have been reported over the years, ranging from 5 k_BT to 230 k_BT (Betz et al. 2009; Brochard and Lennon 1975; Evans 1983; Evans et al. 2008; Park et al. 2010a; Strey et al. 1995; Zilker et al. 1992). A reasonable hypothesis for this disparity is that the bending modulus depends on the length scale of the measurements. On length scales smaller than the mesh size of the spectrin network of $\approx$ 80 nm, the average bending modulus could be due mostly to the cytoplasmic membrane, while the spectrin network would add a contribution at longer length scales (Auth et al. 2007). When measuring RBC elasticity on small length scales, values for κ of 5 k_BT (Zilker et al. 1992) and 7 k_BT (Park et al. 2010a) have been reported in contrast to the much larger values for length scales of the entire RBC (Betz et al. 2009; Evans 1983; Evans et al. 2008; Strey et al. 1995). According to the above hypothesis, this would imply a substantial bending modulus for the spectrin network. It may be noted, however, that this is inconsistent

Table 1. A summary of values reported for the bending rigidity, κ, of discocytic red blood cells from the literature and RBC_{cm} membranes. (Errors are expressed as standard deviations.)

Technique	κ (k_BT)	Lengthscale (μm)	Reference
Chronological Literature	RBC		
Flickering Analysis	3-9	>0.6	[Brochard and Lennon 1975]
Micropipette Aspiration Buckling	43	>7	[Evans 1983]
Reflection Interference Microscopy	5±1.5	>0.25	[Zilker et al. 1992]
Reflection Interference Microscopy	97±37	>2	[Strey et al. 1995]
Diffraction Phase Microscopy	16±0.3	>0.1	[Popescu et al. 2006]
Reanalysis of [Popescu et al. 2006]	14,25	>0.1	[Auth et al. 2007]
Flickering Analysis	210	>0.7	[Evans et al. 2008]
Optical Tweezer	68±0.68	>7	[Betz et al. 2009]
Flickering Analysis	67±13	>1.5	[Yoon et al. 2009]
Diffraction Phase Microscopy	7±3	>0.1	[Park et al. 2010a]
Diffraction Phase Microscopy	5±2	>0.1	[Park et al. 2011]
	RBC_{cm}		
Diffuse X-ray Scattering	2 - 6	< 0.08	[Himbert et al. 2022]
Neutron Spin Echo	4 - 7	< 0.08	[Himbert et al. 2022]
Molecular Dynamics	4	< 0.08	[Himbert et al. 2022]

with a report that the bending modulus of the spectrin network is very small (Gov et al. 2003). Supposing that the bending modulus of the cytoplasmic membrane is only about 5 k_BT, it is noteworthy that this κ is significantly smaller than bending rigidities reported for single-component lipid bilayers of similar thickness that have values of κ typically between 15 k_BT and 50 k_BT (Chakraborty et al. 2020; Dimova 2014; Eid et al. 2020, 2021; Hu et al. 2013; Kelley et al. 2020b; Kučerka et al. 2006; Liu and Nagle 2004; Nagao et al. 2017; Nagle et al. 2015; Pabst et al. 2010; Pan et al. 2009a; Rheinstädter et al. 2006; Tristram-Nagle and Nagle 2007).

An important question concerns the effect of cholesterol on the membrane's mechanical properties. Intriguingly, the cytoplasmic membrane contains a considerable amount of cholesterol (50 mol%), which is conventionally thought to stiffen lipid membranes, although that depends on the type of lipid (Pan et al. 2009a). Table 2 lists literature values of κ and K_A for different lipid-cholesterol mixtures. One may intuitively speculate that cholesterol's rigid nature increases the membrane's rigidity, which can indeed be observed in a fully saturated lipid bilayer. The addition of cholesterol in DMPC bilayers (30 mol%) led to a five-fold increase in κ in XDS experiments. However, the effect is typically significantly reduced when there is a single mono-unsaturated chain (1.5-fold increase in κ in SOPC bilayers) and vanishes when both chains are unsaturated (see table entries for DOPC) and with increasing tail length (see table entry for di22:1 PC). This contrasts with the area compression modulus, which increases in the presence of cholesterol in all reported lipid mixtures (see Table 2). In contrast to these findings, an increase in κ with increasing cholesterol content has recently been reported in DOPC membranes from NSE (Chakraborty et al. 2020), as also listed in Table 2.

Biological membranes are typically composed of hundreds of different lipid species, which raises the question of how mechanical properties scale in multi-component lipid mixtures. A striking result was published by Brüning et al. 2013), who reported a low $\kappa = 1.8 \pm 0.3$ k_BT in a DMPC:DOPC mixture (molar ratio 1:1). This is surprising since κ of DMPC and DOPC have been independently reported to be >10 k_BT (Brüning et al. 2013; Liu and Nagle 2004; Pan et al. 2009b; Rawicz et al. 2000b) and the result thus contradicts the intuitive hypothesis of an averaged bending modulus in lipid mixtures. The same change was observed when doping DMPC with lipid 1,2-dioleoyl-3-trimethylammonium-propane (DOTAB) (Brüning et al. 2013). Several studies have confirmed the trend of a lowered κ in lipid mixtures (Kelley et al. 2020a; Kumagai et al. 2019; Peralta et al. 2018), though not consistently (Kumagai et al. 2019, Sorre et al. 2009), and a full picture is still lacking.

Measurements of the Bending Rigidity of the RBC_{cm}

The measurement of κ on nanoscales is based on the observation of the underlying thermal membrane undulations to determine the height-height correlation function. As shown in Figure 5A, (x, y) refer to coordinates in the membrane plane while z is directed along the membrane normal. $u(r)$ is defined as the vertical membrane displacement relative to a flat membrane (Figure 5B). This time-dependent membrane height profile can directly be calculated from MD computer simulations and κ be determined by calculating relaxation and time-dependent autocorrelation

Table 2. A summary of values reported for the bending rigidity κ and area compression modulus, K_A, for different lipid bilayers containing cholesterol (errors represent the standard error of the mean).

Lipid	Method	κ (k_BT)	K_A (m_N/m)	References
DMPC				
0 mol% Cholesterol	XDS	13±1	-	(Pan et al. 2008b)
30 mol% Cholesterol	XDS	65±7	-	
SOPC				
0 mol% Cholesterol	XDS	22±2	-	(Pan et al. 2009b)
30 mol% Cholesterol	XDS	33±3	-	
50 mol% Cholesterol	XDS	33±3	-	
SOPC				
0 mol% Cholesterol	MA	-	290±17	(Rawicz et al. 2008)
50 mol% Cholesterol	MA	-	1130±110	
DOPC				
0 mol% Cholesterol	XDS	18±2	-	(Pan et al. 2009b)
30 mol% Cholesterol	XDS	18±2	-	
50 mol% Cholesterol	XDS	18±2	-	
DOPC				
0 mol% Cholesterol	NT	26±3	-	(Gracia et al. 2010)
30 mol% Cholesterol	NT	23±3	-	
DOPC				
0 mol% Cholesterol	MA	16±2	-	(Sorre et al. 2009)
30 mol% Cholesterol	MA	15±4		
DOPC				
30 mol% Cholesterol	NT	20±6	-	(Tian et al. 2009)
DOPC				
0 mol% Cholesterol	MA	-	310±20	(Rawicz et al. 2008)
50 mol% Cholesterol	MA	-	870±20	
DOPC				
0 mol% Cholesterol	NSE	19.05 ± 0.65	230±8	(Chakraborty et al. 2020)
10 mol% Cholesterol	NSE	22.46 ± 1.77	325±26	
20 mol% Cholesterol	NSE	30.34 ± 2.47	471±38	
diC 22:1 PC				
0 mol% Cholesterol	XDS	29±3	-	(Pan et al. 2009b)
30 mol% Cholesterol	XDS	30±3	-	

functions. Bending rigidity is determined in XDS experiments from the ensemble-averaged mean-square fluctuations in the Fourier coefficients of the height-height correlation functions (Nagle 2021). Stacked membranes, typically supported by a silicon wafer, are used to measure X-ray diffuse scattering, as shown in Figure 5C. Those membrane stacks are created by producing ghosts (empty RBCs) and applying and annealing them on the wafer, as outlined in (Himbert et al. 2017). A potential asymmetry of the membranes is averaged out by a random stacking of the membranes on the silicon wafer. The relaxation of undulation modes of liposomes is measured using the neutron spin echo technique (shown in part D). We point to the very insightful discussion of the differences between the techniques by John Nagle in (Nagle 2021).

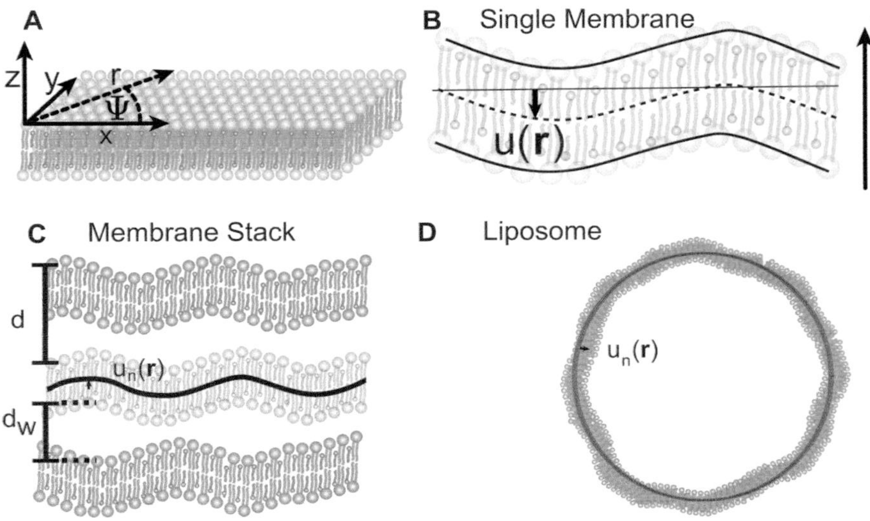

Figure 5. **(A)** The z-axis is normal to the membrane surface (out-of-plane direction). In-plane coordinates are described by the two-dimensional vector $\vec{r} = (x, y) = (r\cos(\Psi), r\sin(\Psi))$. **(B)** Membrane out-of-plane fluctuations are described by local deviations $u(\vec{r})$ relative to a flat membrane. Membrane stacks are usually used to measure X-ray diffuse scattering (XDS) **(C)** with the lamellar spacing d and thickness of the water layer between membranes d_w. The neutron spin echo technique measures the relaxation of undulation modes in liposomes (in **D**).

In MD simulations, the lipid composition of the RBC_{cm} model can be matched to results from mass spectrometry on extracted lipids from native RBCs (Stefanoni et al. 2020) to create realistic and relevant model systems. The cholesterol-to-lipid ratio in these models was ≈ 1. The membrane models contained ≈ 5,000 molecules forming a patch of ≈ 30 nm×30 nm. The membrane composition was recreated in coarse-grained MD simulations by mapping experimental lipids to the molecules available in the MARTINI force field. Figure 6A and B show a three-dimensional rendering of the model (side- and top-view). The undulation spectrum was determined and is shown in Figure 6C. $\vec{Q}$ with the in-plane component $Q_{||}$ was used to distinguish between the Fourier space of the membrane and the X-ray and neutron scattering vector $\vec{q}$. The spectrum has a $Q_{||}^4$ dependency in the low-$Q_{||}$ regime ($Q_{||} < 0.2$ Å^{-1}) in good agreement with the Helfrich–Canham (HC) theory. This theory models the membrane as an elastic sheet and is only valid for length scales much larger than the membrane thickness, i.e., small $Q_{||}$ (Brown 2008; Fowler et al. 2016). The spectrum consequently deviates from the $Q_{||}^4$ dependency for $Q_{||} > 0.2$ Å^{-1} due to molecular tilt that is characterized by the tilt modulus K_t (May et al. 2007). Corresponding fits for values of $Q_{||} < 0.2$ Å^{-1} are displayed as a red solid line from which the bending modulus and tilt modulus were determined to be $\kappa = 4.2 \pm 0.8\ k_BT$ and $K_t = 3.63 \pm 1$ mN/m.

The geometry of X-ray scattering experiments on stacks of membranes is depicted in Figure 6D. The most intense scattering is specular ($q_{||} = 0$ Å^{-1}). As shown in Figure 6E this includes peaks due to the average lamellar repeat distance in the stack of membranes, and it includes the sharp line of reflectivity

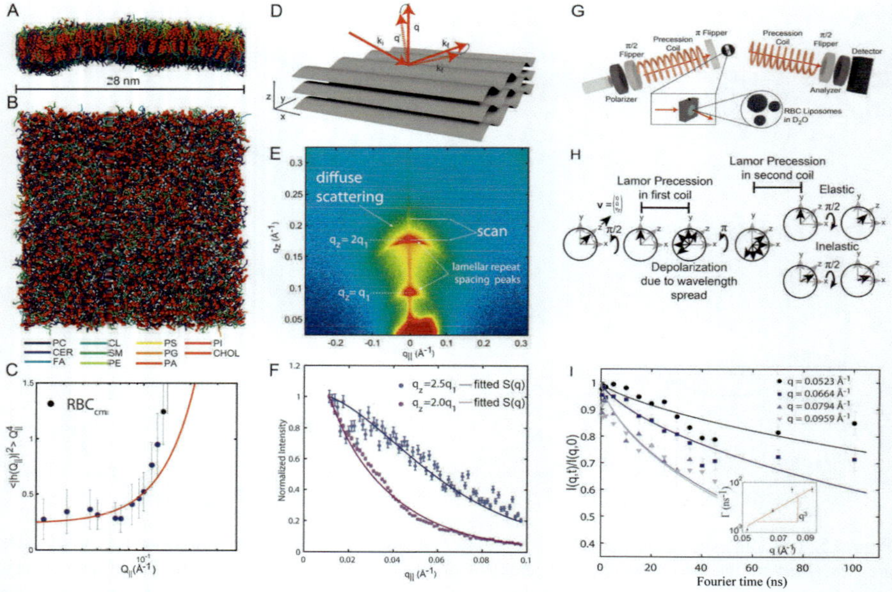

Figure 6. Renderings of the RBC$_{cm}$ simulations after 5 μs (side-view **A** and top-view **B**). **(C)** Fluctuation spectrum with fits the fluctuation model. **(D)** Schematic of a stack of fluctuating membranes and the geometry of specular ($q_{||} = 0$ Å^{-1}) and off-specular ($q_{||} \neq 0$ Å^{-1}) X-ray scattering. **(E)** q-space X-ray intensity map of a solid supported RBC$_{cm}$ membrane stack. Two orders of lamellar peaks surrounded by diffuse x-ray intensity are visible. **(F)** Off-specular intensities at $q_z = 2q_1$ and $q_z = 2.5q_1$ with fits of the structure factor $S(q_z, q_{||})$. **(G)** Experimental setup of the NSE experiment. **(I)** Intermediate scattering function I(q, t)/I(q, 0) and fits. The inset shows the determined decay constant Γ as a function of the scattering vector q.

from the silicon substrate. Only two lamellar repeat spacing peaks are visible for the RBC$_{cm}$, indicating a high degree of structural disorder within membranes. The first order peak was observed at $q_1 = 0.084$Å^{-1} corresponding to a membrane d-spacing of $d = 74.8$ Å. Most importantly, for elastic properties, a cloud of diffuse off-specular scattering was observed. Figure 6F displays the $q_{||}$ dependence for $q_z = 2q_1$ and $q_z = 2.5q_1$.

This off-specular scattering ($q_{||} \neq 0$ Å^{-1}) is due to thermal fluctuations of membrane undulation modes and compression modes of the stack of membranes. The energy of these fluctuations is given by smectic liquid crystal elasticity theory as (Helfrich 1978; Rheinstädter et al. 2006)

$$H = \int_A d^2r \sum_{n=1}^{N-1} \frac{1}{2} \left(\kappa \left(\nabla_{||}^2 u_n\right)^2 + B(u_{n+1} - u_n)^2 \right), \tag{3}$$

where u_n is the locally varying displacement. κ is the bending modulus, B is the compression modulus, N is the number of membranes and d is the lamellar spacing between neighboring membranes. Given values of κ and B, this model predicts the structure factor $S(q_z, q_{||})$ (Lyatskaya et al. 2000), which contains the important $q_{||}$ dependence of the diffuse scattering intensity. The best values were obtained by varying κ and B to provide the best fit of this model to the off-specular diffuse intensity (Himbert et al. 2022). Values of $\kappa = 2\ k_BT$ and $B = 2 \cdot 10^{-7}\ k_BT/\text{Å}^4$ were obtained. This value of κ is the smallest of the range of values shown in Table 1. It is known that the true value of κ is generally larger because XDS is also affected by the shorter range tilt degree of freedom that adds terms to Equation (3). The larger value of κ ($\kappa = 6\ k_BT$) at the end of the range shown in Table 1 was, thus, estimated by including contributions from tilt.

Membrane dynamics in unilamellar RBC_{cm} liposomes were measured by NSE (Himbert et al. 2022). Those liposomes were produced through the rehydration of annealed RBC membranes from ghosts. The basic setup of the experiment is shown in Figure 6G. Nanometer-sized RBC_{cm} liposomes were immersed in D_2O and mounted in a custom-made titanium chamber (Himbert et al. 2022). The precession of the neutron spin in a well-defined magnetic field is used to determine the energy transfer between neutrons and membranes on length scales of $\approx$ 10 nm (Figure 6H). The data of the intermediate scattering were fitted to the Zilman and Granek (Zilman and Granek 1996) theory,

$$I(q, t)/I(q, 0) = \exp\left[-(\Gamma t)^{2/3}\right], \tag{4}$$

as shown by solid lines in function in Figure 6 I. The inset shows the decay constant Γ versus q. These values of Γ were then used to obtain the effective bending modulus $\tilde{\kappa}$ using

$$\Gamma(q) = 0.025 \left(\frac{k_BT}{\tilde{\kappa}}\right)^{1/2} \left(\frac{k_BT}{\eta}\right) q^3, \tag{5}$$

where k_BT is thermal energy, and η is the solvent viscosity.

The theory of Watson and Brown (Watson et al. 2011) relates the effective bending modulus $\tilde{\kappa}$ to the customary bending modulus κ by the formula:

$$\tilde{\kappa} = \kappa + h^2 K_A, \tag{6}$$

that also involves the area compressibility modulus K_A and the distance h of the neutral surface of each monolayer from the bilayer midplane. To obtain κ, it is necessary to eliminate K_A. The range of values of $\kappa = 4\ k_BT$ to $\kappa = 7\ k_BT$ listed in Table 1 is the result of using a model that is appropriate for bilayers with a high concentration of cholesterol (Nagle 2021; Nagle et al. 2021). The high end of this range also included a correction factor for diffusion of the unilamellar liposomes in Equation (4), whereas the low end of this range did not attempt to compensate for liposome diffusion.

Implications

The bending rigidity of intact red blood cells has been reported many times, as shown in Table 1. The disparate experimental results have been appropriately described as puzzling (Auth et al. 2007; Strey et al. 1995). It has been suggested that these apparent controversial results can be explained by the complex interplay between the membrane bilayer and the spectrin network (Strey et al. 1995).

The correlation of the magnitude of κ with the length scale of the experiments has also been proposed (Auth et al. 2007; Lipowsky and Girardet 1990). One would associate a crossover length scale with the 80 nm mesh of the spectrin network. At length scales substantially greater than that, the composite RBC shell is homogeneous and would be characterized by a bending modulus for both the cytoplasmic membrane and the spectrin network. Measurements on the length scale of the whole cell, such as buckling in an aspiration pipette experiment (43 k_BT) (Evans 1983) and deformations induced by optical tweezers ($\kappa = 67\ k_BT$ (Betz et al. 2009)), provide these values. For length scales smaller than the crossover length scale, most of the bending would be situated in the cytoplasmic membrane between the ribs of the spectrin network, and κ values from measurements on those length scales would approach those of just the cytoplasmic membrane, which is likely to be homogeneous down to a length scale of 10 nm.

However, as suggested by Himbert et al. (2022), the literature data do not support this scenario. The smaller values in Table 1 by Brochard et al. ($\kappa = 3–9\ k_BT$ (Brochard and Lennon 1975)), Park et al. ($\kappa = 7\ k_BT$ (Park et al. 2010a)), and Zilker et al. ($\kappa = 5\ k_BT$ (Zilker et al. 1992)) are from measurements with length scales of the order of the wavelength (400 nm) of the optical methods employed. This is not smaller than the spectrin network length scale of 80 nm, so crossover to the value of κ for the cytoplasmic membrane would only be expected to have just begun. Instead, those κ values are much smaller than those obtained from the largest length scale measurements. At even smaller length scales, one would expect a complete crossover

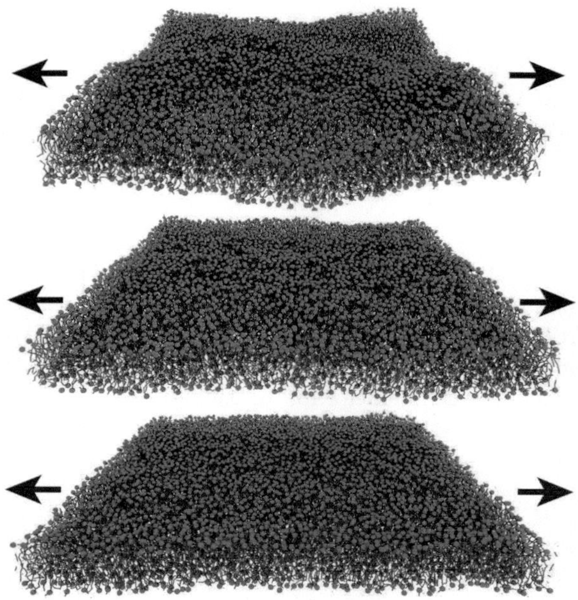

Figure 7. A small bending rigidity κ leads to thermally induced undulations that make the cell's projected area smaller than the local area. These undulations can be pulled out, resulting in an increase in the projected area and leading to a regime of area strain that costs little energy. A small κ of the RBC_{cm} can provide a greater range of mechanical flexibility that might be advantageous for blood flow.

to an even smaller value of κ, which would be just that of the cytoplasmic membrane. Contrarily, the values from MD simulations, XDS, and NSE are roughly equal to the small values obtained at the optical length scales (Brochard and Lennon 1975; Park et al. 2010b, 2011; Zilker et al. 1992). The data, thus, suggest that there is no length scale dependence in κ. This also implies that the contribution of the spectrin network to the RBC bending modulus is negligible, in agreement with (Gov et al. 2003).

The nanoscopic methods obtain values of κ for the cytoplasmic RBC_{cm} in the range of 4 k_BT to 6 k_BT. Even though this is a rather large uncertainty range, it is still significant in that the bending modulus of the RBC_{cm} is relatively small compared to most pure lipid bilayers, such as POPC, for which κ is about 20 k_BT (Kučerka et al. 2006). This may well be attributed to the large lipid diversity in this biological membrane, which is similar to what occurs in HIV mimic membranes (Boscia et al. 2013). The nanoscopic regime is most relevant for cellular processes that take place between the ribs of the spectrin network. The passive transport of small molecules is especially intrinsically related to the membrane's properties on small-length scales (Angles et al. 2021; Ghysels et al. 2019). Red blood cells are required to exchange oxygen and carbon dioxide across the membrane efficiently. One may speculate that such permeability is enhanced in a softer membrane, and a standard measure of softness is having a smaller bending modulus. As such, a smaller bending modulus of the RBC_{cm} would generally indicate a physiological advantage.

Another possible advantage of a small bending modulus of the cytoplasmic membrane might be the reduction in energy cost for the process of squeezing the RBC through small capillaries. This hypothesis is based on the possibility that such mechanical processes might require local area changes in the cytoplasmic membrane. These changes could be associated with changes in the local area of the spectrin network if necessary. Even if the spectrin network is rigid with respect to local area changes, changes in its local curvature would necessarily alter the local area of the attached cytoplasmic membrane.

The free energy for area change can be considered in terms of the area compressibility modulus K_A, which is typically a fairly stiff modulus of around 250 mN/m. This modulus is associated with area changes per molecule in a flat membrane and the work done to change the molecular packing. However, in the flaccid, low surface tension regime, the membrane exhibits thermally induced undulations that make the cell's projected area smaller than the local area (Rawicz et al. 2000a). Small increases in tension smooth out these undulations, increasing projected area corresponding to a much smaller apparent K_A than the one usually reported. Indeed, an apparent K_A of only 15 μN/m has been reported for the RBC (Park et al. 2010a). This implies the existence of a regime of area strain that costs very little energy. The extent to which the area can change in this low-cost regime varies nearly inversely with the bending modulus κ. This regime has been measured to extend to an area increase of about 2% for a lipid bilayer with $\kappa = 10\ k_BT$ (Rawicz et al. 2000a). The smaller κ of the cytoplasmic membrane in RBCs thus increases the low-cost regime, providing a greater range of mechanical flexibility that might be advantageous for blood flow. This concept is illustrated in Figure 7.

Another interesting question concerns the role of cholesterol in membrane stiffness. While an increase in stiffness has often been observed in single-component model membranes, the data in Table 2 suggest that the presence of cholesterol increases κ in saturated bilayers made of DMPC, while this effect was not always observed in unsaturated bilayers of SOPC, DOPC, and diC 22:1 PC. However, the values of κ in Table 2 for unsaturated bilayers are all around 20–30 k_BT. This makes the small bending rigidity of the cytoplasmic membrane in RBCs (4–7 k_BT) even more puzzling, considering their high lipid-to-cholesterol ratio of 1:1. However, it has been shown that the stiffening effect of cholesterol decreases with increasing unsaturation (Pan et al. 2009a), and it disappears in DOPC and diC22:1PC, which have just one unsaturated double bond in each chain. Extrapolating this trend suggests that cholesterol might even decrease the bending modulus of membranes with a significant concentration of lipids with multiple double bonds. The softness of the cytoplasmic membrane in RBCs may thus be partially explained by the presence of lipids with higher degrees of tail unsaturation. While it is difficult to produce RBC membranes without unsaturated lipids to test this hypothesis, MD simulations provide an ideal tool for manipulating membrane composition.

Figure 8 shows snapshots of MD simulations of RBC cytoplasmic membranes with the original composition (Figure 8A) and membranes from which polyunsaturated lipids (Figure 8B) or cholesterol molecules (Figure 8C) have been removed. Values for κ were determined using the same analysis outlined above. The removal of polyunsaturated lipids (while maintaining a 50 mol% cholesterol content) and the removal of cholesterol (while maintaining the lipid composition) both led to

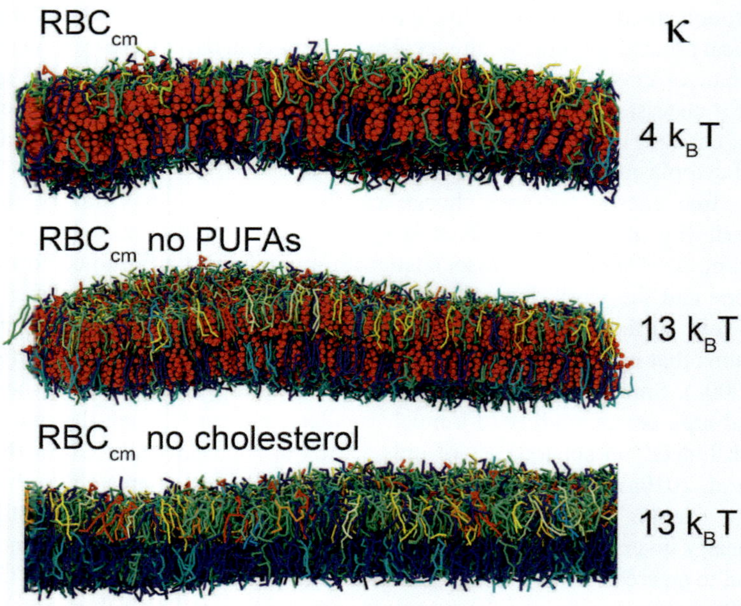

Figure 8. Snapshots of MD simulations of RBC_{cm} mimicking the composition of RBC_{cm} determined by mass spectroscopy (**A**) and RBC_{cm} models where any polyunsaturated lipid molecules were removed (in **B**) and where cholesterol was removed (in **C**). κ was found to increase in both cases from 4 k_BT to 13 k_BT, respectively. Adapted from Himbert et al. (2022).

a significant increase in bending rigidity, from 4 k_BT to 13 k_BT and from 4 k_BT to 7 k_BT, respectively. While the exact molecular mechanism is yet to be determined, these results unequivocally identify the interaction between polyunsaturated lipids and cholesterol as the main cause of the significant softness of the cytoplasmic membrane in RBCs and potentially in other biological membranes. This finding also raises further questions about the applicability of results obtained from simpler artificial model membranes to processes in biological membranes.

Summary & Conclusion

Values for κ for the cytoplasmic component of the RBC between 4 k_BT and 6 k_BT have been reported from nanoscopic techniques such as X-ray diffuse scattering, neutron spin echo spectrometry, and Molecular Dynamics simulations. The cytoplasmic membrane of the RBC is thus significantly softer on the appropriate length scale for biological processes than synthetic bilayers, and much smaller than the value at the length scale of the entire RBC due to its spectrin network. This value of κ is also softer than values reported for most single-component lipid bilayers despite a high cholesterol concentration of approximately 50 mol%. The probable reason for this extreme softness is likely the interaction between the polyunsaturated lipids and cholesterol molecules. The biological significance of a soft RBC membrane is twofold. It can be suggested that nature has designed the RBC to be soft in regions involved in the permeability of gas molecules and also to accommodate possible local area changes. The softness of the cytoplasmic membrane would also create a regime of area strain that costs little energy. This mechanism would facilitate area and RBC shape changes by smoothing out the membrane undulations and can potentially explain the high deformability of red blood cells.

Acknowledgments

We thank all the authors of the original studies, in particular Dr. John Nagle (Carnegie Mellon), who developed the formalism to determine κ from the analysis of X-ray diffuse scattering and was significantly involved in the explanation to explain the extreme softness of the RBC_{cm} and its biological implications.

References

Aiwei, T., B.R. Capraro, C. Esposito et al. 2009. Bending stiffness depends on curvature of ternary lipid mixture tubular membranes. Biophysical Journal 97(6): 1636–1646.

Akari, K., F.G. Dupuy, Z. Arsov et al. 2019. Elastic behavior of model membranes with antimicrobial peptides depends on lipid specificity and d-enantiomers. Soft Matter 15(8): 1860–1868.

Almizraq, R., J.D.R. Tchir, J.L. Holovati et al. 2013. Storage of red blood cells affects membrane composition, microvesiculation, and in vitro quality. Transfusion 53(10): 2258–2267.

Angles, G., A. Hail, R.J. Dotson et al. 2021. Atomistic simulations modify interpretation of spin-label oximetry data. part 1: Intensified water–lipid interfacial resistances. Applied Magnetic Resonance 52(10): 1261–1289.

Auth, T., S.A. Safran and N.S. Gov. 2007. Fluctuations of coupled fluid and solid membranes with application to red blood cells. Physical Review E 76(5): 051910.

Betz, T., M. Lenz, J.-F. Joanny et al. 2009. ATP-dependent mechanics of red blood cells. Proceedings of the National Academy of Sciences 106(36): 15320–15325.

Boscia, A.L., K. Akabori, Z. Benamram et al. 2013. Membrane structure correlates to function of llp2 on the cytoplasmic tail of hiv-1 gp41 protein. Biophysical Journal 105(3): 657–666.

Bouvrais, H. 2012. Bending rigidities of lipid bilayers: Their determination and main inputs in biophysical studies. In Advances in Planar Lipid Bilayers and Liposomes 15: 1–75. Elsevier.

Brochard, F. and J.F. Lennon. 1975. Frequency spectrum of the flicker phenomenon in erythrocytes. Journal de Physique 36(11): 1035–1047.

Brown, L.H F. 2008. Elastic modeling of biomembranes and lipid bilayers. Annual Reviews Physical Chemistry 59: 685–712.

Brüning, B., R. Stehle and P. Falus. 2013. Influence of charge density on bilayer bending rigidity in lipid vesicles: A combined dynamic light scattering and neutron spin-echo study. The European Physical Journal E 36(7): 1–8.

Campillo, C., P. Sens, D. Köster et al. 2013. Unexpected membrane dynamics unveiled by membrane nanotube extrusion. Biophysical Journal 104(6): 1248–1256.

Chakraborty, S., M. Doktorova, T.R. Molugu et al. 2020. How cholesterol stiffens unsaturated lipid membranes. Proceedings of the National Academy of Sciences 117(36): 21896–21905.

Ciana, A., C. Achilli and G. Minetti. 2014. Membrane rafts of the human red blood cell. Molecular Membrane Biology 31(2-3): 47–57.

Dimova, R. 2014. Recent developments in the field of bending rigidity measurements on membranes. Advances in Colloid and Interface Science 208: 225–234.

Dodge, J.T. and G. B. Phillips. 1967. Composition of phospholipids and of phospholipid fatty acids and aldehydes in human red cells. Journal of Lipid Research 8(6): 667–675.

Eid, J., H. Greige-Gerges, L. Monticelli et al. 2021. Elastic moduli of lipid membranes: Reproducibility of AFM measures. Chemistry and Physics of Lipids 234: 105011.

Eid, J., H. Razmazma, A. Jraij et al. 2020. On calculating the bending modulus of lipid bilayer membranes from buckling simulations. The Journal of Physical Chemistry B 124(29): 6299–6311.

Evans, A.E., R. Waugh and L. Melnik. 1976. Elastic area compressibility modulus of red cell membrane. Biophysical Journal 16(6): 585–595.

Evans, E.A. 1983. Bending elastic modulus of red blood cell membrane derived from buckling instability in micropipet aspiration tests. Biophysical Journal 43(1): 27–30.

Evans, E. and W. Rawicz. 1990. Entropy-driven tension and bending elasticity in condensed-fluid membranes. Physical Review Letters 64(17): 2094.

Evans, J., W. Gratzer, N. Mohandas et al. 2008. Fluctuations of the red blood cell membrane: relation to mechanical properties and lack of atp dependence. Biophysical Journal 94(10): 4134–4144.

Falck, E., M. Patra, M. Karttunen et al. 2004. Lessons of slicing membranes: interplay of packing, free area, and lateral diffusion in phospholipid/cholesterol bilayers. Biophysical Journal 87(2): 1076–1091.

Fowler, W.F., J. Hélie, A. Duncan et al. 2016. Membrane stiffness is modified by integral membrane proteins. Soft Matter 12(37): 7792–7803.

Ghysels, A., A. Krämer, R.M. Venable et al. 2019. Permeability of membranes in the liquid ordered and liquid disordered phases. Nature Communications 10(1): 1–12.

Gov, N., A.G. Zilman and S. Safran. 2003. Cytoskeleton confinement and tension of red blood cell membranes. Physical Review Letters 90(22): 228101.

Gracia, R.S., N. Bezlyepkina, R.L. Knorr et al. 2010. Effect of cholesterol on the rigidity of saturated and unsaturated membranes: fluctuation and electrodeformation analysis of giant vesicles. Soft Matter 6(7): 1472–1482.

Helfrich, W 1978. Steric interaction of fluid membranes in multilayer systems. Zeitschrift für Naturforschung A 33(3): 305–315.

Himbert, S., R.J. Alsop, M. Rose et al. 2017. The molecular structure of human red blood cell membranes from highly oriented, solid supported multi-lamellar membranes. Scientific Reports 7: 39661, Jan 2017.

Himbert, S., M.J. Blacker, A. Kihm et al. 2020. Hybrid erythrocyte liposomes: Functionalized red blood cell membranes for molecule encapsulation. Advanced Biosystems 4(3): 1900185.

Himbert, S., A. D'Alessandro, S.M. Qadri et al. 2022. The bending rigidity of the red blood cell cytoplasmic membrane. PLoS One 17(8): e0269619.

Himbert, S., A. D'Alessandro, S.M. Qadri et al. 2022. The bending rigidity of the red blood cell cytoplasmic membrane. PLoS One 17(8): e0269619.

Hu, M., P. Diggins IV and M. Deserno. 2013. Determining the bending modulus of a lipid membrane by simulating buckling. The Journal of Chemical Physics 138(21): 214110.

Ingólfsson, H.I., M.N. Melo, F.J. Van Eerden et al. 2014. Lipid organization of the plasma membrane. Journal of the American Chemical Society 136(41): 14554–14559.

Kelley, E.G., P.D. Butler, R. Ashkar et al. 2020a. Scaling relationships for the elastic moduli and viscosity of mixed lipid membranes. Proceedings of the National Academy of Sciences 117(38): 23365–23373.

Kelley, E.G., P.D. Butler, R. Ashkar et al. 2020b. Scaling relationships for the elastic moduli and viscosity of mixed lipid membranes. Proceedings of the National Academy of Sciences 117(38): 23365–23373.

Kirchhausen, T. 2012. Bending membranes. Nature Cell Biology 14(9): 906–908.

Kučerka, N., S. Tristram-Nagle and J.F. Nagle. 2006. Structure of fully hydrated fluid phase lipid bilayers with monounsaturated chains. The Journal of Membrane Biology 208(3): 193–202.

Kučerka, N., D. Marquardt, T.A. Harroun et al. 2010. Cholesterol in bilayers with PUFA chains: Doping with DMPC or POPC results in sterol reorientation and membrane-domain formation. Biochemistry 49: 7485.

Li, Y., R. Lipowsky and R. Dimova. 2011. Membrane nanotubes induced by aqueous phase separation and stabilized by spontaneous curvature. Proceedings of the National Academy of Sciences 108(12): 4731–4736.

Lipowsky, R. and M. Girardet. 1990. Shape fluctuations of polymerized or solid like membranes. Physical Review Letters 65(23): 2893.

Liu, S.-C., L.H. Derick and J. Palek. 1987. Visualization of the hexagonal lattice in the erythrocyte membrane skeleton. The Journal of Cell Biology 104(3): 527–536.

Liu, Y. and J.F. Nagle. 2004. Diffuse scattering provides material parameters and electron density profiles of biomembranes. Physical Review E 69(4): 040901.

Lorent, J.H., K.R. Levental, L. Ganesan et al. 2020. Plasma membranes are asymmetric in lipid unsaturation, packing and protein shape. Nature Chemical Biology 16(6): 644–652.

Lyatskaya, Y., Y. Liu, S. Tristram-Nagle et al. 2000. Method for obtaining structure and interactions from oriented lipid bilayers. Physical Review E 63: 011907.

May, E.R., A. Narang and D.I. Kopelevich. 2007. Role of molecular tilt in thermal fluctuations of lipid membranes. Physical Review E 76(2): 021913.

McMahon, H.T. and E. Boucrot. 2015. Membrane curvature at a glance. Journal of Cell Science 128(6): 1065–1070.

McMahon, H.T. and J.L. Gallop. 2005. Membrane curvature and mechanisms of dynamic cell membrane remodelling. Nature 438(7068): 590–596.

Mikhalyov, I. and A. Samsonov. 2011. Lipid raft detecting in membranes of live erythrocytes. Biochimica et Biophysica Acta (BBA)-Biomembranes 1808(7): 1930–1939.

Mohandas, N. and P.G. Gallagher. 2008. Red cell membrane: past, present, and future. Blood 112(10): 3939–3948.

Muller, A.W. and S. Frings. 2009. Tier-und Humanphysiologie: Eine Einführung. Springer- Verlag.

Nagle, J.F., E.A. Evans, P. Bassereau et al. 2021. A needless but interesting controversy. Proceedings of the National Academy of Sciences 118(20). ISSN 0027-8424. doi:10.1073/pnas.2025011118. URL https://www. pnas.org/content/118/20/e2025011118.

Nagle, J.F., M.S. Jablin and S. Tristram-Nagle. 2015. What are the true values of the bending modulus of simple lipid bilayers? Chemistry and Physics of Lipids 185: 3–10.

Nagle, J.F. 2021. Measuring the bending modulus of lipid bilayers with cholesterol. Physical Review E 104(4): 044405.

Nagao, M., E.G. Kelley, R. Ashk par et al. 2017. Probing elastic and viscous properties of phospholipid bilayers using neutron spin echo spectroscopy. The Journal of Physical Chemistry Letters 8(19): 4679–4684.

Nickels, J.D., M.D. Smith, R.J. Alsop et al. 2019. Lipid rafts: buffers of cell membrane physical properties. The Journal of Physical Chemistry B 123(9): 2050–2056.

Pabst, G., N. Kučerka, M.-P. Nieh et al. 2010. Applications of neutron and X-ray scattering to the study of biologically relevant model membranes. Chemistry and Physics of Lipids 163(6): 460–479.

Pan, J., S. Tristram-Nagle and J.F. Nagle. 2009a. Effect of cholesterol on structural and mechanical properties of membranes depends on lipid chain saturation. Physical Review E 80(2): 021931.

Pan, J., S. Tristram-Nagle and J.F. Nagle. 2009b. Effect of cholesterol on structural and mechanical properties of membranes depends on lipid chain saturation. Physical Review E 80(2): 021931.

Pan, J., T.T. Mills, S. Tristram-Nagle et al. 2008a. Cholesterol perturbs lipid bilayers nonuniversally. Physical Review Letters 100: 198103.

Pan, J., T.T. Mills, S. Tristram-Nagle et al. 2008b. Cholesterol perturbs lipid bilayers nonuniversally. Physical Review Letters 100(19): 198103.

Park, Y., C.A. Best and T. Kuriabova. 2011. Measurement of the nonlinear elasticity of red blood cell membranes. Physical Review E 83(5): 051925.

Park, Y., C.A. Best, K. Badizadegan et al. 2010a. Measurement of red blood cell mechanics during morphological changes. Proceedings of the National Academy of Sciences 107(15): 6731–6736.

Park, Y., C.A. Best, K. Badizadegan et al. 2010b. Measurement of red blood cell mechanics during morphological changes. Proceedings of the National Academy of Sciences 107(15): 6731–6736.

Peralta, M.F., H. Smith, D. Moody et al. 2018. Effect of anti-leishmania drugs on the structural and elastic properties of ultradeformable lipid membranes. The Journal of Physical Chemistry B 122(29): 7332–7339.

Phillips, R., J. Kondev, J. Theriot et al. 2012. Physical Biology of the Cell. Garland Science.

Popescu, G., T. Ikeda, K. Goda et al. 2006. Optical measurement of cell membrane tension. Physical Review Letters 97(21): 218101.

Radmacher, M. 2002. 4—Measuring the elastic properties of living cells by the atomic force micro- scope. Methods in Cell Biology 68(1): 67–90.

Rappolt, M. and G. Pabst. 2008. Flexibility and structure of fluid bilayer interfaces. Structure and dynamics of membranous interfaces, pages 45–81.

Rawicz, W., B.A. Smith, T.J. McIntosh et al. 2008. Elasticity, strength, and water permeability of bilayers that contain raft microdomain-forming lipids. Biophysical Journal 94(12): 4725–4736.

Rawicz, W., K.C. Olbrich, T. McIntosh et al. 2000a. Effect of chain length and unsaturation on elasticity of lipid bilayers. Biophysical Journal 79(1): 328–339.

Rawicz, W., K.C. Olbrich, T. McIntosh et al. 2000b. Effect of chain length and unsaturation on elasticity of lipid bilayers. Biophysical Journal 79(1): 328–339.

Reisz, J.A., C. Zheng, A. D'Alessandro et al. 2019. Untargeted and semi- targeted lipid analysis of biological samples using mass spectrometry-based metabolomics. In High-Throughput Metabolomics, pages 121–135. Springer.

Rheinstädter, M.C. and O.G. Mouritsen. 2013. Small-scale structures in fluid cholesterol-lipid bilayers. Curr. Opin. Colloid Interface Sci. 18: 440–447.

Rheinstädter, M.C., W. Häussler and T. Salditt. 2006. Dispersion relation of lipid membrane shape fluctuations by neutron spin-echo spectrometry. Physical Review Letters 97: 048103.

Simons, K. and M.J. Gerl. 2010. Revitalizing membrane rafts: new tools and insights. Nat. Rev. Mol. Cell Biol. 11(10): 688–699.

Singer, S.J. and G.L. Nicolson. 1972. The fluid mosaic model of the structure of cell membranes. Science 175: 720–731.

Sorre, B., A. Callan-Jones, J.-B. Manneville et al. 2009. Curvature-driven lipid sorting needs proximity to a demixing point and is aided by proteins. Proceedings of the National Academy of Sciences 106(14): 5622–5626.

Stefanoni, D., H.K.H. Shin, J.H. Baek et al. 2020. Red blood cell metabolism in rhesus macaques and humans: comparative biology of blood storage. Haematologica 105(8): 2174.

Strey, H., M. Peterson and E. Sackmann. 1995. Measurement of erythrocyte membrane elasticity by flicker eigenmode decomposition. Biophysical Journal 69(2): 478–488.

Tristram-Nagle, S. and J.F. Nagle. 2007. Hiv-1 fusion peptide decreases bending energy and promotes curved fusion intermediates. Biophysical Journal 93(6): 2048–2055.

Tristram-Nagle, S., D.J. Kim, N. Akhunzada et al. 2010. Structure and water permeability of fully hydrated diphytanoylpc. Chemistry and Physics of Lipids 163(6): 630–637.

Ukideve, A., Z. Zhao, A. Fehnel et al. 2020. Erythrocyte-driven immunization via biomimicry of their natural antigen-presenting function. Proceedings of the National Academy of Sciences 117(30): 17727–17736.

Van Meer, G., D.R. Voelker and G.W. Feigenson. 2008. Membrane lipids: where they are and how they behave. Nature Reviews Molecular Cell Biology 9(2): 112–124.

Watson, M.C., Y. Peng, Y. Zheng et al. 2011. The intermediate scattering function for lipid bilayer membranes: From nanometers to microns. The Journal of Chemical Physics 135(19): 194701.

Yoon, Y.-Z., H. Hong, A. Brown et al. 2009. Flickering analysis of erythrocyte mechanical properties: dependence on oxygenation level, cell shape, and hydration level. Biophysical Journal 97(6): 1606–1615.

Zilker, A., M. Ziegler and E. Sackmann. 1992. Spectral analysis of erythrocyte flickering in the 0.3–4-μm^{-1} regime by microinterferometry combined with fast image processing. Physical Review A 46(12): 7998.

Zilman, A.G. and R. Granek. 1996. Undulations and dynamic structure factor of membranes. Physical Review Letters 77: 4788–4791.

10

Shape and Function of the Golgi Apparatus

Felix Campelo

Introduction

Historical Introduction

The Golgi apparatus, also referred to as the Golgi complex and henceforth referred to simply as the "Golgi" from now on (Fabene and Bentivoglio 1998) is an organelle of the endomembrane system that was first discovered in cells of the spinal ganglia by the Italian pathologist Camillo Golgi in 1898 (Bentivoglio and Mazzarello 1998). By using a variant of his previously reported chemical staining technique for nervous tissues—nowadays known as the Golgi stain—Golgi identified an *apparato reticolare interno* (internal reticular apparatus) of unknown function. The debate about whether this reticular apparatus was indeed a bona fide cellular structure and not an artifact of the staining technique remained opened until the 1950s (Farquhar and Palade 1981; 1998) when the Golgi was visualized in unstained cells using recently developed electron microscopy (EM) techniques (Bentivoglio and Mazzarello 1998; Dalton and Felix 1954).

During the following decade, researchers primarily focused on providing a morphological characterization of the Golgi structure in various cell types. Those studies highlighted the ubiquity of the organelle and suggested the idea that its function could be linked to the secretion of proteins, as the Golgi was seen to be enlarged in specialized secretory cells and to be spatially associated with secretory storage granules (Farquhar and Palade 1981; 1998; Palade 1992). However, it was not until the 1960s that—thanks to advances in biochemistry techniques—the secretory and processing functions of the Golgi were experimentally delineated (Farquhar and Palade 1998). On the one hand, the use of cell fractionation assays led to the

ICFO-Institut de Ciencies Fotoniques, The Barcelona Institute of Science and Technology, 08860, Castelldefels (Barcelona), Spain.
Email: felix.campelo@icfo.eu

discovery of specific enzymatic activity—that of the enzyme galactosyltransferase—in intact Golgi fractions isolated from rat or bovine liver (Fleischer et al. 1969; James Morre et al. 1969). On the other hand, by using EM autoradiography, George Palade and colleagues were able to biochemically monitor in cells of the exocrine pancreas the vectorial transport of secretory cargo proteins along what we now know as the secretory pathway, that is, from the endoplasmic reticulum (ER) to the Golgi and secretory granules for secretion (Palade 1975; 1992). Collectively, the scientific discoveries and observations made during those years offered compelling evidence for the functional significance of the Golgi in protein processing and protein secretion, ultimately contributing to the emergence of modern cell biology as a new field of biological research (Sabatini 2008).

The Functions of the Golgi

As aforementioned, the Golgi plays a key role in cell physiology by controlling a number of fundamental biological functions (Emr et al. 2009). The main, well-established canonical functions of the Golgi are the following:

- ***Sorting and Secretion***: The Golgi serves as a sorting station for exporting and transporting different lipids and proteins to other intracellular destinations and/or for secretion outside of the cell. Newly synthesized proteins enter the Golgi from the ER in a COPII-dependent transport process (Lorente-Rodríguez and Barlowe 2011). Once at the Golgi, these proteins are enzymatically modified (e.g., by the addition/removal of sugars; see the following item for details). Then, these proteins are sorted and loaded into transport carriers (vesicles, tubules, or other pleiomorphic structures) that take them to their next destination. The Golgi is responsible for protein and lipid sorting based on their destination—including the plasma membrane, lysosomes, secretory granules and even to be sent back to the ER or the ER-Golgi intermediate compartment (ERGIC) (Boncompain and Weigel 2018; De Matteis and Luini 2008; Ford et al. 2021; Guo et al. 2014a; Lorente-Rodríguez and Barlowe 2011; Malhotra and Campelo 2011; Ramazanov et al. 2021; Spang 2015; Stalder and Gershlick 2020; von Blume and Hausser 2019; Wakana and Campelo 2021)—as well as based on cargo characteristics and the physiological and regulatory needs of the cell (Wakana and Campelo 2021).
- ***Protein Processing and Glycosylation***: The Golgi is a hub for the enzymatic modification of proteins and lipids. After their arrival to the Golgi, newly synthesized proteins undergo a series of post-translational modifications, such as glycosylation, phosphorylation, and sulfation (Freeze and Ng 2011; Stanley 2011; Ungar 2009). Among these, protein glycosylation is the most abundant protein modification occurring at the Golgi, and as such, it is one of the main functions of this organelle. Besides being essential for protein function, these modifications help determine their final destination. Importantly, a number of mechanisms control the correct localization of the proteins responsible for these enzymatic activities, such as glycosylation enzymes (Banfield 2011; Lujan and Campelo 2021; Stanley 2011; Welch and Munro 2019).

- ***Lipid Metabolism and Homeostasis***: The control of lipid metabolism is a crucial functional layer within the diverse set of Golgi functions. Lipids play a central role in controlling the shape and dynamics of the Golgi membranes and therefore in the functional organization of Golgi functions (Bankaitis et al. 2012a; Wang et al. 2019). A number of lipid-metabolizing enzymes (such as kinases, phosphatases, phospholipases, phospholipid acyl-transferases, and flippases) are present at the Golgi membranes, implying that some lipid species, such as some sphingolipids and glycosphingolipids, are generated in the Golgi membranes (D'Angelo et al. 2019; Huitema et al. 2004; Parashuraman and D'Angelo 2019). In addition to these, lipid transfer proteins also localize to particular Golgi subregions, such as membrane contact sites (MCS) (Bohnert and Schuldiner 2018; Eisenberg-Bord et al. 2016; Lujan et al. 2021; Phillips and Voeltz 2016; Reinisch and Prinz 2021; Scorrano et al. 2019; Wu et al. 2018). In particular, ER-Golgi MCS have been identified as platforms for the fast and direct transport of lipids between these two organelles and hence stand up as key organizers of lipid homeostasis in the early secretory pathway (Masone et al. 2019; Mesmin et al. 2019; Venditti et al. 2020; Wakana and Campelo 2021). These contact sites are established by the ER-localized vesicle-associated membrane protein-associated proteins (VAPs) that tether the ER with Golgi-associated lipid transfer proteins, such as ceramide transferase 1 (CERT), Nir2, four-phosphate adapter protein 1 (FAPP1), and oxysterol-binding protein (OSBP) (Venditti et al. 2020). Therefore, ER-Golgi MCS maintain and regulate the levels of a number of lipids, including phosphoinositides (phosphatidylinositol and phosphatidylinositol-4-phosphate), sphingolipids (sphingomyelin, ceramide, and glucosylceramide), and cholesterol.

Besides these "canonical functions", the Golgi is involved in other cellular functions. It can store calcium ions, which are important for cell signaling and many cellular processes, such as enzyme activation, prevention of cellular damage, and protein sorting and secretion (Missiaen et al. 2007; Pizzo et al. 2011; Ramazanov et al. 2023). In sperm cells, the Golgi is involved in the formation of an organelle called the acrosome, which localizes to the leading edge of the sperm head and carries enzymes required for breaking down the glycoprotein layer that surrounds the egg and therefore facilitates fertilization (Deneke and Pauli 2021; Khawar et al. 2019). Finally, in recent years, additional functions of the Golgi have been proposed, such as cell cycle regulation (Lu and Pfeffer 2013; Tang and Wang 2013), apoptosis and cell signaling (He et al. 2020; Ignashkova et al. 2017; Li et al. 2016), immune response (Tao et al. 2020), quality control and degradation (Eisenberg-Lerner et al. 2020; Schmidt et al. 2019; Yamaguchi et al. 2016), and mechano-sensing (Romani et al. 2019; Seetharaman and Etienne-Manneville 2018; Théry and Pende 2019). In summary, the diverse functions of the Golgi highlight the need for further research to understand better its biology, architecture, and mechanisms of action, making it a crucial target for future research in fundamental cell biology.

The Shape and Organization of the Golgi

A Hierarchical View of Golgi Architecture

The characteristic Golgi morphology is that of a stack of pancakes, which is seen in most organisms and cell types at interphase. The Golgi is disassembled into small vesicular structures in early mitosis, thereby facilitating the inheritance of Golgi fragments to the two daughter cells during cell division. After mitosis, the Golgi is reassembled into a fully functional structure. In this chapter, we will focus on describing the shape of the Golgi at interphase (see, e.g., Yanzhuang Wang and Seemann 2011 and Wei and Seemann 2017 for reviews on Golgi assembly and disassembly during cell cycle progression). So, what is this shape? The Golgi is morphologically organized in a hierarchical fashion at different levels (Figure 1).

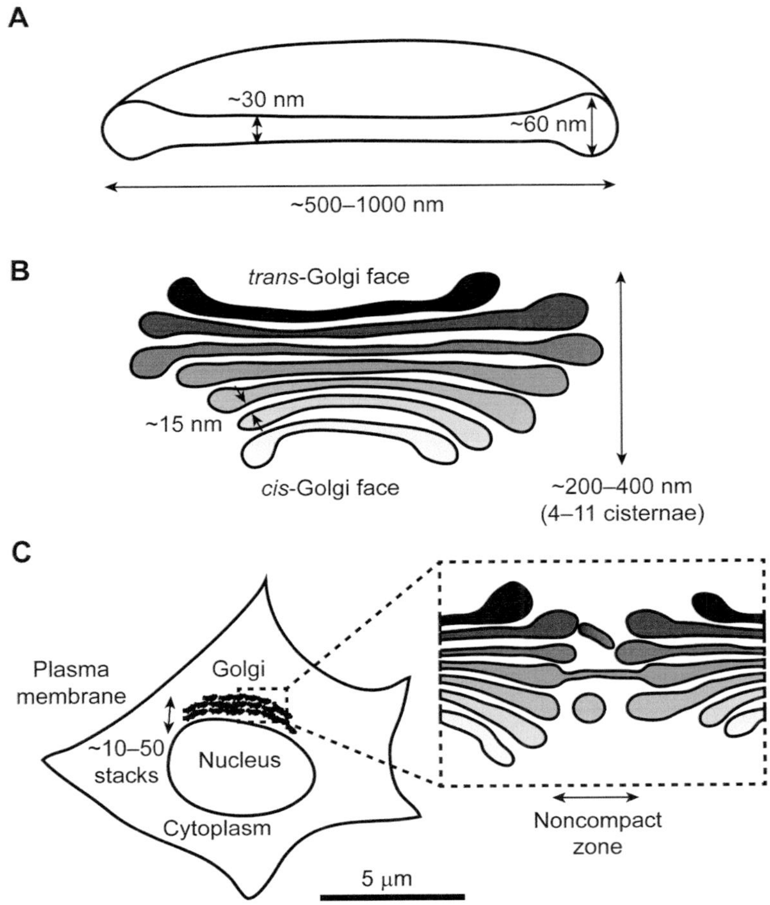

Figure 1. Hierarchical view of Golgi architecture. Schematic view of Golgi morphology at interphase, with typical numbers and sizes indicated. The structure shows a Golgi cisterna (A), a Golgi stack (B), and the Golgi ribbon shown in a cellular context (C).

The fundamental Golgi unit is a Golgi cisterna (first level of hierarchical organization), a flattened, disk-like, membrane-bound structure (Bykov et al. 2017; Marsh 2005). The rims of a cisterna are swollen, giving rise to a discocyte shape similar to that of healthy red blood cells (Figure 1A). Interestingly, it appears that the size of Golgi cisternae is relatively well-conserved among different organisms and cell types. The diameter of a Golgi cisternae is typically in the range of ~ 500–1,000 nm, the lumenal distance between the two parallel membranes is ~ 30 nm, and the diameter of the swollen rims is at least of the order of ~ 60 nm (Figure 1A) (Bykov et al. 2017; Donohoe et al. 2013; Ferraro et al. 2014; Kang and Staehelin 2008; Klumperman 2011; Ladinsky et al. 1999; Mogelsvang et al. 2003; Tie et al. 2022; Trucco et al. 2004; Weidman et al. 1993; Witte et al. 2011). With these characteristic geometrical descriptors, the typical membrane surface area of the single Golgi cisternae can be estimated to be ~ 2–8 μm^2, whereas its volume is ~ 0.025–0.1 μm^3.

The second level of hierarchical organization is that of the Golgi stack. In most organisms, a variable number of individual cisternae (typically ~ 4–11) are stacked together in the axial direction (Figure 1B) (Bykov et al. 2017; Klumperman 2011; Marsh 2005; Rambourg and Clermont 1997). While this stacked organization is widely conserved among species, there are exceptions, such as the yeast *Saccharomyces cerevisiae* (Beznoussenko et al. 2016), in which individual Golgi cisternae are not stacked to one another but remain distributed all over the cytoplasm. In *Drosophila melanogaster*, Golgi cisternae are stacked, but the stacks consist of a smaller number of cisternae (2–3 cisternae per stack) (Kondylis and Rabouille 2009). The typical axial length of the Golgi stack lies between ~ 200–400 nm, whereas the inter-cisternae distance (the distance between the two proximal membranes of consecutive cisternae) is of the order of ~ 15 nm (Figure 1B) (Bykov et al. 2017; Donohoe et al. 2013; Ferraro et al. 2014; Kang and Staehelin 2008; Klumperman 2011; Ladinsky et al. 1999; Mogelsvang et al. 2003; Tie et al. 2022; Trucco et al. 2004; Weidman et al. 1993; Witte et al. 2011).

The third level of hierarchical organization of the Golgi is that of the Golgi ribbon. Most metazoan cells present multiple Golgi stacks that are laterally interconnected in a membrane tubule and microtubule-dependent manner, forming the Golgi ribbon, which is located at the perinuclear area of the cell (Figure 1C) (Klumperman 2011). The lateral interconnection of the Golgi ribbon forms the so-called noncompact zone, and it fundamentally changes how the stacks can communicate with one another. For instance, the number of individual Golgi stacks is ~ 10–50 in nocodazole-treated cells in which the Golgi ribbon in mammalian tissue cultured cells is broken apart into individual Golgi mini-stacks. Interestingly, this final level of organization is absent in yeast species, such as *Pichia pastoris*, as well as in *Drosophila melanogaster*. In the cells of these organisms, individual Golgi stacks are dispersed throughout the cytoplasm (Klumperman 2011; Kondylis and Rabouille 2009; Papanikou and Glick 2009). Noteworthy, it has come to light recently that the Golgi ribbon is a specialized feature found in metazoans. The authors proposed that tethering proteins such as GORASPs and Golgin-45 contributed to the appearance of this ribbon-like organization (Benvenuto et al. 2023). How the appearance of a centralized Golgi ribbon structure during evolution provided these organisms with a functional advantage remains unclear. Nevertheless, the early appearance of the ribbon structure during embryogenesis suggests that the ribbon adaptation may serve a developmental role (Benvenuto et al. 2023). Total Golgi membrane area and lumenal volume have been measured in professional secretory cells (mouse ß cells), where the membrane surface area was estimated to be ~ 260–380 μm^2 and the enclosed volume ~ 3.7–5.8 μm^3 (Noske et al. 2008).

An important property of the Golgi is that not all cisternae are equivalent but have different biochemical identities. This means that the lipidome and proteome vary substantially from cisternae to cisternae. In particular, resident Golgi proteins, such as glycosylation enzymes, are distributed anisotropically across the stack, following a spatial distribution that locates the enzymes in consecutive cisternae matching the order of enzymatic action. This polarized arrangement of Golgi residents occurs in the axial direction, making the Golgi stack a vectorial arrangement of cisternae from the *cis*-Golgi face (where newly synthesized cargo proteins arrive from the ER) through the medial Golgi cisternae toward the *trans*-Golgi cisternae and *trans*-Golgi network (TGN), located at the *trans*-face of the Golgi stack. Hence, early-acting enzymes populate the *cis*-face of the stack, whereas late-acting enzymes do it at the *trans*-face (Lujan and Campelo 2021; Pothukuchi et al. 2019; Stanley 2011). Importantly, theoretical work has shown that Golgi compartmentalization in polarized cisternae permits efficient glycan production regardless of the stochasticity and promiscuity of Golgi glycosylation enzymes (Jaiman and Thattai 2020) and that cell surface glycan complexity controls the number of Golgi cisternae (Yadav et al. 2022). By contrast to the axial polarization, the Golgi ribbon does not seem to be polarized in the lateral direction, meaning that individual Golgi stacks appear to be biochemically and functionally indistinguishable from one another. Interestingly, the establishment of planar cell polarity—that is, the polarization of the cell along the surface layer orthogonal to the apical-basal axis—requires the segregation and therefore differential sorting of planar cell polarity cargo proteins Vangl2 and Frizzled6 (Guo et al. 2014b; Guo et al. 2013; Ma et al. 2018). Whether this segregation leads to laterally distinct composition of different Golgi stacks remains unclear.

A final note on the lateral heterogeneity of Golgi membranes: It has been proposed that highly dynamic, nanoscopic membrane regions—or nanodomains, with sizes $\lesssim$ 20 nm—laterally compartmentalize the membranes of the different Golgi cisternae (Duran et al. 2012; Emr et al. 2009; Ernst et al. 2018; Patterson et al. 2008), thereby adding level of hierarchical organization (level 0, according to the nomenclature used in this chapter). Experimental access to the intricacies of these nanodomains has proven to be challenging (see also our discussion in the "Golgi Shaping by Local Curvature Generation" section below). However, we expect that current advances in chemical biology and fluorescence microscopy techniques that allow for the visualization of endogenous proteins and lipids in live cells at the nanometer and millisecond spatial and temporal scales, respectively, will shed light on the existence and physiological relevance of these nanodomains in the years to come (Holcman et al. 2018; Manzo and Garcia-Parajo 2015; Nixon-Abell et al. 2016; Stockhammer and Bottanelli 2020).

Why is it Important to Understand Golgi Shape?

In the previous section, we have mostly described a static picture of the Golgi shape. However, the Golgi is a highly dynamic organelle that undergoes continuous membrane remodeling (e.g., by fission and fusion of transport intermediates) to accommodate the changing demands of the cell. It is critical to understand the relationship between the Golgi shape and its functions, as this helps us understand how this organelle is regulated, how it communicates with other organelles, and how it responds to various cellular stimuli. Moreover, the structural complexity of the Golgi renders this organelle an appealing model for investigating fundamental biological and biophysical questions related to intracellular organization and transport (Sens and Rao 2013). It is, therefore, critical to understand the molecular factors and biochemical/biophysical mechanisms that contribute to Golgi membrane shaping and relate Golgi shape and function.

Molecular Control of Golgi Shape

The architecture of the Golgi membranes and the overall shape of the Golgi are highly regulated and adapted to meet the physiological needs of the cell. A variety of factors (such as lipids and proteins) and mechanisms (biochemical or biophysical) are involved in Golgi shaping. In the following, we discuss our current knowledge of these topics.

Lipid Factors Involved in Golgi Shaping

The Golgi membranes contain a particular lipidome that distinguishes its membranes from those of other cellular organelles (Van Meer et al. 2008). Such specific lipidome contributes to shaping the Golgi proteome by recruiting and modulating the activity of Golgi proteins that control the function of the Golgi. The mammalian Golgi is known for its specialization in sphingolipid synthesis and produces various sphingolipids, such as sphingomyelin (SM), glucosylceramide (GlcCer), lactosylceramide (LacCer), and higher-order glycosphingolipids (GSLs), which are primarily destined to be exported to the plasma membrane (Anthony H. Futerman and Riezman 2005; Van Meer et al. 2008). The levels of cholesterol are also enriched at the Golgi membranes as compared to the ER membranes (Van Meer 1998; Van Meer et al. 2008). In addition, the Golgi is a place for synthesizing phospholipids, such as phosphatidylcholine (PC) and phosphatidylethanolamine (PE). These major cellular membrane lipid components are important for the local production of lipid second messengers, such as diacylglycerol (DAG) or phosphatidic acid (PA) (Carrasco and Mérida 2007; Stace and Ktistakis 2006; Van Blitterswijk et al. 2003). Furthermore, the Golgi membranes are also enriched in other signaling lipids, such as certain phosphoinositides and mostly phosphatidylinositol-4-phosphate (PI(4)P), and possibly also (albeit to a much lower extent) phosphatidylinositol-(4,5)-bisphosphate (PI(4,5)P_2) (Balla and Balla 2006; Cockcroft and Lev 2020; De Matteis et al. 2005; Posor et al. 2022).

Different factors help create and maintain the specific lipid composition of the Golgi membranes. Firstly, the Golgi localization of lipid-modifying proteins and lipid metabolic enzymes regulates the synthesis and remodeling of specific lipids, allowing for the generation and maintenance of the Golgi-specific lipid profile. Among these proteins, we can find the following:

- ***Phospholipases***: These are enzymes that cleave phospholipids. There are several types of phospholipases that act on different phospholipids and cleave them at specific sites, resulting in the release of various lipid molecules. Examples include phospholipases C (PLC), D (PLD), and also A (PLA). On the one hand, PLCs cleave the headgroup of phosphoinositide molecules between the glycerol backbone and the phosphodiester phosphate to generate DAG and free inositol phosphate. Several PLC isoforms (PLCß3, PLCγ1, and PLCε) have been identified to localize at the TGN and suggested to play potential roles in membrane trafficking processes and generate DAG at the ER-Golgi interface (Díaz Añel 2007; Sicart et al. 2015; Zhang et al. 2013). Despite these findings, the precise regulatory mechanisms governing these pathways and their specific contributions to DAG production at the Golgi under physiological conditions remain incompletely understood. Further investigations are warranted to elucidate the intricate interplay between PLC isoforms, PI4P metabolism, and DAG generation within the Golgi apparatus. On the other hand, PLD catalyzes the reaction that cleaves the phosphodiester bond in a phospholipid (usually PC), generating phosphatidic acid (PA) and a lipid headgroup (e.g., choline when PC is the substrate) (Freyberg et al. 2003; Riebeling et al. 2009). The two isoforms of human PLD, PLD1 and PLD2, partially localize to Golgi membranes, contributing to DAG homeostasis and Golgi shaping and function (Freyberg et al. 2001; Freyberg et al. 2002). Finally, different PLA isoforms localize to Golgi membranes, where they cleave a phospholipid at the glycerol position to release a free fatty acid and a lysophospholipid. In particular, cPLA(2) (Bechler et al. 2012; Kita et al. 2019; Stewart et al. 2002) and PLA2G6 (Ben-Tekaya et al. 2010; Lin et al. 2018; Schmidt et al. 2010) partially locate in different regions of the Golgi and regulate Golgi shaping.
- ***Lipid Transferases***: These enzymes transfer specific functional groups from a donor lipid molecule to an acceptor lipid molecule. Hence, the reaction catalyzed by these enzymes has two substrates (the donor and the acceptor) and two products. Of special interest in Golgi lipidomics is a particular type of lipid transferase: SM synthase (SMS) (Huitema et al. 2004; Subathra et al. 2011; Tafesse et al. 2007). SMS uses ceramide and PC as substrates and cleaves PC (similar to what PLC does; see above) to yield a DAG molecule and a phosphorylcholine functional group, which is immediately transferred to ceramide to convert to SM. Hence, the SMS-mediated reaction can be

summarized as ceramide + PC ⇌ SM + DAG. Notice the double arrow in the reaction, which means that SMS can catalyze the reaction in the two directions, although the forward direction generally appears more efficient. Humans express two SMS isoforms, SMS1 and SMS2, with the former mostly localized at the *trans*-Golgi membranes and the latter at the PM (Huitema et al. 2004). The catalytic activity of SMS faces the lumenal side of the Golgi membranes, therefore generating SM specifically in the lumenal leaflet. The generation of SM and DAG by the SMS-mediated reaction at the Golgi membranes has been shown to play important roles in shaping the Golgi membranes (Campelo et al. 2017; van Galen et al. 2014) and in controlling its secretory (Campelo et al. 2017; Capasso et al. 2017; Duran et al. 2012; Lujan et al. 2022) and processing (van Galen et al. 2014) functions.

- ***Sphingomyelinases*:** These enzymes cleave SM to produce ceramide and a free headgroup. Examples include neutral sphingomyelinase and acid sphingomyelinase. Although they mostly localize to the cell surface, neutral sphingomyelinase SMPD3 has been reported to localize to the Golgi, where it maintains sphingolipid and DAG homeostasis and, as a consequence, sustains protein secretion (Stoffel et al. 2016).
- ***Glycosyltransferases*:** These are enzymes that catalyze the addition of sugar residues to lipids, which is an important event in glycolipid synthesis. Examples include glucosylceramide synthase (D'Angelo et al. 2013; Futerman and Pagano 1991; Tu and Banfield 2010). Glycosphingolipids (GSLs) are a particularly important (and highly represented) class of glycolipids that are formed by the covalent linkage (a ß-glycosidic bond) between a glycan and a ceramide. Similar to SM, GSLs are mostly present in the lumenal leaflet of the Golgi membranes because the catalytic activity of most glycosyltransferases faces the Golgi lumen (Rini et al. 2022). Because many different glycans can be linked to different ceramides, the metabolism of GSLs is highly complex (D'Angelo, Capasso, et al. 2013). Although not essential for cell survival, GSLs play crucial roles in organizing the Golgi membranes. For instance, the levels of gangliosides—a type of GSL found in high levels in the nervous system—particularly GD3 have been linked to Golgi morphology, protein secretion, and regulation of Golgi function in apoptosis under stress (Hicks and Machamer 2005).
- ***Lipid Kinases and Phosphatases*:** These are enzymes that add or remove, respectively, phosphate groups from lipids. Of special importance in Golgi lipid homeostasis are phosphoinositide kinases and phosphatases (De Matteis et al. 2005; Graham and Burd 2011). There are several isoforms of these enzymes, including phosphatidylinositol-4-phosphate 5-kinases (PIP5Ks), which generates PI(4,5)P_2 from PI(4)P; phosphatidylinositol 4-kinases (PI4Ks) generates PI(4)P from PI; phosphatidylinositol 4-phosphate phosphatases (Sac1), which dephosphorylates PI(4)P to produce PI; and phosphatidylinositol 4,5-bisphosphate phosphatases (INPP5), which dephosphorylate PI(4,5)P_2 to produce PI(4)P. At the Golgi membranes, the main phosphoinositide is PI(4)

P, which is produced by the Golgi-localized PI4KIIIß, PI4KIIα, and PI4KIIß (Balla and Balla 2006; Posor et al. 2022). Regarding INPP5 phosphatases, it has been reported that OCRL-1 and INPP5E dephosphorylate PI(4,5)P2 to form PI4P at the Golgi membranes (Bielas et al. 2009; Choudhury et al. 2009; Kong et al. 2000). The Sac1 phosphatase localizes to the ER side of ER-Golgi MCS, and only upon specific conditions, such as serum deprivation (Blagoveshchenskaya et al. 2008) it translocates to the Golgi membranes. These enzymes are involved in the regulation of a variety of cellular processes, including Golgi structure and function, vesicle trafficking, and membrane signaling (Bankaitis et al. 2012b; Campelo et al. 2017; Capasso et al. 2017; Cheong et al. 2010; De Matteis et al. 2013; De Matteis et al. 2005; von Blume and Hausser 2019; Wakana and Campelo 2021).

- ***Acyltransferases*:** Lipid acyltransferases, such as lysophosphatidic acid acyltransferase (LPAAT), catalyze the linkage of an acyl chain to a lipid molecule. LPAAT3 (LPAAT-γ/AGPAT3) regulates Golgi tubulation, and in the absence of this protein, the Golgi appears fragmented (Schmidt and Brown 2009). Of interest, the Golgi fission-inducing protein CtBP1/BARS, which was initially thought to have an intrinsic LPAAT activity, regulates Golgi shaping by an extrinsic association with an LPAAT activity (Gallop et al. 2005; Pagliuso et al. 2016; Spanfò et al. 1999; Valente et al. 2013; Yang et al. 2005; Zhukovsky et al. 2019).

Secondly, the non-homogeneous exchange of lipids between the compartments of the secretory pathway by vesicular trafficking also contributes to outlining the specific profile of the Golgi lipidome. This is because lipids undergo lateral partitioning and differential sorting at different steps of the secretory pathway, implying a different lipid composition in transport intermediates as compared to that in the bulk donor organelle membrane (Brügger et al. 2000; Deng et al. 2016; Klemm et al. 2009; Lev 2006; Melero et al. 2018; Nickel et al. 1998; van Meer et al. 1987). Consequently, lipid compositional gradients along the organelles of the secretory pathway arise. Regarding the Golgi membranes, preferential sorting of lipids into membrane-bound transport intermediates (or exclusion by retention in the Golgi membranes) leads to selective depletion (or enrichment) of specific lipid species in the Golgi. Of course, this mechanism parallels lipid enzymes at the Golgi. A clear example is the case of SM, the levels of which follow a gradient along the secretory pathway, being low at the ER (~3 mol %), increasing at the Golgi membranes (~ 8 mol %), and reaching a maximum at the PM (~ 16 mol %) (Van Meer 1998). Because SM is produced *de novo* at the *trans*-Golgi membranes by SMS (see above), its levels are expected to be "high" there. In addition, SM is preferentially enriched into specific transport carriers at the TGN for export to the cell surface (Deng et al. 2016; Duran et al. 2012; Klemm et al. 2009), which creates a flux of SM out of the Golgi (negative flux at the Golgi), thereby partially counterbalancing the positive flux of SM emanating from SMS-mediated synthesis. Whether there are feedback mechanisms between synthesis and export to maintain SM homeostasis at the Golgi membranes remains to be fully ascertained. On the one hand, it is known that when SM homeostasis is disrupted by adding a short-chain ceramide analog, Golgi export is largely reduced (Capasso et al. 2017; Duran et al. 2012; Rosenwald and Pagano 1993). However, whether abrogation of SM export from the Golgi (with the concomitant increase in SM levels) feeds back to inhibit SMS-mediated SM production to maintain SM homeostasis remains unclear.

Lastly, maintaining the Golgi lipid composition is also controlled by the vectorial, non-vesicular transport of particular lipid species to and from the Golgi membranes. Such non-vesicular lipid transport is mediated by specialized lipid transfer proteins that shuttle lipid molecules from a donor to an acceptor membrane. Because this is a non-vesicular lipid transport process, it requires spatial proximity between the donor and acceptor membranes so the transfer proteins can channel or drive the passage of the amphipathic lipids (Reinisch and Prinz 2021). This interorganellar lipid transfer occurs at specialized membrane regions—named membrane contact sites (MCS)—where the two organelles are in close proximity (~ 30 nm separation) (Bohnert and Schuldiner 2018; Eisenberg-Bord et al. 2016; Phillips and Voeltz 2016; Scorrano et al. 2019; Wu et al. 2018). In particular, the Golgi membranes form MCS with the ER, in so-called ER-Golgi MCS, which host a number of lipid transfer proteins involved in the regulation of the lipid composition in both organelles (Lujan et al. 2021; Mesmin et al. 2019; Venditti et al. 2020). The main ER-Golgi MCS proteins involved in lipid homeostasis are briefly summarized as follows:

- ***VAP-A/B (Vesicle-Associated Membrane Protein-Associated Proteins A/B).*** These integral transmembrane proteins are inserted into the ER membrane and serve as tethers to establish MCS with other organelles. They do that by VAP major sperm domain (MSP), which binds FFAT motif-containing proteins, such as CERT or OSBP (see below).
- ***CERT (Ceramide Transfer Protein)*****:** CERT was the first lipid transfer protein found to non-vesicularly shuttle lipids between two distinct membranes (Hanada et al. 2003). CERT contains a pleckstrin homology (PH) domain for binding to PI(4)P and hence targets this protein to the *trans*-Golgi/TGN membranes; an FFAT domain for interaction with VAPs; and a StART domain for the directional transfer of ceramide from the ER to the Golgi.
- ***OSBP (Oxysterol-Binding Protein)*****:** OSBP is a protein of the Osh family of sterol sensor proteins. Similar to CERT, OSBP contains a PH domain that binds PI(4)P and also ADP-ribosylation factor 1 (Arf1)-GTP, thus targeting OSBP to the late Golgi cisternae; and an FFAT domain for binding the ER in *trans* through VAPs. In addition, OSBP contains an OSBP-related domain (ORD), which acts as a bidirectional lipid transfer domain that catalyzes the transport of cholesterol from the ER to the trans-Golgi membranes and the reciprocal transport of PI(4)P from the Golgi to the ER membranes (Mesmin et al. 2013). Mechanistically, there is experimental evidence suggesting that OSBP can function as a ferry-bridge protein, meaning that it combines its tethering function by linking the ER and Golgi membranes together with its lipid transfer function by shuttling lipids

between these two membranes (Antonny et al. 2018). Importantly, this ferry-bridge mode of action has also been proposed for other proteins with a similar domain structure, such as CERT (Antonny et al. 2018). Besides OSBP, other OSBP-related protein (ORP) family members, such as ORP4L or ORP9, have also been reported to function at ER-Golgi MCS (David et al. 2021; Venditti et al. 2020).

- ***FAPP1/2 (Four-Phosphate-Adapter-Protein 1/2).*** FAPP1 regulates PI(4)P levels by working in conjunction with Sac1 and VAPs at ER-Golgi MCS (Venditti et al. 2019). Depletion of FAPP1 increases Golgi PI(4)P, and FAPP1 acts as a negative regulator of carrier biogenesis for secretion (Venditti et al. 2019). FAPP2 is a lipid transfer protein that specifically binds and transfers glucosylceramide (D'Angelo et al. 2007; D'Angelo, Uemura, et al. 2013). Although it is not a bona fide ER-Golgi MCS component, FAPP2 contains a PH domain for binding PI(4)P and the small G protein Arf1-GTP at the trans-Golgi membranes and an FFAT domain that potentially leads to interaction with VAPs at the ER membrane (Mikitova and Levine 2012). Finally, FAPP2 contains a glycolipid transfer protein homology (GLTPH) domain for the non-vesicular trafficking of glucosylceramide not from the ER but from the *cis*-Golgi membranes, where this lipid is synthesized, directly toward the *trans*-Golgi membranes (Mesmin et al. 2019).
- ***Nir2*:** Nir2 is a peripheral protein that contains a phosphatidylinositol (PI) transfer protein (PITP) domain that has been proposed to allow Nir2 to shuttle PI from the ER to the Golgi membranes. Nir2 also contains an FFAT domain for ER-Golgi MCS localization through the interaction with the ER-localized VAPs (Litvak et al. 2005; Peretti et al. 2008).
- ***Sac1*:** At the ER side, this lipid phosphatase dephosphorylates PI(4)P that is shuttled from the Golgi via OSBP to produce PI, thereby establishing a PI/PI4P cycle at the ER-Golgi interface (Antonny, Bigay, and Mesmin 2018).
- ***SCAP (Sterol Regulatory Element-Binding Protein (SREBP) Cleavage-Activating Protein)*:** When cholesterol levels are low, SCAP escorts the transcription factor SREBP from the ER to the Golgi. At the Golgi, SREBP is cleaved so its transcriptionally active domain can translocate into the nucleus to upregulate the expression of proteins involved in the production and uptake of cholesterol (Brown et al. 2018). However, when cells are fed with cholesterol, SCAP interacts with the Sac1-VAP-OSBP complex at ER-Golgi MCS to regulate PI(4)P turnover and the formation of transport carriers at the TGN (Wakana et al. 2021; Wakana and Campelo 2021).

When taken together, these factors ensure the specific lipid composition of the Golgi membranes, which in turn plays a critical role in the recruitment and modulation of peripheral membrane proteins and the regulation of the shape and function of the Golgi.

Protein Factors Involved in Golgi Shaping

The Golgi has a well-determined proteome, which is variable across the stack (from *cis-* to *trans-*) and is a fundamental driver of the shape and function of this organelle. The main proteins of the Golgi are logically related to its functions. Besides lipid metabolism-related proteins outlined in the previous section, the Golgi hosts a number of processing enzymes, structural proteins, as well as trafficking-related proteins.

- ***Enzymes and Transporters*****:** The Golgi is highly enriched in enzymes responsible for protein maturation (e.g., glycosylation) (Stanley 2011). Moreover, a number of nucleotide sugar transporters are also inserted in the Golgi membranes, thereby facilitating the overall glycosylation process. In addition, Golgi-localized ion transporters, such as SPCA1 calcium pump, and proton pumps, such as the vacuolar [H+]ATPase (V-ATPase),regulate the ionic concentration and pH in the lumen of Golgi cisternae and, as a consequence, its functions (Forgac 2007; Kellokumpu 2019).
- ***Trafficking-Related Proteins*****:** The protein machinery involved in the different steps required for trafficking is vast. These include sorting receptors, such as KDEL receptor (Pelham 1991; Popoff et al. 2011), mannose 6-phosphate receptor (Braulke and Bonifacino 2009), Cab45 (von Blume et al. 2012), TGN46 (Lujan et al. 2022), or Syndecan-1 (Sundberg et al. 2019); budding effectors, such as COPI, COPII, and clathrin coat components (Kirchhausen 2000), or caveolins (Parton et al. 2021); fission factors, such as protein kinase D (Campelo and Malhotra 2012; Malhotra and Campelo 2011; Wakana and Campelo 2021), CtBP1/BARS (C-terminal-binding protein 1/brefeldin A ADP-ribosylation substrate) (Bonazzi et al. 2005; Valente et al. 2013; Weigert et al. 1999), or dynamin (Jones et al. 1998); tethering factors, such as golgins (Munro 2011; Short et al. 2005; Witkos and Lowe 2017) or the DCGE sub-family of tethers (Dsl1, COG, GARP and exocyst complexes) (Launay et al. 2023; Sztul and Lupashin 2009); fusion factors, such as SNARE proteins (Malsam and Söllner 2011); and general regulators, such as Arf proteins (D'Souza-Schorey and Chavrier 2006) and Rab GTPases (Novick and Zerial 1997; Pfeffer 2017; Zerial and Stenmark 1993), and their effectors, such as GTPase-activating proteins (GAPs) and guanine nucleotide exchange factors (GEFs) (Grosshans et al. 2006; Thomas and Fromme 2020). In particular, Arf proteins, including Golgi-localized isoforms Arf1, Arf3, Arf4, and Arf5 have been shown to play different roles in recruiting different coats (clathrin or COPI) as well as been involved in TGN and *cis*-Golgi tubulation (Adarska et al. 2021; Bottanelli et al. 2017; Wong-Dilworth et al. 2023).
- ***Structural Proteins*****:** These include proteins involved in Golgi stacking and Golgi ribbon linking, such as Golgi matrix proteins (including golgins, see above) and GORASPs, which have been involved in maintenance of the overall Golgi ribbon structure by acting as "Golgi glue" proteins (Grond et al. 2020; Lee et al. 2014; Xiang and Wang 2010); as well as proteins involved in Golgi positioning (Yadav and Linstedt 2011). Of special importance for the

maintenance of the Golgi structure is the cytoskeleton (Klumperman 2011; Lowe 2011). Microtubules help position the Golgi near the centrosome microtubule-organizing center (Yadav and Linstedt 2011) and for the maintenance and re-assembly after mitosis of the Golgi structure (Miller et al. 2009). When microtubules are depolymerized or destabilized, the Golgi tends to disperse throughout the cytoplasm, and the Golgi ribbon disassembles into so-called Golgi mini-stacks (Thyberg and Moskalewski 1999). Actin filaments also play a role in Golgi positioning and organization. Actin filaments interact with Golgi membranes and regulate secretory cargo sorting and transport carrier formation at the Golgi (Curwin et al. 2012; Egea et al. 2013; von Blume et al. 2009). In addition, actin-associated proteins, such as molecular motors, act in conjunction with the cytoskeleton to provide structural integrity of the Golgi (Egea et al. 2013; Thyberg and Moskalewski 1999). Moreover, the PI(4)P-binding, TGN-localized protein GOLPH3 has been shown to link the Golgi membranes to the actin cytoskeleton mediated by binding to the myosin motor Myo-18B (Dippold et al. 2009) as well as to have an intrinsic curvature generation capability (Rahajeng et al. 2019).

Importantly, in the last decades, a large number of pharmacological reagents and chemical inhibitors have been established as Golgi-disturbing agents. These are valuable tools to help us investigate how different biochemical pathways affect the shape and function of the Golgi. Very common examples are chemicals such as brefeldin A, a fungal metabolite that inhibits the secretion of proteins from the Golgi apparatus and causes the redistribution of Golgi components to the ER and the endosomes; the ionophore monensin, an antibiotic that perturbs the Golgi pH, disrupts the transport of proteins from the Golgi apparatus to the lysosomes and causes the Golgi cisternae to become swollen and fragmented; and the microtubule depolymerizing agent nocodazole, which dissolve the Golgi ribbon in so-called Golgi mini-stacks (Bui et al. 2023; Dinter and Berger 1998). The list of Golgi-perturbing agents is long, so in order to avoid as much as possible from being repetitive, we defer to the excellent reviews by Dinter and Berger (Dinter and Berger 1998) and by Bui et al. (2023) for comprehensive lists of small molecule pharmacological agents that affect Golgi shape and function.

Physical Control of Golgi Shape

Because the Golgi is an organelle built from membranes, its morphology is linked to that of its membranes. Hence, to understand the biophysical mechanisms of Golgi shaping, it is fundamental to understand the mechanism of membrane deformation (Bassereau et al. 2018; Evans and Skalak 1979; Nelson et al. 1989; Seifert 1997). From a physics standpoint, membranes can be thought of as elastic quasi-two-dimensional structures because their thickness (~ 4 nm) is generally small as compared to their lateral dimensions (from ~ 50–100 nm for small transport carriers to ~ 1 μm for individual Golgi cisternae). This implies that, in many circumstances, membranes can be effectively treated as two-dimensional elastic surfaces embedded in a three-dimensional space (Helfrich 1973). In addition, in most physiological conditions, biological membranes behave as fluids in the in-plane directions (Evans and Skalak 1979). Therefore, in the following, we briefly describe the mechanics of two-dimensional fluid membranes.

Membrane Mechanics

To describe the elastic deformations relevant for membrane shaping purposes, we will follow Wolfgang Helfrich's seminal paper on lipid bilayer elasticity (Helfrich 1973). Helfrich rationalized that there are three classes of elastic strains that are important in biological membrane mechanics.

The first deformation is the stretching-compression of the membrane. The amphipathic nature of membrane lipids makes them, on the one hand, self-assemble into bilayer structures, which seal forming closed structures due to the energetic hydrophobic penalty of free edges (Frolov et al. 2011). On the other hand, it imposes a preferred area per lipid molecule within the bilayer configuration (the area per phospholipid is typically ~ 0.6 nm^2; Klauda et al. 2010), which results from the balance between the attractive forces at the polar-nonpolar interface and the repulsive forces between the lipid headgroups as well as between the acyl chains (Israelachvili 2010). This force balance tightly controls the area per lipid (Chapman 1981), which means that the local elastic stretching modulus of a lipid monolayer is high, implying that we can consider, for most practical situations, that lipid monolayers are non-stretchable surfaces. A quantity usually referred to when dealing with biological membranes is the membrane tension or the in-plane tension. The membrane tension is the mechanical force (stress) required to change the apparent (or projected) surface area of the membrane (Barnoy et al. 2023; Sens and Plastino 2015) without necessarily changing the true area of the membrane (that is, the number of lipids multiplied by the average area per lipid). Hence, if we define the effective area of the membrane, A as the projected area of the membrane in the membrane plane, then the membrane tension, σ, is the thermodynamic conjugate variable to the membrane projected area (Safran 2018). Therefore, the infinitesimal work, dF_t, performed against membrane tension required to change the projected area by an infinitesimal amount, dA, is:

$$dF_t = \sigma \, dA$$

For an in-depth discussion of the difference between the local stretching-compression deformation, change in the in-plane area per lipid, and the effective stretching-compression change in the projected area of the membrane, see Barnoy et al. (2023).

The second deformation arises from the tilting and splay of lipid molecules (that is, the transverse shear strain of the elastic material) without creating a net membrane bending. Because this deformation arises from the changes in the internal structure of the membrane (Campelo et al. 2014; Campelo et al. 2008), we will not be discussing it here.

Finally, the third type of deformation that elastic fluid surfaces can sustain is bending deformation, i.e., the generation of membrane curvature (Kozlov and Taraska 2023; Zimmerberg and Kozlov 2006). Intuitively, the curvature of a surface is a

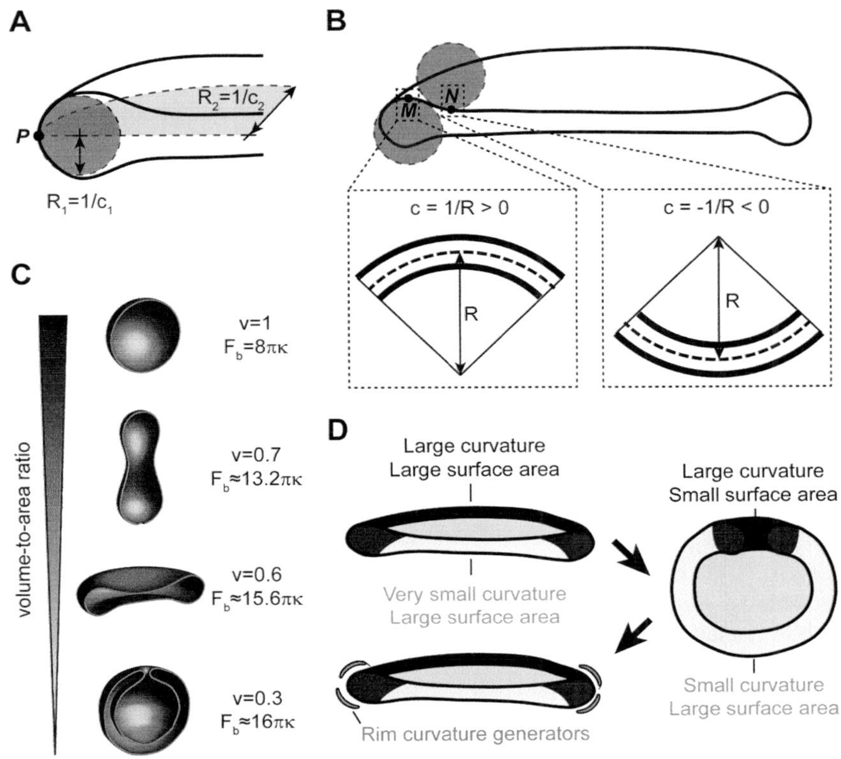

Figure 2. Physical control of Golgi shape. (A) Description of the local curvatures and radii of curvature of a Golgi-like surface at a point P. (B) Definition of positive and negative local curvatures. (C) Theoretically, estimated shapes and bending energies that minimize the bending energy for a given volume-to-area ratio. (D) Proposed model for the control of Golgi cisternae morphology was (adapted from (Campelo et al. (2017)). A flat cisterna (left) has two major regions of different curvatures and surface areas (the large, highly- curved rim, and the large flatter central part). A curled cisterna (right) also has two major regions of different curvatures and surface areas (the small but highly- curved rim, and the large, mildly curved central part). In the absence of rim curvature generators, the curled shape is energetically favorable, whereas when curvature generators stabilize the cisternal rim, the flat cisterna is energetically favorable.

measure of how much the surface deviates from a flat plane. Mathematically, the local curvature of a surface along a given direction is the inverse of the radius of curvature of that surface along that specific direction (Figure 2A). The radius of curvature is the radius of the circumference that locally describes the surface at that point and along that direction. In addition, the curvature is defined with a sign (positive or negative). The most used sign convention in membrane biophysics defines positive (negative) curvatures for those surfaces bulging toward (away from) the lipid headgroup region (Figure 2B). An isotropic two-dimensional surface can be locally described by two principal curvatures (the curvatures along the two principal directions), c_1 and c_2 (Figure 2A). By convenience, one usually works with the so-called total curvature, J, which is the sum of the two principal curvatures, $J = c_1+c_2$; and the Gaussian curvature, K, which is the product of the two principal curvatures, $K = c_1 c_2$, as they are invariant quantities that do not depend on the choice of the coordinate system (Do Carmo 2016). The so-called mean curvature, H, is often used instead of the total curvature. The mean curvature is simply half of the total curvature, $H = J/2$. Because the energetics of membrane bending deformations has been exhaustively reviewed in the past (Campelo et al. 2014; Deserno 2015; Seifert 1997; Zimmerberg and Kozlov 2006), here we will briefly describe the main idea behind the Helfrich bending energy, which states that an infinitesimal thermodynamic work, dF_b , needs to be performed to change the total curvature by an infinitesimal amount, dJ. The proportionality is given by a stress, in this case, the bending moment, τ , multiplied by the surface area of the membrane region, A:

$$dF_b = A\,\tau\,dJ$$

In the Helfrich model, which is valid for small curvatures, the bending moment depends linearly on the total curvature of the system:

$$\tau = \kappa\,(J - J_s)$$

Here, is the membrane bending modulus (an elastic parameter that defines how "bendable" a membrane is), and J_s is the spontaneous curvature of the membrane, which is related to the membrane stress (torque) in a flat state (see (Kozlov 2018) for a detailed discussion about the differences between the spontaneous curvature and the intrinsic curvature, this latter being the curvature the membrane adopts in a relaxed state). Hence, the total thermodynamic work that needs to be performed to bend a piece of the membrane of area A from a reference state in which it adopts the spontaneous curvature, J_s, to a state of curvature J is given by $F_b = \int_{Js}^{J} dF_b$, which adopts the classical expression of the Helfrich bending energy (Helfrich 1973):

$$F_b = \frac{1}{2}\kappa\,(J - J_s)^2 A.$$

It is important to highlight that, as we stated above, here we consider the membrane as a two-dimensional surface; hence, the actual lipid deformations resulting from membrane bending (such as splay of the lipid tails or stretching-compression of the polar-head region) are not explicitly considered. Moreover, we have only focused on the energy associated with changes in the total curvature while disregarding the Gaussian curvature term. This latter is important only in cases when there are topological changes (such as fission, fusion, and pore formation/sealing) or local changes in the elastic modulus (e.g., due to the existence of different membrane nanodomains), and we will not discuss here the Gaussian curvature energy (see, e.g., Seifert (1997) for details).

Volume-to-Area-Driven Golgi Shaping

By using the Helfrich energy (see previous section), it is possible to find the shapes of closed vesicles (model lipid bilayers) with a fixed surface area and fixed enclosed volume by minimizing the Helfrich bending energy with the volume and area constraints. The problem of finding the stationary shapes of lipid vesicles using different curvature models (such as those including spontaneous curvature) was extensively studied in the last decade of the 20th century (Seifert 1997). Given that the Helfrich energy is scale-invariant, a parameter that can be used to describe vesicle shapes is the so-called reduced volume, *v*, which is the dimensionless volume-to-area ratio of the vesicle normalized to belong to the [0,1] interval, and is given by the formula $v = 6\sqrt{\pi}\ V/A^{3/2}$. In the simplest case where the spontaneous curvature is zero and a spherical topology, there are three branches of local minima of the bending energy, corresponding to three qualitatively different shapes: prolates (for large values of the reduced volume, $v > 0.65$), oblates or discocytes—the characteristic shape of a Golgi cisterna – (for intermediate values, $0.59 < v < 0.65$), and stomatocytes (for small values, $v < 0.59$) (Figure 2C) (Seifert 1997). The energy required to bend a flat membrane surface into such shapes is equal to for a sphere ($v = 1$)—independently of its size —and increases as the reduced volume decreases, reaching a maximum energy of 16 $\pi\kappa$ for the limiting case of a stomatocyte of minimal reduced volume ($v \rightarrow 0$) (Figure 2C). As we discussed above, typical Golgi cisternae have very small volume-to-area ratios, typically $v < 0.1$, which would mean that, according to a simple model that describe Golgi cisternal shape by its surface-to-volume ratio, they should acquire a stomatocyte (cup-like) shape. Even when considering a non-vanishing, homogenous spontaneous curvature, Golgi-like oblate shapes are rarely found at very low reduced volumes (Seifert et al. 1991), suggesting that there needs to be additional factors governing the shape of this organelle, such as inhomogeneous curvature generation.

Golgi Shaping by Local Curvature Generation

A number of mechanisms for the generation of membrane curvature have been proposed (Day and Stachowiak 2020; Derganc et al. 2013; Kozlov et al. 2014a; Kozlov and Taraska 2023; McMahon and Gallop 2005; Zimmerberg and Kozlov 2006). Although both lipids and proteins can bend membranes, proteins are generally thought to be more effective in generating large membrane curvatures (Kozlov and Taraska 2023; Zimmerberg and Kozlov 2006), as the ones required to stabilize the highly-curved rims of Golgi cisternae.

- ***Lipids:*** They can generate or sustain Golgi curvature in different ways. First, as explained above, certain lipids recruit specific peripheral membrane proteins involved in curvature generation, sensing, or stabilization. For example, on the cytoplasmic leaflet of the Golgi membranes, DAG recruits the fission factor PKD (Baron and Malhotra 2002; Malhotra and Campelo 2011; Wakana and Campelo 2021); and PI(4)P recruits lipid transfer proteins, such as CERT or OSBP (Mesmin et al. 2019; Wakana and Campelo 2021), as well as the BAR (Bin/Amphiphysin/RVS) domain-containing proteins arfaptin1 and arfaptin2 (Cruz-Garcia et al. 2013). Second, certain lipids, such as SM and cholesterol, can segregate into small and highly dynamic liquid-ordered nanodomains (Campelo et al. 2017; Duran et al. 2012; Levental and Veatch 2016; Lingwood and Simons 2010). However, the size and lifetime of such nanodomains are largely unknown (Sezgin et al. 2017). Such nanodomains are expected to be more rigid (i.e., have a larger bending rigidity) than the surrounding membrane (Dimova 2014) and, therefore, be less prone to localize to or sustain regions of high curvature. However, theoretical modeling suggests that spontaneous partitioning of such nanodomains to the central flat regions of Golgi cisternae is not sufficient to stabilize cisternal morphology (Campelo et al. 2017). Third, non-vesicular transfer of lipids by lipid transfer proteins can, on the one hand, alter the volume-to-area ratio and hence the shape of a closed membrane-bound Golgi cisterna (see above); on the other hand, it creates an area imbalance between the two lipid monolayers, which results in overall bilayer bending (Kozlov and Taraska 2023; Seifert 1997; Varma and Deserno 2022). And fourth, lipids themselves can bend the membrane. This is because some lipids have a non-cylindrical effective molecular shape (e.g., a conical shape), which creates a monolayer lateral stress profile that leads to a non-vanishing monolayer spontaneous curvature (Campelo and Kozlov 2014; Campelo et al. 2008; Illya et al. 2005; Marsh 2007). Examples of such lipids are DAG or ceramide, which are conical lipids and therefore generate negative membrane curvature (bulging away from the headgroup region), and lipids such as lysolipids, which are inverted conical lipids and generate positive membrane curvature (Zimmerberg and Kozlov 2006). It is important to stress that these lipids only bend the bilayer if they are asymmetrically distributed between the two leaflets of the Golgi membrane (if, e.g., equal amounts of DAG are present in the two lipid monolayers, each monolayer would tend to bend in the opposite direction so the net bilayer curvature cancels out, to a first approximation). Therefore, although cellular membranes are asymmetric (Lorent et al. 2020), lipids, such as DAG, that have a high flip-flop rate (spontaneous exchange rate between the two monolayers of a bilayer) are not expected to be strong inducers of membrane curvature (Campelo et al. 2017).

- *Proteins*: They can be strong inducers of membrane curvature, which they accomplish in different ways, which we briefly summarize here (see (Kozlov and Taraska 2023) for a comprehensive review on this topic). First, the hydrophobic insertion mechanism consists of the shallow insertion of hydrophobic protein regions into the membrane bilayer. Shallow hydrophobic insertions expand the surface area of each of the two monolayers to a different extent, leading to area asymmetry and membrane bending when the two monolayers are laterally coupled (Campelo et al. 2008). Furthermore, insertions locally modify the lipid material around the insertion, leading to wedging and overall membrane bending (Campelo et al. 2008). At the Golgi, small G proteins such as Arf1 can induce curvature in such a way (Beck et al. 2008). Arf1 is necessary for COPI carrier formation at the *cis*-Golgi as well as for TGN export in PKD-dependent (Pusapati et al. 2010; Wakana and Campelo 2021) as well as in other PKD-independent routes (Bottanelli et al. 2017; Wong-Dilworth et al. 2023), but its role in overall Golgi morphology has not been fully resolved. Second, the scaffolding mechanism is based on the fact that membrane proteins with a pre-defined curved shape can impose their shape into the membrane by strong lipid-protein interactions (Kozlov et al. 2014b; McMahon and Gallop 2005). Examples of this mechanism for Golgi proteins are classical coats, such as the COPI and clathrin coats, that impose their polymerized spherical curvature to the underlying membrane (Kirchhausen 2000; Kozlov and Taraska 2023), as well as BAR-domain-containing proteins, such as arfaptins (Cruz-Garcia et al. 2013; Gehart et al. 2012; Man et al. 2011). Third, the cytoskeleton can also impose membrane curvature using molecular motor-mediated forces applied to the membrane (Miserey-Lenkei et al. 2010; Wakana et al. 2013), and by direct force application and modulation of the membrane tension (Carlsson 2018). Fourth, the crowding mechanism is based on the idea that membrane-associated proteins locally enriched on a membrane region can exert laterally repulsive (or attractive) forces at a distance from the membrane. Such forces result in positive (or negative) membrane bending when the proteins are somehow anchored to the membrane and, therefore, maintain their relative position with respect to the membrane plane (Day and Stachowiak 2020; Kozlov et al. 2014b; Kozlov and Taraska 2023; Stachowiak et al. 2012). Noteworthy, the anisotropic lateral distribution of the Golgi trafficking machinery (enriched at cisternal rims) with respect to Golgi enzymes (enriched at the central, flat part of the cisterna) could contribute to providing the driving force to generate the curvature of discocyte-shaped Golgi cisterna (Campelo et al. 2017; Ernst et al. 2018; Tie et al. 2018). According to this view, Golgi enzymes with small cytosolic tails and large lumenal domains would stabilize the central part of the cisternae against being positively curved, while effectors of the trafficking machinery, such as golgins, Rabs, SNAREs, and coats, which possess large cytosolic domains and small lumenal tails, would maintain the large positive curvature of cisternal rims. Crowding-generating forces can arise from entropic or electrostatic repulsion between membrane-anchored polypeptides (Stachowiak et al. 2012), as well as from multivalent protein-protein interactions, such as those arising in intrinsically-disordered proteins (Yuan et al. 2021), which are in turn prone to form biomolecular condensates (Gouveia et al. 2022; Mittag and Pappu 2022; Shin and Brangwynne 2017). Examples of those proteins in the Golgi are beginning to be described. In the Golgi lumen, proteins such as chromogranins (Parchure et al. 2022) and TGN46 (Lujan et al. 2022) have been shown to possess the capability to form biomolecular condensates by their intrinsically disordered regions. On the cytosolic side of the Golgi, a number of Golgins and Golgi-associated proteins that do not have intrinsically disordered regions have been shown to phase separate and contribute to maintaining the overall Golgi matrix morphology (Jimenez and Perez 2023; Rebane et al. 2020; Rothman 2019; Ziltener et al. 2020). Based on these latter results, a model in which the Golgi structure is templated by a series of Golgin condensates overall creates a two-dimensional "cocoon" that holds the polarized architecture of the Golgi (Rebane et al. 2020).

Theoretical analysis of Golgi morphology has revealed that a way to stabilize the flat cisternal morphology with physiological volume-to-area ratios is by localizing a certain amount of curvature inducers at the rims of the cisternae (Campelo et al. 2017). As aforementioned, for a homogeneous cisterna with a small volume-to-area ratio, the elastic shape that minimizes the overall energy of the system is that of a stomatocyte (or curled cisterna). The reason for that is the high energetic penalty that the large cisternal rim region (of large curvature) has on the bending energy. The bending energy needed to generate a flat cisternal structure grows approximately linearly with the reduction of the reduced volume, v, whereas the curled stomatocyte shape has an almost constant energy ($F_b \approx 16\ \pi\kappa$) regardless of the reduced volume, (Seifert 1997) (see Figure 2C). Hence, it was proposed that if there were proteins generating and maintaining the curvature of the rim, then the flat cisterna morphology would be stabilized (Campelo et al. 2017). Based on these computations, the Golgi membrane morphology was suggested to be maintained by membrane curvature generators via a positive feedback loop where the highly bent rims of the Golgi cisternae provide optimal nucleation sites for the budding of transport carriers and the concomitant stabilization of the rim curvature (Campelo et al. 2017) (Figure 2D). Interestingly, there is experimental evidence showing that in conditions where the flat cisterna curl into a stomatocyte-like shape—by inhibition of SM metabolism and transient activation of PKD (Campelo et al. 2017; Capasso et al. 2017; Duran et al. 2012; van Galen et al. 2014) curvature-generating proteins such as clathrin are released from the Golgi membranes (Campelo et al. 2017). It will be important in the future to further establish which curvature-generating proteins are preferentially enriched at the rims of the Golgi (Ernst et al. 2018; Tie et al. 2022) and elucidate their specific roles in Golgi morphology.

Intermembrane Interactions at the Golgi Membranes

Golgi membranes face the crowded environment of the cytoplasm. On the one hand, as mentioned above, the Golgi cisternae in most eukaryotes are stacked together, making them somehow spatially confined by proximity between cisternal membranes. On the other hand, ER membranes can be in close apposition to *trans*-Golgi membranes, forming ER-Golgi MCS, which similarly leads to Golgi membrane confinement. Such spatial confinement produced by intermembrane proximity may contribute to Golgi shaping by altering its mechanical and thermodynamic state. First, confinement means that thermal undulations of Golgi membranes are reduced, which can alter membrane tension and bending rigidity (Helfrich 1985; Helfrich and Servuss 1984), protein diffusion (Brown 2003), and also promote protein nano-clustering mediated by Casimir-like long-range attractive forces (Bitbol et al. 2010; MacHta et al. 2012; Pezeshkian et al. 2017). It is tempting to speculate that the localization of ER-Golgi MCS at specific subdomains of the *trans*-Golgi membranes can lead to an inhomogeneous suppression of thermal undulations, possibly leading to lateral segregation of Golgi functions based on altered mechanical membrane properties (Lujan et al. 2021). Interestingly, highly-curved transport carriers seem to form at regions of the TGN that are close but not overlapping with the ER-Golgi MCS regions (Wakana et al. 2021). However, it is still unclear if the possible modulation of thermal fluctuation at MCS plays a significant role in Golgi shaping.

Second, direct or indirect protein-protein or protein-membrane interactions have been suggested to mediate Golgi stacking by effectively creating adhesion forces between subsequent Golgi cisternae (Klumperman 2011; Lee et al. 2014; Rothman 2019; Warren and Malhotra 1998). Theoretical work by Derganc et al. considered the physical factors that regulate Golgi stacking (Derganc et al. 2006). They showed that the balance between adhesion energy and bending energy sets the number of cisternae in a flat stack (Derganc et al. 2006). The physical mechanisms by which different golgins, GORASP proteins, and other unknown factors control cisternal stacking and ribbon formation remain incompletely understood (Grond et al. 2020; Lee et al. 2014; Rothman 2019). Finally, another mechanism that could help stabilize flat Golgi cisternae is by means of intralumenal protein bridges, in analogy to the protein bridges that maintain the morphology of peripheral ER sheets (Shibata et al. 2009). Such bridges would consist of transmembrane proteins located on the flat central part of the cisterna, where they would dimerize in *trans*- with a partner protein on the opposite membrane within the same cisterna, thereby forming a lumenal bridge that connects the two parallel membranes. However, no clear candidates for Golgi intralumenal bridges have yet been identified.

Conclusion

In conclusion, the Golgi apparatus is a complex and dynamic organelle that plays a critical role in several cellular processes. Golgi morphology and functions are tightly linked, and understanding this relationship is essential for understanding how this organelle works at the cellular level. As discussed in this chapter, the Golgi typically consists of a series of flattened, stacked cisternae with numerous tubular connections between them. This unique shape provides a large surface area for various enzymatic reactions (such as protein glycosylation) to efficiently and timely occur within its relatively small lumenal volume. The flattened cisternal morphology comprises two regions of divergent curvatures: the central flat part of the cisterna and the highly-curved cisternal rims. The flat cisternal morphology ensures that the rims occupy a relatively large fraction of the entire cisterna surface area, which facilitates the efficient remodeling of the membrane to produce transport carriers that bud off from the Golgi to shuttle cargo proteins and lipids to their respective cellular destinations, such as the plasma membrane or the endo-lysosomal system. Furthermore, it has been suggested that the interconnected tubular networks that form between the Golgi cisternae to establish the Golgi ribbon structure play a vital role in intra-Golgi transport. These tubules facilitate the rapid movement of proteins and lipids between the various cisternae compartments, allowing for coordinated processing and efficient cargo transport. Therefore, the unique membrane shape of the Golgi apparatus is intricately linked to its essential functions in the modification, sorting, and transport of cellular molecules. The Golgi interactions with other organelles further highlight its central role in intracellular communication and coordination.

Understanding the dynamic interplay between Golgi membrane shaping and its diverse cellular functions poses a formidable challenge in the field. The intricate balance between maintaining the Golgi architecture and ensuring the polarized distribution of components, alongside the constant remodeling required for efficient cargo processing and trafficking, demands a comprehensive exploration. Unraveling the molecular intricacies that govern Golgi membrane dynamics and their functional implications in a (patho-) physiological context remains a key hurdle in advancing our comprehension of the intracellular organization of the secretory pathway.

To overcome these challenges and achieve the breakthrough of a better understanding of how Golgi membrane dynamics impacts function, a multi-faceted approach integrating advanced imaging techniques, quantitative analyses, and theoretical and computational biophysics tools will be imperative. Comprehensive studies combining structural biology, lipidomics, and proteomics will also be essential to unveil the intricate molecular networks orchestrating Golgi's shape and function. Collaborative efforts across disciplines, integrating experimental data with computational modeling and novel deep learning approaches, can provide a holistic understanding of the regulatory mechanisms governing Golgi dynamics. Additionally, leveraging emerging technologies such as live-cell super-resolution microscopy and intracellular single-particle tracking will be pivotal in capturing real-time and single-molecule insights into Golgi membrane dynamics with unprecedented spatial and temporal resolution. Only through such an interdisciplinary and technology-driven approach can we aspire to unlock the mysteries surrounding the shape and function of the Golgi apparatus.

Acknowledgments

I thank members of the Single Molecule Biophotonics group at ICFO and a large number of colleagues and collaborators for valuable discussions. I am also very grateful to David Gershlick, Misha Kozlov, Lei Lu, Manuel Muñiz, and Yuichi Wakana for critical reading of this chapter and for providing valuable comments. I acknowledge support from the Government of Spain (RYC-2017–22227, PID2019-106232RB-I00/10.13039/501100011033/110198RB-I00; PID2022-138282NB-I00 project funded by the MCIN/AEI/10.13039/501100011033/FEDER, UE; Severo Ochoa CEX2019-000910-S), Fundació Privada Cellex, Fundació Privada Mir-Puig, and Generalitat de Catalunya (CERCA, AGAUR).

References

Adarska, P., L. Wong-Dilworth and F. Bottanelli. 2021. ARF GTPases and their ubiquitous role in intracellular trafficking beyond the Golgi. Frontiers in Cell and Developmental Biology 9 (July). Frontiers Media S.A.: 679046. doi:10.3389/FCELL.2021.679046/BIBTEX.

Antonny, B., J. Bigay and B. Mesmin. 2018. The Oxysterol-Binding Protein Cycle: Burning off PI(4) P to Transport Cholesterol. Annual Review of Biochemistry 87 (1). Annual Reviews: 809–837. doi:10.1146/annurev-biochem-061516-044924.

Balla, A. and T. Balla. 2006. Phosphatidylinositol 4-Kinases: Old Enzymes with Emerging Functions. Trends in Cell Biology. Elsevier Current Trends. doi:10.1016/j.tcb.2006.05.003.

Banfield, D.K. 2011. Mechanisms of protein retention in the Golgi. Cold Spring Harbor Perspectives in Biology 3(8): 1–14. doi:10.1101/cshperspect.a005264.

Bankaitis, V.A., R. Garcia-Mata and C.J. Mousley. 2012a. Golgi membrane dynamics and lipid metabolism. Current Biology 22 (10):R414-24. doi:10.1016/j.cub.2012.03.004.

Bankaitis, V.A., R. Garcia-Mata and C.J. Mousley. 2012b. Golgi membrane dynamics and lipid metabolism. Current Biology 22 (10). Cell Press:R414–R424. doi:10.1016/j.cub.2012.03.004.

Barnoy, A., A.K. Tsaturyan and M.M. Kozlov. 2023. Mechanism of tension propagation in cell membranes. BioRxiv, March. Cold Spring Harbor Laboratory, 2023.03.22.533804. doi:10.1101/2023.03.22.533804.

Baron, C.L. and V. Malhotra. 2002. Role of diacylglycerol in PKD Recruitment to the TGN and protein transport to the plasma membrane. Science 295 (5553). American Association for the Advancement of Science: 325–328. doi:10.1126/science.1066759.

Bassereau, P., R. Jin, T. Baumgart et al. 2018. The 2018 biomembrane curvature and remodeling roadmap. Journal of Physics D: Applied Physics. Vol. 51. Institute of Physics Publishing. doi:10.1088/1361-6463/aacb98.

Bechler, M.E., P. De Figueiredo and W.J. Brown. 2012. A PLA1-2 punch regulates the Golgi complex. Trends in Cell Biology 22(2): 116–124. doi:10.1016/j.tcb.2011.10.003.

Beck, R., Z. Sun, F. Adolf et al. 2008. Membrane curvature induced by Arf1-GTP is essential for vesicle formation. Proceedings of the National Academy of Sciences of the United States of America 105(33): 11731–11736. doi:10.1073/pnas.0805182105.

Ben-Tekaya, H., R.A. Kahn and H.P. Hauri. 2010. ADP Ribosylation Factors 1 and 4 and Group VIA Phospholipase A 2 regulate morphology and intraorganellar traffic in the endoplasmic reticulum-Golgi intermediate compartment. Molecular Biology of the Cell 21(23): 4130–4140. doi:10.1091/mbc.E10-01-0022.

Bentivoglio, M. and P. Mazzarello. 1998. One hundred years of the Golgi apparatus: History of a disputed cell organelle. Italian Journal of Neurological Sciences 19 (4). Ital J Neurol Sci.: 241–247. doi:10.1007/BF02427612.

Benvenuto, G., S. Leone, E. Astoricchio et al. 2023. Evolution of the ribbon-like organization of the Golgi apparatus in animal cells. BioRxiv, May. Cold Spring Harbor Laboratory, 2023.02.16.528797. doi:10.1101/2023.02.16.528797.

Beznoussenko, G.V., A. Ragnini-Wilson, C. Wilson et al. 2016. Three-dimensional and immune electron microscopic analysis of the secretory pathway in *Saccharomyces Cerevisiae*. Histochemistry and Cell Biology 146(5). Springer Verlag:515–527. doi:10.1007/s00418-016-1483-y.

Bielas, S.L., J.L. Silhavy, F. Brancati et al. 2009. Mutations in INPP5E, Encoding Inositol Polyphosphate-5-Phosphatase E, Link Phosphatidyl Inositol Signaling to the Ciliopathies. Nature Genetics 2009 41: 9 41 (9). Nature Publishing Group:1032–1036. doi:10.1038/ng.423.

Bitbol, A.F., P.G. Dommersnes and J.B. Fournier. 2010. Fluctuations of the Casimir-like Force between Two Membrane Inclusions. Physical Review E - Statistical, Nonlinear, and Soft Matter Physics 81 (5). American Physical Society:050903. doi:10.1103/PhysRevE.81.050903.

Blagoveshchenskaya, A., Y.C. Fei, H.M. Rohde et al. 2008. Integration of Golgi trafficking and growth factor signaling by the lipid phosphatase SAC1. Journal of Cell Biology 180(4): 803–812. doi:10.1083/jcb.200708109.

Bohnert, M. and M. Schuldiner. 2018. Stepping outside the comfort zone of membrane contact site research. Nature Reviews Molecular Cell Biology 19(8). Nature Publishing Group: 483–484. doi:10.1038/s41580-018-0022-1.

Bonazzi, M., S. Spanò, G. Turacchio et al. 2005. CtBP3/BARS drives membrane fission in dynamin-independent transport pathways. Nature Cell Biology 7(6). Nature Publishing Group: 570–580. doi:10.1038/ncb1260.

Boncompain, G. and A.V. Weigel. 2018. Transport and sorting in the Golgi complex: multiple mechanisms sort diverse cargo. Current Opinion in Cell Biology. Elsevier Ltd. doi:10.1016/j.ceb.2018.03.002.

Bottanelli, F., N. Kilian, A.M. Ernst et al. 2017. A novel physiological role for ARF1 in the formation of bidirectional tubules from the Golgi. Molecular Biology of the Cell 28(12). American Society for Cell Biology:1676–1687. doi:10.1091/mbc.E16-12-0863.

Braulke, T. and J.S. Bonifacino. 2009. Sorting of Lysosomal Proteins. Biochimica et Biophysica Acta - Molecular Cell Research. doi:10.1016/j.bbamcr.2008.10.016.

Brown, F. 2003. Regulation of protein mobility via thermal membrane undulations. Biophysical Journal 84 (2 I). The Biophysical Society:842–853. doi:10.1016/S0006-3495(03)74903-0.

Brown, M.S., A. Radhakrishnan and J.L. Goldstein. 2018. Retrospective on cholesterol homeostasis: the central role of scap. Annual Review of Biochemistry 87(1): 783–807. doi:10.1146/annurev-biochem-062917-011852.

Brügger, B., R. Sandhoff, S. Wegehingel et al. 2000. Evidence for segregation of sphingomyelin and cholesterol during formation of COPI-Coated Vesicles. Journal of Cell Biology 151(3): 507–517. doi:10.1083/jcb.151.3.507.

Bui, S., D. Stark, J. Li et al. 2023. Common markers and small molecule inhibitors in Golgi studies. Methods in Molecular Biology 2557: 453–493. doi:10.1007/978-1-0716-2639-9_27.

Bykov, Y.S., M. Schaffer, S.O. Dodonova et al. 2017. The structure of the COPI coat determined within the cell. ELife 6 (November). doi:10.7554/eLife.32493.

Campelo, F., C. Arnarez, S.J. Marrink et al. 2014. Helfrich model of membrane bending: from gibbs theory of liquid interfaces to membranes as thick anisotropic elastic layers. Advances in Colloid and Interface Science 208: 25–33. doi:10.1016/j.cis.2014.01.018.

Campelo, F. and M.M. Kozlov. 2014. Sensing membrane stresses by protein insertions. PLoS Computational Biology 10(4). doi:10.1371/journal.pcbi.1003556.

Campelo, F. and V. Malhotra. 2012. Membrane fission: the biogenesis of transport carriers. Annual Review of Biochemistry 81: 407–427. doi:10.1146/annurev-biochem-051710-094912.

Campelo, F., H.T. McMahon and M.M. Kozlov. 2008. The hydrophobic insertion mechanism of membrane curvature generation by proteins. Biophysical Journal 95(5): 2325–2339. doi:10.1529/biophysj.108.133173.

Campelo, F., J. van Galen, G. Turacchio et al. 2017. Sphingomyelin metabolism controls the shape and function of the Golgi Cisternae. ELife 6 (May). doi:10.7554/eLife.24603.

Capasso, S. L. Sticco, R. Rizzo et al. 2017. Sphingolipid metabolic flow controls phosphoinositide turnover at the trans -golgi network. The EMBO Journal 36 (12):1736–1754. doi:10.15252/embj.201696048.

Carlsson, A.E. 2018. Membrane bending by actin polymerization. Current Opinion in Cell Biology 50 (February). Elsevier Current Trends:1–7. doi:10.1016/J.CEB.2017.11.007.

Carrasco, S. and I. Mérida. 2007. Diacylglycerol, when simplicity becomes complex. Trends in Biochemical Sciences 32(1): 27–36. doi:10.1016/j.tibs.2006.11.004.

Chapman, D. 1981. The Hydrophobic Effect: Formation of Micelles and Biological Membranes (2nd Edition). Biochemical Society Transactions. Vol. 9. J. Wiley. doi:10.1042/bst0090178.

Cheong, F.Y., V. Sharma, A. Blagoveshchenskaya et al. 2010. Spatial regulation of Golgi phosphatidylinositol-4-Phosphate is required for enzyme localization and glycosylation fidelity. Traffic 11(9): 1180–1190. doi:10.1111/j.1600-0854.2010.01092.x.

Choudhury, R., C.J. Noakes, E. McKenzie et al. 2009. Differential clathrin binding and subcellular localization of OCRL1 splice isoforms. Journal of Biological Chemistry 284 (15). Elsevier: 9965–9973. doi:10.1074/jbc.M807442200.

Cockcroft, S. and S. Lev. 2020. Mammalian PITPs at the Golgi and ER-Golgi membrane contact sites. Contact 3: 251525642096417. doi:10.1177/2515256420964170.

Cruz-Garcia, D., M. Ortega-Bellido, M. Scarpa et al. 2013. Recruitment of Arfaptins to the Trans-Golgi network by PI(4)P and their involvement in cargo export. EMBO Journal 32 (12). Nature Publishing Group 1717–1729. doi:10.1038/emboj.2013.116.

Curwin, A.J , J. Von Blume and V. Malhotra. 2012. Cofilin-mediated sorting and export of specific cargo from the Golgi apparatus in yeast. Molecular Biology of the Cell. doi:10.1091/mbc.E11-09-0826.

D'Angelo, G., S. Capasso, L. Sticco et al. 2013. Glycosphingolipids: Synthesis and Functions. FEBS Journal 280(24). John Wiley & Sons, Ltd:6338–6353. doi:10.1111/febs.12559.

D'Angelo, G., C.J. Clarke and L.C. Silva. 2019. Meeting Report – the 2019 FEBS special meeting on sphingolipid biology: sphingolipids in physiology and pathology. Journal of Cell Science 132(15). Company of Biologists Ltd. doi:10.1242/jcs.235705.

D'Angelo, G., E. Polishchuk, G. Di Tullio et al. 2007. Glycosphingolipid synthesis requires FAPP2 transfer of glucosylceramide. Nature 449(7158). Nature Publishing Group:62–67. doi:10.1038/nature06097.

D'Angelo, G., T. Uemura, C.C. Chuang et al. 2013. Vesicular and non-vesicular transport feed distinct glycosylation pathways in the Golgi. Nature 501(7465). Nature Publishing Group: 116–120. doi:10.1038/nature12423.

D'Souza-Schorey, C. and P. Chavrier. 2006. ARF proteins: roles in membrane traffic and beyond. Nature Reviews Molecular Cell Biology 7(5): 347–358. doi:10.1038/nrm1910.

Dalton, A.J. and M.D. Felix. 1954. Cytologic and cytochemical characteristics of the Golgi substance of epithelial cells of the epididymis–*in Situ*, in homogenates and after isolation. American Journal of Anatomy 94(2). Am. J. Anat: 171–207. doi:10.1002/aja.1000940202.

David, Y., I.G. Castro and M. Schuldiner. 2021. The fast and the furious: Golgi Contact Sites. Https://Doi.Org/10.1177/25152564211034424 4 (August). SAGE PublicationsSage CA: Los Angeles, CA:251525642110344. doi:10.1177/25152564211034424.

Day, K.J. and J.C. Stachowiak. 2020. Biophysical forces in membrane bending and traffic. Current Opinion in Cell Biology. Elsevier Ltd. doi:10.1016/j.ceb.2020.02.017.

De Matteis, Maria A., C. Wilson and G. D'Angelo. 2013. Phosphatidylinositol-4-Phosphate: The Golgi and beyond. BioEssays 35(7). Bioessays:612–622. doi:10.1002/bies.201200180.

De Matteis, Maria Antonietta, A. Di Campli and A. Godi. 2005. The role of the phosphoinositides at the Golgi complex. Biochimica et Biophysica Acta - Molecular Cell Research 1744 (3 SPEC. ISS.): 396–405. doi:10.1016/j.bbamcr.2005.04.013.

De Matteis, Maria Antonietta and A. Luini. 2008. Exiting the Golgi complex. Nature Reviews Molecular Cell Biology 9(4): 273–284. doi:10.1038/nrm2378.

Deneke, V.E. and A. Pauli. 2021. The fertilization enigma: how sperm and egg fuse. Annual Review of Cell and Developmental Biology 37 (October). Annual Reviews:391–414. doi:10.1146/annurev-cellbio-120219-021751.

Deng, Y., F.E. Rivera-Molina, D.K. Toomre et al. 2016. Sphingomyelin is sorted at the trans Golgi network into a distinct class of secretory vesicle. Proceedings of the National Academy of Sciences of the United States of America 113(24). National Academy of Sciences: 6677–6682. doi:10.1073/pnas.1602875113.

Derganc, J., B. Antonny and A. Čopič. 2013. Membrane bending: the power of protein imbalance. Trends in Biochemical Sciences. Elsevier Current Trends. doi:10.1016/j.tibs.2013.08.006.

Derganc, J., A.A. Mironov and S. Svetina. 2006. Physical factors that affect the number and size of Golgi Cisternae. Traffic 7(1). Blackwell Publishing Ltd:85–96. doi:10.1111/j.1600-0854.2005.00365.x.

Deserno, M. 2015. Fluid lipid membranes: from differential geometry to curvature stresses. Chemistry and Physics of Lipids 185. Chem Phys Lipids: 11–45. doi:10.1016/j.chemphyslip.2014.05.001.

Díaz Añel, A.M. 2007. Phospholipase C B3 is a key component in the Gβγ/PKCη/ PKD-Mediated regulation of trans-golgi network to plasma membrane transport. Biochemical Journal 406(1). Portland Press:157–165. doi:10.1042/BJ20070359.

Dimova, R. 2014. Recent developments in the field of bending rigidity measurements on membranes. Advances in Colloid and Interface Science 208(June): 225–234. doi:10.1016/j.cis.2014.03.003.

Dinter, A. and E.G. Berger. 1998. Golgi-disturbing agents. Histochemistry and Cell Biology 109(5-6). Springer: 571–590. doi:10.1007/s004180050256.

Dippold, H.C., M.M. Ng, S.E. Farber-Katz et al. 2009. GOLPH3 bridges phosphatidylinositol-4-phosphate and actomyosin to stretch and shape the Golgi to promote budding. Cell 139(2): 337–351. doi:10.1016/j.cell.2009.07.052.

Do Carmo, M.P. 2016. Differential Geometry of Curves and Surfaces: Revised and Updated Second Edition. Courier Dover Publications.

Donohoe, B.S., B.H. Kang, M.J. Gerl et al. 2013. Cis-Golgi cisternal assembly and biosynthetic activation occur sequentially in plants and algae. Traffic 14(5): 551–567. doi:10.1111/tra.12052.

Duran, J.M., F. Campelo, J. Van Galen et al. 2012. Sphingomyelin organization is required for vesicle biogenesis at the Golgi complex. EMBO Journal 31(24): 4535–4546. doi:10.1038/emboj.2012.317.

Egea, G., C. Serra-Peinado, L. Salcedo-Sicilia et al. 2013. Actin acting at the Golgi. Histochemistry and Cell Biology 140(3): 347–360. doi:10.1007/s00418-013-1115-8.

Eisenberg-Bord, M., N. Shai, M. Schuldiner et al. 2016. A tether is a tether is a tether: tethering at membrane contact sites. Developmental Cell 39(4). Cell Press: 395–409. doi:10.1016/j.devcel.2016.10.022.

Eisenberg-Lerner, A., R. Benyair, N. Hizkiahou et al. 2020. Golgi organization is regulated by proteasomal degradation. Nature Communications 11(1). Nature Publishing Group:1–14. doi:10.1038/s41467-019-14038-9.

Emr, S., B.S. Glick, A.D. Linstedt et al. 2009. Journeys through the Golgi - Taking Stock in a New Era. Journal of Cell Biology. doi:10.1083/jcb.200909011.

Ernst, A.M., S.A. Syed, O. Zaki et al. 2018. S-palmitoylation sorts membrane cargo for anterograde transport in the Golgi. Developmental Cell 47(4). Cell Press: 479–493.e7. doi:10.1016/j.devcel.2018.10.024.

Evans, E.A. and R. Skalak. 1979. Mechanics and thermodynamics of biomembranes: Part 1. CRC Critical Reviews in Bioengineering 3(3). United States:181–330.

Fabene, P.F. and M. Bentivoglio. 1998. 1898–1998: Camillo Golgi and "the Golgi": One Hundred Years of Terminological Clones. Brain Research Bulletin 47(3). Brain Res Bull: 195–198. https://pubmed.ncbi.nlm.nih.gov/9865849/.

Farquhar, M.G. and G.E. Palade. 1981. The Golgi apparatus (Complex) (1954–1981) from artifact to center stage. Journal of Cell Biology. The Rockefeller University Press. doi:10.1083/jcb.91.3.77s.

Farquhar, Marilyn G. and G.E. Palade. 1998. The Golgi Apparatus: 100 years of progress and controversy. Trends in Cell Biology 8(1): 2–10. doi:10.1016/S0962-8924(97)01187-2.

Ferraro, F., J. Kriston-Vizi, D.J. Metcalf et al. 2014. A two-tier golgi-based control of organelle size underpins the functional plasticity of endothelial cells. Developmental Cell 29(3). Dev Cell:292–304. doi:10.1016/j.devcel.2014.03.021.

Fleischer, B., S. Fleischer and H. Ozawa. 1969. Isolation and characterization of Golgi membranes from bovine liver. The Journal of Cell Biology 43(1). The Rockefeller University Press:59–79. doi:10.1083/jcb.43.1.59.

Ford, C., A. Parchure, J. von Blume et al. 2021. Cargo sorting at the Trans-Golgi network at a Glance. Journal of Cell Science 134(23). J Cell Sci. doi:10.1242/jcs.259110.

Forgac, M. 2007. Vacuolar ATPases: Rotary proton pumps in physiology and pathophysiology. Nature Reviews Molecular Cell Biology 8(11). Nature Publishing Group: 917–929. doi:10.1038/nrm2272.

Freeze, H.H. and B.G. Ng. 2011. Golgi glycosylation and human inherited diseases. Cold Spring Harbor Perspectives in Biology 3(9): 1–22. doi:10.1101/cshperspect.a005371.

Freyberg, Z., D. Sweeney, A. Siddhanta et al. 2001. Intracellular localization of phospholipase D1 in mammalian cells. Molecular Biology of the Cell 12(4): 943–955. doi:10.1091/mbc.12.4.943.

Freyberg, Zachary, S. Bourgoin and D. Shields. 2002. Phospholipase D2 is localized to the rims of the Golgi apparatus in mammalian cells. Molecular Biology of the Cell 13(11). Mol Biol Cell:3930–3942. doi:10.1091/mbc.02-04-0059.

Freyberg, Zachary, A. Siddhanta and D. Shields. 2003. "Slip, Sliding Away": Phospholipase D and the Golgi apparatus. Trends in Cell Biology 13(10). Elsevier Current Trends: 540–546. doi:10.1016/j.tcb.2003.08.004.

Frolov, V.A., A.V. Shnyrova and J. Zimmerberg. 2011. Lipid polymorphisms and membrane shape. Cold Spring Harbor Perspectives in Biology 3(11). Cold Spring Harbor Laboratory Press:a004747. doi:10.1101/cshperspect.a004747.

Futerman, A.H. and R.E. Pagano. 1991. Determination of the intracellular sites and topology of glucosylceramide synthesis in rat liver. Biochemical Journal 280(2). Portland Press: 295–302. doi:10.1042/bj2800295.

Futerman, Anthony H. and H. Riezman. 2005. The ins and outs of sphingolipid synthesis. Trends in Cell Biology 15(6): 312–318. doi:10.1016/j.tcb.2005.04.006.

Gallop, J.L., P.J. Butler and H.T. McMahon. 2005. Endophilin and CtBP/BARS Are Not Acyl transferases in endocytosis or Golgi fission. Nature 438(7068): 675–678. doi:nature04136 [pii]10.1038/nature04136.

Gehart, H., A. Goginashvili, R. Beck et al. 2012. The BAR domain protein arfaptin-1 controls secretory granule biogenesis at the Trans-Golgi network. Developmental Cell 23(4). Elsevier: 756–768. doi:10.1016/j.devcel.2012.07.019.

Gouveia, B., Y. Kim, J.W. Shaevitz et al. 2022. Capillary forces generated by biomolecular condensates. Nature 609: 7926 609 (7926). Nature Publishing Group:255–264. doi:10.1038/s41586-022-05138-6.

Graham, T.R. and C.G. Burd. 2011. Coordination of Golgi functions by phosphatidylinositol 4-kinases. Trends in Cell Biology 21(2). Elsevier Current Trends: 113–121. doi:10.1016/j.tcb.2010.10.002.

Grond, R., T. Veenendaal, J.M. Duran et al. 2020. The function of GORASPs in Golgi apparatus organization *in vivo*. Journal of Cell Biology 219(9). Rockefeller University Press. doi:10.1083/jcb.202004191.

Grosshans, B.L., D. Ortiz and P. Novick. 2006. Rabs and their effectors: achieving specificity in membrane traffic. Proceedings of the National Academy of Sciences of the United States of America. National Academy of Sciences. doi:10.1073/pnas.0601617103.

Guo, Y., D.W. Sirkis and R. Schekman. 2014a. Protein sorting at the Trans-Golgi network. Annual Review of Cell and Developmental Biology 30(August): 169–206. doi:10.1146/annurev-cellbio-100913-013012.

Guo, Y., D.W. Sirkis and R. Schekman. 2014b. Protein sorting at the Trans-Golgi network. Annual Review of Cell and Developmental Biology 30(October): 169–206. doi:10.1146/annurev-cellbio-100913-013012.

Guo, Y., G. Zanetti and R. Schekman. 2013. A novel GTP-binding protein-adaptor protein complex responsible for export of Vangl2 from the trans Golgi network. ELife 2013(2). doi:10.7554/eLife.00160.

Hanada, K., K. Kumagai, S. Yasuda et al. 2003. Molecular machinery for non-vesicular trafficking of ceramide. Nature 426 (6968). Nature Publishing Group: 803–809. doi:10.1038/nature02188.

He, Q., H. Liu, C. Huang et al. 2020. Herpes simplex virus 1-induced blood-brain barrier damage involves apoptosis associated with GM130-mediated Golgi stress. Frontiers in Molecular Neuroscience 13 (January). Frontiers Media S.A. doi:10.3389/fnmol.2020.00002.

Helfrich, W. 1973. Elastic properties of lipid bilayers: theory and possible experiments. Zeitschrift Fur Naturforschung - Section C Journal of Biosciences 28(11-12): 693–703. doi:10.1515/znc-1973-11-1209.

Helfrich, W. 1985. Effect of thermal undulations on the rigidity of fluid membranes and interfaces. Journal de Physique 46(7). Société Française de Physique: 1263–1268. doi:10.1051/JPHYS:019850046070126300.

Helfrich, W. and R.M. Servuss. 1984. Undulations, steric interaction and cohesion of fluid membranes. Il Nuovo Cimento D 1984 3: 1 3(1). Springer: 137–151. doi:10.1007/BF02452208.

Hicks, S.W. and C.E. Machamer. 2005. Golgi structure in stress sensing and apoptosis. Biochimica et Biophysica Acta - Molecular Cell Research 1744 (3 SPEC. ISS.). Elsevier: 406–414. doi:10.1016/j.bbamcr.2005.03.002.

Holcman, D., P. Parutto, J.E. Chambers et al. 2018. Single particle trajectories reveal active endoplasmic reticulum luminal flow. Nature Cell Biology 20(10). Nature Publishing Group:1118–1125. doi:10.1038/s41556-018-0192-2.

Huitema, K., J. van den Dikkenberg, J.F. Brouwers et al. 2004. Identification of a family of animal Sphingomyelin synthases. EMBO J. 23(1): 33–44. doi:10.1038/sj.emboj.76000347600034 [pii].

Ignashkova, T.I., M. Gendarme, K. Peschk et al. 2017. Cell survival and protein secretion associated with Golgi integrity in response to Golgi stress-inducing agents. Traffic 18(8). Traffic:530–544. doi:10.1111/tra.12493.

Illya, G., R. Lipowsky and J.C. Shillcock. 2005. Effect of chain length and asymmetry on material properties of bilayer membranes. J. Chem. Phys. 122(24): 244901. http://www.ncbi.nlm.nih.gov/entrez/query.fcgi?cmd=Retrieve&db=PubMed&dopt=Citation&list_uids=16035810.

Israelachvili, J.N. 2010. Intermolecular and Surface Forces, Third Edition. Intermolecular and Surface Forces, Third Edition, January. Elsevier, 1–674. doi:10.1016/C2009-0-21560-1.

Jaiman, A. and M. Thattai. 2020. Golgi compartments enable controlled biomolecular assembly using promiscuous enzymes. ELife 9 (June). eLife Sciences Publications Ltd: 1–22. doi:10.7554/ELIFE.49573.

James Morre, D., L. Marcel Merlin and T.W. Keenan. 1969. Localization of glycosyl transferase activities in a Golgi apparatus-rich fraction isolated from rat liver. Biochemical and Biophysical Research Communications 37(5). Academic Press: 813–819. doi:10.1016/0006-291X(69)90964-4.

Jimenez, A.J. and F. Perez. 2023. Phase partitioning initiates rab protein membrane domain formation and promote vesicle tethering. BioRxiv, April. Cold Spring Harbor Laboratory, 2023.04.17.537227. doi:10.1101/2023.04.17.537227.

Jones, S.M., K.E. Howell, J.R. Henley et al. 1998. Role of dynamin in the formation of transport vesicles from the trans-golgi network. Science 279(5350): 573–577. doi:10.1126/science.279.5350.573.

Kang, B.H. and L.A. Staehelin. 2008. ER-to-Golgi transport by COPII vesicles in arabidopsis involves a ribosome-excluding scaffold that is transferred with the vesicles to the Golgi matrix. Protoplasma 234(1–4): 51–64. doi:10.1007/s00709-008-0015-6.

Kellokumpu, S. 2019. Golgi PH, ion and redox homeostasis: how much do they really matter? Frontiers in Cell and Developmental Biology 7(JUN). Frontiers Media S.A.:93. doi:10.3389/fcell.2019.00093.

Khawar, M.B., H. Gao and W. Li. 2019. Mechanism of acrosome biogenesis in mammals. Frontiers in Cell and Developmental Biology 7(September). Frontiers Media S.A.:195. doi:10.3389/fcell.2019.00195.

Kirchhausen, T. 2000. Three Ways to Make a Vesicle. Nat. Rev. Mol. Cell Biol. 1(3): 187–198. doi:10.1038/35043117.

Kita, Y., H. Shindou and T. Shimizu. 2019. Cytosolic phospholipase A 2 and Lysophospholipid Acyltransferases. Biochimica et Biophysica Acta - Molecular and Cell Biology of Lipids, Novel functions of phospholipase A2, 1864(6): 838–845. doi:10.1016/j.bbalip.2018.08.006.

Klauda, J.B., R.M. Venable, J.A. Freites et al. 2010. Update of the CHARMM All-atom additive force field for lipids: validation on six lipid types. Journal of Physical Chemistry B 114(23). American Chemical Society: 7830–7843. doi:10.1021/jp101759q.

Klemm, R.W., C.S. Ejsing, M.A. Surma et al. 2009. Segregation of sphingolipids and sterols during formation of secretory vesicles at the trans-golgi network. Journal of Cell Biology 185(4). The Rockefeller University Press: 601–612. doi:10.1083/jcb.200901145.

Klumperman, J. 2011. Architecture of the mammalian Golgi. Cold Spring Harbor Perspectives in Biology 3(7): 1–19. doi:10.1101/cshperspect.a005181.

Kondylis, V. and C. Rabouille. 2009. The Golgi apparatus: lessons from drosophila. FEBS Letters. No longer published by Elsevier. doi:10.1016/j.febslet.2009.09.048.

Kong, A.M., C.J. Speed, C.J. O'Malley et al. 2000. Cloning and characterization of a 72-KDa Inositol-Polyphosphate 5-phosphatase localized to the Golgi network. Journal of Biological Chemistry 275 (31). American Society for Biochemistry and Molecular Biology Inc.: 24052–24064. doi:10.1074/jbc.M000874200.

Kozlov, M.M. 2018. Spontaneous and intrinsic curvature of lipid membranes: back to the origins. physics of biological membranes, December. Springer International Publishing, 287–309. doi: 10.1007/978-3-030-00630-3_12/FIGURES/2.

Kozlov, M.M., F. Campelo, N. Liska et al. 2014a. Mechanisms shaping cell membranes. Current Opinion in Cell Biology 29(1). Elsevier Ltd: 53–60. doi:10.1016/j.ceb.2014.03.006.

Kozlov, M.M., F. Campelo, N. Liska et al. 2014b. Mechanisms shaping cell membranes. Current Opinion in Cell Biology 29(1): 53–60. doi:10.1016/j.ceb.2014.03.006.

Kozlov, M.M. and J.W. Taraska. 2023. Generation of nanoscopic membrane curvature for membrane trafficking. Nature Reviews Molecular Cell Biology 24(1). Nature Publishing Group:63–78. doi:10.1038/s41580-022-00511-9.

Ladinsky, M.S., D.N. Mastronarde, J.R. McIntosh et al. 1999. Golgi structure in three dimensions: functional insights from the normal rat kidney cell. Journal of Cell Biology 144(6): 1135–1149. doi:10.1083/jcb.144.6.1135.

Launay, N., M. Ruiz, L. Planas-Serra et al. 2023. RINT1 deficiency disrupts lipid metabolism and underlies a complex hereditary spastic paraplegia. The Journal of Clinical Investigation 133(14). American Society for Clinical Investigation. doi:10.1172/JCI162836.

Lee, I., N. Tiwari, M.H. Dunlop et al. 2014. Membrane adhesion dictates Golgi stacking and cisternal morphology. Proceedings of the National Academy of Sciences of the United States of America 111(5): 1849–1854. doi:10.1073/pnas.1323895111.

Lev, S. 2006. Lipid homoeostasis and Golgi secretory function. Biochemical Society Transactions 34(3). Portland Press: 363–366. doi:10.1042/BST0340363.

Levental, I. and S.L. Veatch. 2016. The continuing mystery of lipid rafts. Journal of Molecular Biology. Academic Press. doi: 10.1016/j.jmb.2016.08.022.

Li, T., H. You, X. Mo et al. 2016. GOLPH3 mediated Golgi stress response in modulating N2A Cell death upon oxygen-glucose deprivation and reoxygenation injury. Molecular Neurobiology 53(2). Humana Press Inc.:1377–1385. doi:10.1007/s12035-014-9083-0.

Lin, G., P.T. Lee, K. Chen et al. 2018. Phospholipase PLA2G6, a Parkinsonism-Associated Gene, Affects Vps26 and Vps35, Retromer Function, and Ceramide Levels, Similar to α-Synuclein Gain. Cell Metabolism 28(4): 605–618.e6. doi:10.1016/j.cmet.2018.05.019.

Lingwood, D. and K. Simons. 2010. Lipid rafts as a membrane-organizing principle. Science. American Association for the Advancement of Science. doi:10.1126/science.1174621.

Litvak, V., N. Dahan, S. Ramachandran et al. 2005. Maintenance of the diacylglycerol level in the Golgi apparatus by the Nir2 protein is critical for Golgi secretory function. Nature Cell Biology 7(3): 225–234. doi:10.1038/ncb1221.

Lorent, J.H., K.R. Levental, L. Ganesan et al. 2020. Plasma membranes are asymmetric in lipid unsaturation, packing and protein shape. Nature Chemical Biology 16(6). Nature Research: 644–652. doi:10.1038/s41589-020-0529-6.

Lorente-Rodríguez, A. and C. Barlowe. 2011. Entry and exit mechanisms at the cis-face of the Golgi complex. Cold Spring Harbor Perspectives in Biology 3(7). Cold Spring Harbor Laboratory Press: 1–15. doi:10.1101/cshperspect.a005207.

Lowe, M. 2011. Structural organization of the Golgi apparatus. Current Opinion in Cell Biology 23(1). Elsevier Current Trends: 85–93. doi:10.1016/j.ceb.2010.10.004.

Lu, A. and S.R. Pfeffer. 2013. Golgi-associated RhoBTB3 targets cyclin E for ubiquitylation and promotes cell cycle progression. Journal of Cell Biology 203(2). J. Cell Biol.:233–250. doi:10.1083/jcb.201305158.

Lujan, P., J. Angulo-Capel, M. Chabanon et al. 2021. Interorganelle communication and membrane shaping in the early secretory pathway. Current Opinion in Cell Biology. Elsevier Ltd. doi:10.1016/j.ceb.2021.01.010.

Lujan, P. and F. Campelo. 2021. Should i stay or should i go? Golgi Membrane Spatial Organization for Protein Sorting and Retention. Archives of Biochemistry and Biophysics 707 (August). Academic Press:108921. doi:10.1016/j.abb.2021.108921.

Lujan, P., C. Garcia-Cabau, Y. Wakana et al. 2022. Sorting of secretory proteins at the Trans-Golgi network by TGN46. BioRxiv, April. Cold Spring Harbor Laboratory, 2022.04.20.488883. doi:10.1101/2022.04.20.488883.

Ma, T., B. Li, R. Wang et al. 2018. A mechanism for differential sorting of the planar cell polarity proteins Frizzled6 and Vangl2 at the Trans-Golgi network. Journal of Biological Chemistry 293 (22). American Society for Biochemistry and Molecular Biology Inc.:8410–8427. doi:10.1074/jbc.RA118.001906.

MacHta, B.B., S.L. Veatch and J.P. Sethna. 2012. Critical casimir forces in cellular membranes. Physical Review Letters 109(13). Phys. Rev. Lett. doi:10.1103/PHYSREVLETT.109.138101.

Malhotra, V. and F. Campelo. 2011. PKD regulates membrane fission to generate TGN to cell surface transport carriers. Cold Spring Harbor Perspectives in Biology 3(2). Cold Spring Harbor Laboratory Press:1–9. doi:10.1101/cshperspect.a005280.

Malsam, J. and T.H. Söllner. 2011. Organization of SNAREs within the Golgi stack. Cold Spring Harbor Perspectives in Biology 3(10): 1–17. doi:10.1101/cshperspect.a005249.

Man, Z., Y. Kondo, H. Koga et al. 2011. Arfaptins are localized to the Trans-Golgi by interaction with Arl1, but Not Arfs. Journal of Biological Chemistry 286(13): 11569–11578. doi:10.1074/jbc.M110.201442.

Manzo, C. and M.F. Garcia-Parajo. 2015. A review of progress in single particle tracking: from methods to biophysical insights. Reports on Progress in Physics 78(12). IOP Publishing:124601. doi:10.1088/0034-4885/78/12/124601.

Marsh, B.J. 2005. Lessons from tomographic studies of the mammalian Golgi. Biochimica et Biophysica Acta - Molecular Cell Research. Vol. 1744. Elsevier. doi:10.1016/j.bbamcr.2005.04.002.

Marsh, D. 2007. Lateral pressure profile, spontaneous curvature frustration, and the incorporation and conformation of proteins in membranes. Biophysical Journal 93(11): 3884–3899. doi:10.1529/biophysj.107.107938.

Masone, M.C., V. Morra and R. Venditti. 2019. Illuminating the membrane contact sites between the endoplasmic reticulum and the Trans-Golgi Network. FEBS Letters 593(22): 3135–3148. doi:10.1002/1873-3468.13639.

McMahon, H.T. and J.L. Gallop. 2005. Membrane curvature and mechanisms of dynamic cell membrane remodelling. Nature 438(7068): 590–596. doi:10.1038/nature04396.

Melero, A., N. Chiaruttini, T. Karashima et al. 2018. Lysophospholipids facilitate COPII vesicle formation. Current Biology 28(12). Cell Press: 1950–1958.e6. doi:10.1016/j.cub.2018.04.076.

Mesmin, B. J. Bigay, J. Moser Von Filseck et al. 2013. XA four-step cycle driven by PI(4)P hydrolysis directs sterol/PI(4)P exchange by the ER-Golgi tether OSBP. Cell 155(4). Elsevier:830. doi:10.1016/j.cell.2013.09.056.

Mesmin, B., D. Kovacs and G. D'Angelo. 2019. Lipid exchange and signaling at ER–Golgi contact sites. Current Opinion in Cell Biology 57(April). Elsevier Current Trends: 8–15. doi:10.1016/j.ceb.2018.10.002.

Mikitova, V. and T.P. Levine. 2012. Analysis of the key elements of FFAT-like motifs identifies new proteins that potentially bind VAP on the ER, Including Two AKAPs and FAPP2. Edited by Zhe Zhang. PLoS ONE 7 (1). Public Library of Science: e30455. doi:10.1371/journal.pone.0030455.

Miller, P.M., A.W. Folkmann, A.R.R. Maia et al. 2009. Golgi-Derived CLASP-dependent microtubules control Golgi organization and polarized trafficking in motile cells. Nature Cell Biology 11(9). Nature Publishing Group: 1069–1080. doi:10.1038/ncb1920.

Miserey-Lenkei, S., G. Chalancon, S. Bardin et al. 2010. Rab and actomyosin-dependent fission of transport vesicles at the Golgi complex. Nature Cell Biology 12(7): 645–654. doi:10.1038/ncb2067.

Missiaen, L., L. Dode, J. Vanoevelen et al. 2007. Calcium in the Golgi apparatus. Cell Calcium 41(5). Churchill Livingstone: 405–416. doi:10.1016/j.ceca.2006.11.001.

Mittag, T. and R.V. Pappu. 2022. A conceptual framework for understanding phase separation and addressing open questions and challenges. Molecular Cell 82(12). Cell Press: 2201–2214. doi:10 1016/J.MOLCEL.2022.05.018.

Mogelsvang, S., N. Gomez-Ospina, J. Soderholm et al. 2003. Tomographic Evidence for Continuous Turnover of Golgi Cisternae in Pichia Pastoris. Molecular Biology of the Cell 14(6): 2277–2291. doi:10 1091/mbc.E02-10-0697.

Munro, S. 2011. The golgin coiled-coil proteins of the golgi apparatus. Cold Spring Harbor Perspectives in Biology 3(6): 1–14. doi:10.1101/cshperspect.a005256.

Nelson, D., T. Piran and S. Weinberg. 1989. Statistical mechanics of membranes and surfaces. Statistical Mechanics of Membranes and Surfaces. WORLD SCIENTIFIC, 1–278. doi:10.1142/9789814541602.

Nickel, W., B. Brügger and F.T. Wieland. 1998. Protein and lipid sorting between the endoplasmic reticulum and the Golgi complex. Seminars in Cell and Developmental Biology 9(5). Academic Press:493–501. doi:10.1006/scdb.1998.0256.

Nixon-Abell, J., C.J. Obara, A.V. Weigel et al. 2016. Increased spatiotemporal resolution reveals highly dynamic dense tubular matrices in the peripheral ER. Science 354(6311). doi:10.1126/science.aaf3928.

Noske, A.B. A.J. Costin, G.P. Morgan et al. 2008. Expedited approaches to whole cell electron tomography and organelle mark-up *in situ* in high-pressure frozen pancreatic islets. Journal of Structural Biology 161(3). Academic Press: 298–313. doi:10.1016/j.jsb.2007.09.015.

Novick, P. and M. Zerial. 1997. The diversity of rab proteins in vesicle transport. Current Opinion in Cell Biology 9(4). Elsevier Current Trends: 496–504. doi:10.1016/S0955-0674(97)80025-7.

Pagliuso, A., C. Valente, L.L. Giordano et al. 2016. Golgi membrane fission requires the CtBP1-S/BARS-induced activation of lysophosphatidic acid acyltransferase δ. Nature Communications 7(1). Nature Publishing Group: 12148. doi:10.1038/ncomms12148.

Palade, G. 1975. Intracellular aspects of the process of protein synthesis. Science 189(4200): 347–358. http://www.ncbi.nlm.nih.gov/entrez/query.fcgi?cmd=Retrieve&db=PubMed&dopt=Citation&list_uids=1096303.

Palade, G.E. 1992. Intracellular aspects of the process of protein secretion. Physiology or Medicine: 1971–1980 19711980:177.

Papanikou, E. and B.S. Glick. 2009. The yeast Golgi apparatus: insights and mysteries. FEBS Letters. No longer published by Elsevier. doi:10.1016/j.febslet.2009.10.072.

Parashuraman, S. and G. D'Angelo. 2019. Visualizing sphingolipid biosynthesis in cells. Chemistry and Physics of Lipids 218 (January). Elsevier:103–111. doi:10.1016/j.chemphyslip.2018.11.003.

Parchure, A., M. Tian, D. Stalder et al. 2022. Liquid–liquid phase separation facilitates the biogenesis of secretory storage granules. Journal of Cell Biology 221(12). NLM (Medline). doi:10.1083/jcb.202206132.

Parton, R.G., V. Tillu, K.A. McMahon et al. 2021. Key phases in the formation of caveolae. Current Opinion in Cell Biology 71 (August). Elsevier Current Trends:7–14. doi:10.1016/j.ceb.2021.01.009.

Patterson, G.H., K. Hirschberg, R.S. Polishchuk et al. 2008. Transport through the Golgi apparatus by rapid partitioning within a two-phase membrane system. Cell 133(6): 1055–1067. doi:10.1016/j.cell.2008.04.044.

Pelham, H.R.B. 1991. Recycling of proteins between the endoplasmic reticulum and Golgi complex. Current Opinion in Cell Biology 3(4). Elsevier Current Trends: 585–591. doi:10.1016/0955-0674(91)90027-V.

Peretti, D., N. Dahan, E. Shimoni et al. 2008. Coordinated lipid transfer between the endoplasmic reticulum and the Golgi complex requires the VAP proteins and is essential for Golgi-mediated transport. Edited by Vivek Malhotra. Molecular Biology of the Cell 19(9): 3871–3884. doi:10.1091/mbc.E08-05-0498.

Pezeshkian, W., H. Gao, S. Arumugam et al. 2017. Mechanism of Shiga toxin clustering on membranes. ACS Nano 11(1). American Chemical Society: 314–324. doi:10.1021/ACSNANO.6B05706/SUPPL_FILE/NN6B05706_SI_010.AVI.

Pfeffer, S.R. 2017. Rab GTPases: master regulators that establish the secretory and endocytic pathways. Molecular Biology of the Cell. American Society for Cell Biology. doi:10.1091/mbc.E16-10-0737.

Phillips, M.J. and G.K. Voeltz. 2016. Structure and function of ER membrane contact sites with other organelles. Nature Reviews Molecular Cell Biology 17(2). Nature Publishing Group: 69–82. doi:10.1038/nrm.2015.8.

Pizzo, P., V. Lissandron, P. Capitanio et al. 2011. Ca2+ signalling in the Golgi apparatus. Cell Calcium 50 (2). Churchill Livingstone: 184–192. doi:10.1016/j.ceca.2011.01.006.

Popoff, V., F. Adolf, B. Brügge et al. 2011. COPI budding within the Golgi stack. Cold Spring Harbor Perspectives in Biology 3(11). doi:10.1101/cshperspect.a005231.

Posor, Y., W. Jang and V. Haucke. 2022. Phosphoinositides as membrane organizers. Nature Reviews Molecular Cell Biology 23(12). Nature Publishing Group: 797–816. doi:10.1038/s41580-022-00490-x.

Pothukuchi, P., I. Agliarulo, D. Russo et al. 2019. Translation of genome to glycome: role of the Golgi apparatus. FEBS Letters 593(17). Wiley Blackwell: 2390–2411. doi:10.1002/1873-3468.13541.

Pusapati, G.V., D. Krndija, M. Armacki et al. 2010. Role of the second cysteine-rich domain and Pro275 in protein Kinase D2 interaction with ADP-ribosylation factor 1, Trans-Golgi network recruitment, and Protein Transport. Edited by Benjamin S. Glick. Molecular Biology of the Cell 21(6): 1011–1022. doi:10.1091/mbc.E09-09-0814.

Rahajeng, J., R.S. Kuna, S.L. Makowski et al. 2019. Efficient Golgi forward trafficking requires GOLPH3-Driven, PI4P-Dependent Membrane Curvature. Developmental Cell 50(5). Cell Press: 573-585.e5. doi:10.1016/j.devcel.2019.05.038.

Ramazanov, B.R., R. Di Martino, A. Kumar et al. 2023. Calcium flux through ER-TGN contact sites facilitates cargo export. BioRxiv, January. Cold Spring Harbor Laboratory, 2022.12.19.521097. doi:10.1101/2022.12.19.521097.

Ramazanov, B.R., M.L. Tran and J. von Blume. 2021. Sending out molecules from the TGN. Current Opinion in Cell Biology. Elsevier Ltd. doi:10.1016/j.ceb.2021.02.005.

Rambourg, A. and Y. Clermont. 1997. Three-dimensional structure of the Golgi apparatus in mammalian cells. The Golgi Apparatus. Birkhäuser, Basel, 37–61. doi:10.1007/978-3-0348-8876-9_2.

Rebane, A.A., P. Ziltener, L.C. LaMonica et al. 2020. Liquid–liquid phase separation of the Golgi matrix protein GM130. FEBS Letters 594(7). John Wiley & Sons, Ltd:1132–1144. doi:10.1002/1873-3468.13715.

Reinisch, K.M. and W.A. Prinz. 2021. Mechanisms of nonvesicular lipid transport. Journal of Cell Biology 220(3). NLM (Medline). doi:10.1083/JCB.202012058.

Riebeling, C., A.J. Morris and D. Shields. 2009. Phospholipase D in the Golgi apparatus. Biochimica et Biophysica Acta - Molecular and Cell Biology of Lipids 1791(9). Elsevier B.V.:876–880. doi:10.1016/j.bbalip.2009.04.003.

Rini, J.M., K.W. Moremen, B.G. Davis et al. 2022. Glycosyltransferases and glycan-processing enzymes. Essentials of Glycobiology. Cold Spring Harbor Laboratory Press. doi:10.1101/GLYCOBIOLOGY.4E.6.

Romani, P., I. Brian, G. Santinon et al. 2019. Extracellular matrix mechanical cues regulate lipid metabolism through lipin-1 and SREBP. Nature Cell Biology 21(3). Nature Publishing Group: 338–347. doi:10.1038/s41556-018-0270-5.

Rosenwald, A.G. and R.E. Pagano. 1993. Inhibition of glycoprotein traffic through the secretory pathway by ceramide. Journal of Biological Chemistry 268(7): 4577–4579. doi:10.1016/s0021-9258(18)53432-3.

Rothman, J.E. 2019. Jim's View: is the Golgi stack a phase-separated liquid crystal? FEBS Letters 593(19). John Wiley & Sons, Ltd: 2701–2705. doi:10.1002/1873-3468.13609.

Sabatini, D.D. 2008. George Palade 1912–2008. Nature Cell Biology 10(12). Nature Publishing Group: 1374–1374. doi:10.1038/ncb1208-1374.

Safran, S.A. 2018. Statistical thermodynamics of surfaces, interfaces, and membranes. Statistical Thermodynamics of Surfaces, Interfaces, and Membranes, January. CRC Press, 1–266. doi:10.1201/9780429497131.

Schmidt, J.A. and W.J. Brown. 2009. Lysophosphatidic Acid acyltransferase 3 regulates Golgi complex structure and function. Journal of Cell Biology 186(2): 211–218. doi:10.1083/jcb.200904147.

Schmidt, J.A., D.N. Kalkofen, K.W. Donovan et al. 2010. A role for Phospholipase A2 activity in membrane tubule formation and TGN trafficking. Traffic 11(12): 1530–1536. doi:10.1111/j.1600-0854.2010.01115.x.

Schmidt, O., Y. Weyer, V. Baumann et al. 2019. Endosome and Golgi-associated degradation (EGAD) of membrane proteins regulates sphingolipid metabolism. The EMBO Journal 38(15). EMBO J. doi:10.15252/embj.2018101433.

Scorrano, L., M.A. De Matteis, S. Emr et al. 2019. Coming together to define membrane contact sites. Nature Communications 10(1). Nature Publishing Group:1287. doi:10.1038/s41467-019-09253-3.

Seetharaman, S. and S. Etienne-Manneville. 2018. Integrin diversity brings specificity in mechanotransduction. Biology of the Cell 110(3). John Wiley & Sons, Ltd: 49–64. doi:10.1111/boc.201700060.

Seifert, U. 1997. Configurations of Fluid Membranes and Vesicles. Advances in Physics. doi:10.1080/00018739700101488.

Seifert, U., K. Berndl and R. Lipowsky. 1991. Shape transformations of vesicles: phase diagram for spontaneous- curvature and bilayer-coupling models. Physical Review A 44 (2). American Physical Society:1182–1202. doi:10.1103/PhysRevA.44.1182.

Sens, P. and J. Plastino. 2015. Membrane tension and cytoskeleton organization in cell motility. Journal of Physics Condensed Matter 27(27). J. Phys. Condens Matter. doi:10.1088/0953-8984/27/27/273103.

Sens, P. and M. Rao. 2013. (Re)Modeling the Golgi. In Methods in Cell Biology 118: 299–310. doi:10.1016/B978-0-12-417164-0.00018-5.

Sezgin, E., I. Levental, S. Mayor et al. 2017. The mystery of membrane organization: composition, regulation and roles of lipid rafts. Nature Reviews Molecular Cell Biology 2017 18:6 18 (6). Nature Publishing Group:361–374. doi:10.1038/NRM.2017.16.

Shibata, Y., J. Hu, M.M. Kozlov et al. 2009. Mechanisms shaping the membranes of cellular organelles. Annual Review of Cell and Developmental Biology 25(1). Annual Reviews: 329–354. doi:10.1146/annurev.cellbio.042308.113324.

Shin, Y. and C.P. Brangwynne. 2017. Liquid phase condensation in cell physiology and disease. Science 357(6357). American Association for the Advancement of Science. doi:10.1126/science.aaf4382.

Short, B., A. Haas and F.A. Barr. 2005. Golgins and GTPases, giving identity and structure to the Golgi apparatus. Biochimica et Biophysica Acta - Molecular Cell Research. Elsevier. doi:10.1016/j.bbamcr.2005.02.001.

Sicart, A., M. Katan, G. Egea et al. 2015. PLCγ1 Participates in protein transport and diacylglycerol production triggered by cargo arrival at the Golgi. Traffic 16(3). Blackwell Munksgaard: 250–266. doi:10.1111/tra.12246.

Spanfò, S., M.G. Silletta, A. Colanzi et al. 1999. Molecular cloning and functional characterization of brefeldin A-ADP-Ribosylated substrate: a novel protein involved in the maintenance of the Golgi structure. Journal of Biological Chemistry 274(25). Elsevier: 17705–17710. doi:10.1074/jbc.274.25.17705.

Spang, A. 2015. The road not taken: less traveled roads from the TGN to the plasma membrane. Membranes 5(1). MDPI AG:84–98. doi:10.3390/membranes5010084.

Stace, C.L. and N.T. Ktistakis. 2006. Phosphatidic Acid- and Phosphatidylserine-Binding proteins. Biochimica et Biophysica Acta - Molecular and Cell Biology of Lipids. doi:10.1016/j.bbalip.2006.03.006.

Stachowiak, J.C., E.M. Schmid, C.J. Ryan et al. 2012. Membrane bending by protein-protein crowding. Nature Cell Biology 14(9): 944–949. doi:10.1038/ncb2561.

Stalder, D. and D.C. Gershlick. 2020. Direct trafficking pathways from the Golgi apparatus to the plasma membrane. Seminars in Cell and Developmental Biology 107 (November). Elsevier Ltd:112–125. doi:10.1016/j.semcdb.2020.04.001.

Stanley, P. 2011. Golgi Glycosylation. Cold Spring Harbor Perspectives in Biology 3(4): 1–13. doi:10.1101/cshperspect.a005199.

Stewart, A., M. Ghosh, D.M. Spencer et al. 2002. Enzymatic properties of human cytosolic phospholipase A2γ. Journal of Biological Chemistry 277(33): 29526–29536. doi:10.1074/jbc.M204856200.

Stockhammer, A. and F. Bottanelli. 2020. Appreciating the small things in life: STED microscopy in living cells. Journal of Physics D: Applied Physics 54(3). IOP Publishing:11. doi:10.1088/1361-6463/abac81.

Stoffel, W., I. Hammels, B. Jenke et al. 2016. Neutral sphingomyelinase (SMPD3) deficiency disrupts the Golgi secretory pathway and causes growth inhibition. Cell Death and Disease 7(11). Nature Publishing Group: e2488–e2488. doi:10.1038/cddis.2016.385.

Subathra, M., A. Qureshi and C. Luberto. 2011. Sphingomyelin synthases regulate protein trafficking and secretion. PLoS ONE 6(9). doi:10.1371/journal.pone.0023644.

Sundberg, E.L., Y. Deng and C.G. Burd. 2019. Syndecan-1 mediates sorting of soluble lipoprotein lipase with Sphingomyelin-rich membrane in the Golgi apparatus. Developmental Cell 51(3). Cell Press:387-398.e4. doi:10.1016/j.devcel.2019.08.014.

Sztul, E. and V. Lupashin. 2009. Role of vesicle tethering factors in the ER-Golgi membrane traffic. FEBS Letters 583(23). NIH Public Access: 3770. doi:10.1016/J.FEBSLET.2009.10.083.

Tafesse, F.G., K. Huitema, M. Hermansson et al. 2007. Both sphingomyelin synthases SMS1 and SMS2 are required for sphingomyelin homeostasis and growth in human HeLa cells. Journal of Biological Chemistry 282(24): 17537–17547. doi:10.1074/jbc.M702423200.

Tang, D. and Y. Wang. 2013. Cell cycle regulation of Golgi membrane dynamics. Trends in Cell Biology 23(6). Elsevier Current Trends:296–304. doi:10.1016/j.tcb.2013.01.008.

Tao, Y., Y. Yang, R. Zhou et al. 2020. Golgi apparatus: an emerging platform for innate immunity. Trends in Cell Biology 30(6). Elsevier Current Trends: 467–477. doi:10.1016/j.tcb.2020.02.008.

Théry, M. and M. Pende. 2019. Golgi mechanics controls lipid metabolism. Nature Cell Biology 21(3). Nature Publishing Group: 301–302. doi:10.1038/s41556-019-0289-2.

Thomas, L.L. and J.C. Fromme. 2020. Extensive GTPase crosstalk regulates Golgi trafficking and maturation. Current Opinion in Cell Biology 65 (August). Elsevier Current Trends: 1–7. doi:10.1016/j.ceb.2020.01.014.

Thyberg, J. and S. Moskalewski. 1999. Role of microtubules in the organization of the Golgi complex. Experimental Cell Research 246(2). Academic Press:263–279. doi:10.1006/excr.1998.4326.

Tie, H.C., A. Ludwig, S. Sandin et al. 2018. The spatial separation of processing and transport functions to the Interior and Periphery of the Golgi Stack. ELife 7 (November). eLife Sciences Publications Ltd. doi:10.7554/eLife.41301.

Tie, H.C., D. Mahajan and L. Lu. 2022. Visualizing intra-golgi localization and transport by side-averaging Golgi ministacks. Journal of Cell Biology 221(6). Rockefeller University Press. doi:10.1083/jcb.202109114.

Trucco, A., R.S. Polischuck, O. Martella et al. 2004. Secretory traffic triggers the formation of tubular continuities across Golgi sub-compartments. Nature Cell Biology 6(11): 1071–1081. doi:10.1038/ncb1180.

Tu, L. and D.K. Banfield. 2010. Localization of Golgi-Resident glycosyltransferases. Cellular and Molecular Life Sciences. Vol. 67. Springer. doi:10.1007/s00018-009-0126-z.

Ungar, D. 2009. Golgi linked protein glycosylation and associated diseases. Seminars in Cell and Developmental Biology 20(7). Academic Press: 762–769. doi:10.1016/j.semcdb.2009.03.004.

Valente, C., A. Luini and D. Corda. 2013. Components of the CtBP1/BARS-dependent fission machinery. Histochemistry and Cell Biology. Springer Verlag. doi:10.1007/s00418-013-1138-1.

Van Blitterswijk, W.J., A.H. Van Der Luit, R.J. Veldman et al. 2003. Ceramide: second messenger or modulator of membrane structure and dynamics? Biochemical Journal 369(2): 199–211. doi:10.1042/BJ20021528.

van Galen, J., F. Campelo, E. Martínez-Alonso et al. 2014. Sphingomyelin homeostasis is required to form functional enzymatic domains at the Trans-Golgi network. Journal of Cell Biology 206(5):609–618. doi:10.1083/jcb.201405009.

van Meer, G., E.H. Stelzer, R.W. Wijnaendts-van-Resandt et al. 1987. Sorting of sphingolipids in epithelial (Madin-Darby Canine Kidney) Cells. The Journal of Cell Biology 105(4). The Rockefeller University Press:1623–1635. doi:10.1083/jcb.105.4.1623.

Van Meer, Gerrit. 1998. Lipids of the Golgi Membrane. Trends in Cell Biology. doi:10.1016/S0962-8924(97)01196-3.

Van Meer, Gerrit, D.R. Voelker and G.W. Feigenson. 2008. Membrane lipids: where they are and how they behave. Nature Reviews Molecular Cell Biology 9(2). Nature Publishing Group:112–124. doi:10.1038/nrm2330.

Varma, M. and M. Deserno. 2022. Distribution of cholesterol in asymmetric membranes driven by composition and differential stress. Biophysical Journal 121(20). Cell Press:4001–4018. doi:10.1016/J.BPJ.2022.07.032.

Venditti, R., M.C. Masone and M.A. De Matteis. 2020. ER-Golgi membrane contact sites. Biochemical Society Transactions. Portland Press Ltd. doi:10.1042/BST20190537.

Venditti, R. M.C. Masone, L.R. Rega et al. 2019. The activity of Sac1 across ER–TGN contact sites requires the four-phosphate-adaptor-protein-1. Journal of Cell Biology 218(3): 783–797. doi:10.1083/jcb.201812021.

von Blume, J., A.M. Alleaume, C. Kienzle et al. 2012. Cab45 is required for Ca2+;-dependent secretory cargo sorting at the Trans-Golgi network. Journal of Cell Biology 199(7): 1057–1066. doi:10.1083/jcb.201207180.

von Blume, J., J.M. Duran, E. Forlanelli et al. 2009. Actin remodeling by ADF/Cofilin is required for cargo sorting at the Trans-Golgi network. Journal of Cell Biology 187(7): 1055–1069. doi:10.1083/jcb.200908040.

von Blume, J. and A. Hausser. 2019. Lipid-dependent coupling of secretory cargo sorting and trafficking at the Trans-Golgi network. FEBS Letters. John Wiley & Sons, Ltd. doi:10.1002/1873-3468.13552.

Wakana, Y. and F. Campelo. 2021. The PKD-dependent biogenesis of TGN-to-Plasma membrane transport carriers. Cells 10 (7). Multidisciplinary Digital Publishing Institute: 1618. doi:10.3390/cells10071618.

Wakana, Y., K. Hayashi, T. Nemoto et al. 2021. The ER cholesterol sensor SCAP promotes CARTS biogenesis at ER–Golgi membrane contact sites. Journal of Cell Biology 220(1). NLM (Medline). doi:10.1083/jcb.202002150.

Wakana, Y., J. Villeneuve, J. van Galen et al. 2013. Kinesin-5/Eg5 is important for transport of CARTS from the Trans-Golgi network to the cell surface. Journal of Cell Biology 202(2): 241–250. doi:10.1083/jcb.201303163.

Wang, Yanzhuang and J. Seemann. 2011. Golgi biogenesis. Cold Spring Harbor Perspectives in Biology 3(10): 1–12. doi:10.1101/cshperspect.a005330.

Wang, Yaxi, C.J. Mousley, M.G. Lete and V.A. Bankaitis. 2019. An equal opportunity collaboration between lipid metabolism and proteins in the control of membrane trafficking in the Trans-Golgi and endosomal systems. Current Opinion in Cell Biology 59(August): 58–72. doi:10.1016/j.ceb.2019.03.012.

Warren, G. and V. Malhotra. 1998. The organisation of the Golgi apparatus. Current Opinion in Cell Biology. doi:10.1016/S0955-0674(98)80064-1.

Wei, J.H. and J. Seemann. 2017. Golgi ribbon disassembly during mitosis, differentiation and disease progression. Current Opinion in Cell Biology 47(August). Elsevier Current Trends: 43–51. doi:10.1016/J.CEB.2017.03.008.

Weidman, P., R. Roth and J. Heuser. 1993. Golgi membrane dynamics imaged by freeze-etch electron microscopy: views of different membrane coatings involved in tubulation versus vesiculation. Cell 75(1): 123–133. doi:10.1016/S0092-8674(05)80089-3.

Weigert, R., M.G. Silletta, S. Spanò et al. 1999. CtBP/BARS induces fission of golgi membranes by acylating lysophosphatidic acid. Nature 402(6760): 429–433. doi:10.1038/46587.

Welch, L.G. and S. Munro. 2019. A tale of short tails, through thick and thin: investigating the sorting mechanisms of Golgi enzymes. FEBS Letters 593(17). John Wiley & Sons, Ltd: 2452–2465. doi:10.1002/1873-3468.13553.

Witkos, T.M. and M. Lowe. 2017. Recognition and tethering of transport vesicles at the Golgi apparatus. Current Opinion in Cell Biology. Elsevier Ltd. doi:10.1016/j.ceb.2017.02.003.

Witte, K., A.L. Schuh, J. Hegermann, et al. 2011. TFG-1 function in protein secretion and oncogenesis. Nature Cell Biology 13(5): 550–558. doi:10.1038/ncb2225.

Wong-Dilworth, L., C. Rodilla-Ramirez, E. Fox et al. 2023. STED imaging of endogenously tagged ARF GTPases reveals their distinct nanoscale localizations. Journal of Cell Biology 222(7). The Rockefeller University Press. doi:10.1083/JCB.202205107.

Wu, H., P. Carvalho and G.K. Voeltz. 2018. Here, there, and everywhere: the importance of ER membrane contact sites. Science 361 (6401). American Association for the Advancement of Science:eaan5835. doi:10.1126/science.aan5835.

Xiang, Y. and Y. Wang. 2010. GRASP55 and GRASP65 play complementary and essential roles in golgi cisternal stacking. Journal of Cell Biology 188 (2):237–251. doi:10.1083/jcb.200907132.

Yadav, A., Q. Vagne, P. Sens et al. 2022. Glycan processing in the Golgi: optimal information coding and constraints on cisternal number and enzyme specificity. ELife 11 (February). eLife Sciences Publications Ltd. doi:10.7554/ELIFE.76757.

Yadav, S. and A.D. Linstedt. 2011. Golgi positioning. Cold Spring Harbor Perspectives in Biology 3(5): 1–17. doi:10.1101/cshperspect.a005322.

Yamaguchi, H., S. Arakawa, T. Kanaseki et al. 2016. Golgi membrane-associated degradation pathway in yeast and mammals. The EMBO Journal 35(18): 1991–2007. doi:10.15252/embj.201593191.

Yang, J.S., S.Y. Lee, S. Spanò et al. 2005. A role for BARS at the fission step of COPI vesicle formation from Golgi membrane. EMBO Journal 24(23): 4133–4143. doi:10.1038/sj.emboj.7600873.

Yuan, F., H. Alimohamadi, B. Bakka et al. 2021. Membrane bending by protein phase separation. Proceedings of the National Academy of Sciences of the United States of America 118(11). National Academy of Sciences: e2017435118. doi:10.1073/pnas.2017435118.

Zerial, M. and H. Stenmark. 1993. Rab GTPases in vesicular transport. Current Opinion in Cell Biology 5(4). Elsevier Current Trends: 613–620. doi:10.1016/0955-0674(93)90130-I.

Zhang, L., S. Malik, J. Pang et al. 2013. Phospholipase Cε hydrolyzes perinuclear phosphatidylinositol 4-phosphate to regulate cardiac hypertrophy. Cell 153(1): 216–227. doi:10.1016/J.CELL.2013.02.047.

Zhukovsky, M.A., A. Filograna, A. Luini et al. 2019. The Structure and Function of Acylglycerophosphate Acyltransferase 4/Lysophosephatidic Acid Acyltransferase Delta (AGPAT4/LPAATδ). Frontiers in Cell and Developmental Biology. Frontiers Media S.A. doi:10.3389/fcell.2019.00147.

Ziltener, P., A.A. Rebane, M. Graham et al. 2020. The golgin family exhibits a propensity to form condensates in living cells. FEBS Letters 594(19). John Wiley & Sons, Ltd:3086–3094. doi:10.1002/1873-3468.13884.

Zimmerberg, J. and M.M. Kozlov. 2006. How proteins produce cellular membrane curvature. Nature Reviews Molecular Cell Biology 7(1): 9–19. doi:10.1038/nrm1784.

11

Inositol Lipids Are Key to Shaping Membranes and to Define the Cellular Lipid Landscape

Tamas Balla,[1,2,*] *Yeun Ju Kim,*[2] *Amrita Mandal,*[2] *Gergo Gulyas,*[2] *Yang Niu,*[2] *Alejandro Alvarez-Prats*[2] and *Joshua Pemberton*[2]

Phosphoinositides: A Unique Class of Regulatory Phospholipids

Inositol-containing phospholipids were first described in lipid extracts from the brain and were subsequently identified as phosphatidylinositol (PI) and its phosphorylated derivatives, the phosphoinositides (PPIns). The unique feature of these lipids was their rapid labeling with ^{32}P-phosphate and their accelerated ^{32}P-labeling after stimulation with secretagogues, suggesting their high metabolic turnover rates (these historical aspects are detailed in (Balla 2013). Seven isomers of PPIns have been identified, bearing phosphates at one or more of the 3-, 4-, or 5-position of the inositol ring, in addition to the phosphate that forms the phosphodiester group that links the inositol ring to the diacylglycerol (DAG) backbone (Figure 1A). PPIn lipids are synthesized from PI through sequential phosphorylation events by PI- and PPIn kinase enzymes that come in multiple forms and show a high degree of specificity in their substrate recognition and the position at which they phosphorylate the inositol ring (Burke et al. 2022). A similarly large number of PPIn phosphatases exist that remove phosphates at specific positions, which are also highly selective for the specific PPIn isomers that can be used as substrates (Balla 2013; Sasaki et al. 2009). Several human diseases are associated with mutations in PPIn lipid kinases as well as phosphatases (Billcliff and Lowe 2014; Burke et al. 2022; Dickson and Hille 2019).

[1] Bldg 35A, Rm 2D842, 35 Convent Drive, Bethesda, MD 20892.

[2] Section on Molecular Signal Transduction, Program for Developmental Neuroscience, *Eunice Kennedy Shriver* National Institute of Child Health and Human Development, National Institutes of Health, Bethesda, MD 20892.

* Corresponding author: ballat@mail.nih.gov

The first well-established role of PPIns and PI 4,5-bisphosphate [PI(4,5)P_2], in particular, was as the substrate for conserved phospholipase C (PLC) enzymes that produce the essential second messengers, inositol 1,4,5-trisphosphate (InsP_3) and diacylglycerol (DAG) upon receptor activation (Berridge 1984). Many cell surface receptors, including G protein-coupled receptors (GPCRs), as well as receptor tyrosine kinases (RTKs), use PLC-dependent PI(4,5)P_2 hydrolysis as a major signal transduction pathway to induce elevations to intracellular Ca^{2+} concentrations and activate isoforms of protein kinase C (PKC). However, PPIns were also soon recognized as important regulatory or signaling lipids within both the plasma membrane (PM) and on the cytosolic surface of intracellular organelles. For example, in addition to being consumed by PLC activation, PI(4,5)P_2 was identified as a critical factor for membrane attachment of clathrin-adaptor proteins, such as AP2 (Beck and Keen 1991) or AP180 (Ford et al. 2001), and is directly involved in the allosteric regulation of numerous ion channels and transporters (Hilgemann and Ball 1996; Huang et al. 1998) in the PM. The subsequent discovery of the 3-phosphorylation of inositol lipids has further consolidated this notion, as PI 3,4,5-trisphosphate [PI(3,4,5)P_3] was shown to directly mediate the actions of several RTKs, including insulin receptors and several oncogenes (Cantley et al. 1991; Ruderman et al. 1990). Even the mono-phosphorylated PPIn lipids, PI 3-phosphate [PI(3)P] and PI 4-phosphate [PI(4)P], are critical regulators of membrane dynamics within the early endosome (Schu et al. 1993; Simonsen et al. 1998) and Golgi compartment (Godi et al. 2004; Wang et al. 2003), respectively. These actions were found to be independent of the role of these lipids as precursors for the generation of other phosphorylated PPIns species. All these developments have led to the notion that the PPIns distribution, which works with diverse families of small GTP-binding Rab proteins, serves as an essential determinant defining the unique identity of organellar membranes.

More recently, yet another fundamental role for PPIn lipids, especially PI(4)P, has been demonstrated across eukaryotic model systems, namely, the involvement of PPIn species in the dynamic control of non-vesicular lipid transport between the ER and other organelle membranes. The role of PI(4)P for supporting lipid-exchange is now well-established for cholesterol (Cho) (Mesmin et al. 2013) and ceramide (Cer) (Hanada et al. 2003) transport between the ER and the Golgi, as well as part of phosphatidylserine (PS) transport between the ER and the PM (Chung et al. 2015; Moser von Filseck et al. 2015). The pivotal role of PI(4)P in these non-vesicular lipid transport processes also helped to explain why PI 4-kinases (PI4Ks) serve as essential host factors that are used by several varieties of picornaviruses to generate an optimized membrane compartment, with a specialized lipid composition, that supports viral replication (Hsu et al. 2010; Reiss et al. 2011; Tai et al. 2009; Vaillancourt et al. 2009).

Phosphoinositides and Subcellular Lipid Distribution

PPIns have a pivotal role in establishing and maintaining the unique lipid composition of organelle membranes. As part of these integrated actions, PPIns can alter the local biophysical properties of membranes and facilitate direct interactions with unique protein effectors that can both assemble to impact membrane shape or nucleate the

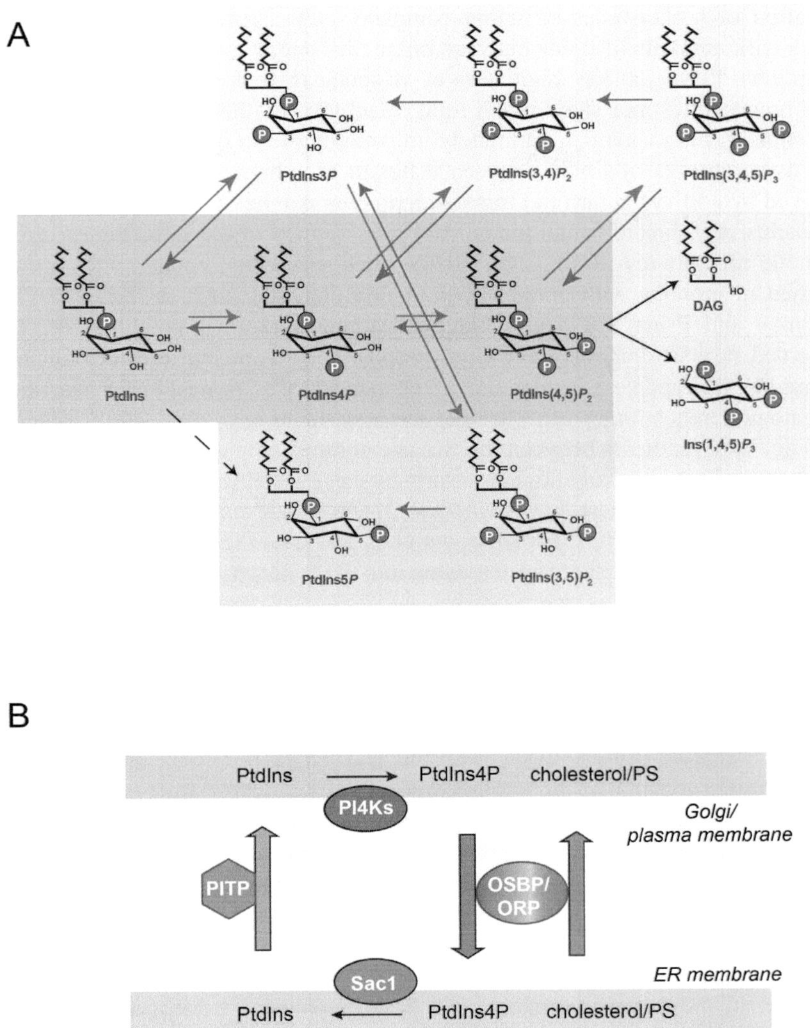

Figure 1. (A) Phosphoinositides (PPIn) are produced by phosphorylation of phosphatidylinositol (PtdIns or PI) at specific positions on the inositol ring. In PI, the myo-inositol headgroup is linked to the diacylglycerol (DAG) backbone by a phosphodiester phosphate that is connected to position 1 of the inositol ring. The acyl-side chains esterifying the sn1 and sn2 of DAG (shown truncated here) usually are stearoyl and arachidonoyl-enriched in PI. Kinase and phosphatase reactions are shown by arrows. Note that there is often more than one isoform of the enzymes that execute either the specific phosphorylation (kinase) or de-phosphorylation (phosphatase) steps. For simplicity, the full enzyme names are not listed here. **(B)** The principle of PI(4)P-driven non-vesicular lipid transport. A PI(4)P gradient is set up between the ER and membranes of other organelles that are in proximity at membrane contact sites (only Golgi and PM are listed for simplicity). This gradient results from PI4K activity in the target membrane and the PI(4) P phosphatase activity of Sac1 in the ER. A critical feature of this transport system to ensure directional transport of Cho or other lipids (here PS is listed) is the ability of the lipid transport protein (OSBP or its related protein, ORPs) to bind either PI(4)P or the counter-transported lipid in their cargo binding pocket. Supply of PI (PtdIns) from the ER, presumably by PI transfer proteins (PITPs), is an important component of this transport system. See text for further details.

formation of higher-order signaling complexes. Furthermore, vesicular transport moves bulk amounts of lipids between organelles and, as will be discussed in later paragraphs, PPIns are key regulators of vesicular trafficking pathways. However, these processes do not show much lipid selectivity, as the entire membrane, with its complex lipid makeup, is ultimately transitioning from one organelle to another. Selective transport of lipids between different membranes is now thought to be achieved by non-vesicular lipid transfer, a process that has been primarily, although not exclusively, linked to unique contact sites formed at the interface of different organelle membranes (Prinz et al. 2020). Such membrane contact sites are often enriched in proteins with unique hydrophobic domains, such as the SMP/TULIP (Toulmay and Prinz 2012) or Chorein-N domains (Kumar et al. 2018). These conserved protein domains have been proposed to adopt hydrophobic tunnels of various lengths capable of lipid transfer. Yet, in most cases, these transfer mechanisms show no preference for a particular lipid and may function by dissipating, rather than building, lipid gradients between the two membrane compartments (Leonzino et al. 2021). Overall, the control of lipid transport by these proteins is still largely unknown, but recent studies suggest that they work together with multi-transmembrane lipid scramblases (Li et al. 2021; Ghanbarpour et al. 2021; Guillen-Samander et al. 2022) that may function to facilitate lipid loading into the hydrophobic transfer domains for membrane expansion.

Evidently, continuous lipid transport between membranes may be upheld by concentration gradients that are maintained as the lipid cargo is metabolically converted or flipped/flopped to the other membrane leaflet after being deposited in the target membrane. Several examples exist for such transport-coupled lipid synthetic cascades. Often, when the lipid product is not transported out of the membrane, the enzymatic reaction that drives lipid synthesis cannot proceed because of product feedback inhibition. A good example involves the synthesis of PS by the PSS1 enzyme, which rapidly loses activity when the product, PS, is not transported out from the ER (Sohn et al. 2016). Conversely, the transport process will stop if the lipid is not converted enzymatically at the target membrane, as is the case for the PS transport mechanisms coupled to PS decarboxylation and PE production within the mitochondria (Vance 2015).

None of these mechanisms explain how lipids can be transported against an established concentration gradient. One of the seminal discoveries of the last ten years was the recognition that a PI(4)P gradient can be used to drive the uphill transport of Cho from the ER to the Golgi compartment by the oxysterol-binding protein (OSBP) (Mesmin et al. 2013) (Figure 1B). This discovery was preceded by several earlier studies that hinted at Golgi-enriched PI(4)P as a critical lipid that is directly recognized by the pleckstrin homology (PH) domains of several lipid transport proteins. These included the Cer transfer protein, CERT (Hanada et al. 2003), the Cho transporter, OSBP, as well as some of the yeast OSBP homologs, the Osh proteins (Levine and Munro 1998; Levine and Munro 2002), and the glucosyl-Cer transfer protein, FAPP2 (D'Angelo et al. 2007). Moreover, in addition to regulating protein localization, the activity of PI4K enzymes was found to be important for the transport of Cer by CERT (Toth et al. 2006), Cho by OSBP (Banerji et al. 2010), or glucosyl-Cer by FAPP2 (D'Angelo et al. 2007). However, not until the demonstration that PI(4)P and Cho both bind to and compete for the same lipid-binding pocket in the Osh4 protein (de Saint-Jean et al. 2011) did the possibility arise that PI(4)P might serve as the counter-transport currency that effectively fuels the transport of structural lipids between adjacent membranes (Kim et al. 2013). A seminal paper from the group of Bruno Antonny (Mesmin et al. 2013) provided both the conceptual outline and experimental support for this mechanism, wherein a PI(4)P gradient that is established by Golgi-resident PI 4-kinase enzymes, PI4KB (also called PI4KIIIβ) and PI4K2A (also called PI4KIIα), which function to generate PI(4)P in the TGN membrane, and the ER-localized Sac1 phosphatase, which dephosphorylates PI(4) P to PI in the ER, functions as the driving force for Cho transport out of the ER and into the Golgi membranes by the coordinate transfer activities of OSBP (Figure 1B).

A key element for this system to operate efficiently is the high-affinity binding of PI(4)P to the lipid-binding pocket of OSBP, which prevents the loading of Cho into the ORD lipid-transfer domain at membranes where the local PI(4)P levels are high. However, when the ORD faces the ER, the released PI(4)P is efficiently eliminated by the Sac1 phosphatase to allow the subsequent loading of the Cho synthesized in the ER into the ORD. As Cho is released at the Golgi from the ORD, PI(4)P then binds and prevents the backward transport of Cho. This four-step cycle mechanism ensures a directional transport of Cho if the PI(4)P gradient is maintained. The energy for this transport process is ultimately provided by ATP hydrolysis, which occurs when PI(4)P is generated by the PI4K enzyme (Antonny et al. 2018). This general concept has since been recognized to govern the transport of PS from the ER to the PM by the mammalian ORP5/8 and yeast Osh6 proteins (Chung et al. 2015; Moser von Filseck et al. 2015) as well as the transport of Cho (Radulovic et al. 2022) or PS (Tan and Finkel 2022) from the ER to damaged lysosomes, which is mediated by the mammalian ORP1L and ORP9/10 proteins, respectively. Overall, this summary highlights the pivotal role of PI(4)P and the PI4K enzymes in establishing and maintaining the specific lipid composition of organellar membranes.

Proteins Contain Selective Phosphoinositide-Recognition Modules

While the biophysical properties of the resident lipid species that form membranes are important for allowing membranes to be pliable and assume certain shapes, integral and peripheral protein effectors also play a critical role in recognizing and creating specific membrane geometries. One of the emergent properties of membranes that is critical for the processes of membrane fusion and fission is the curvature. PPIns play important roles in many of these membrane-shaping processes.

Early studies aimed at defining proteins that could enzymatically modify PPIns were focused on the phospholipase C (PLC) enzymes that generate essential second messengers by hydrolyzing PI(4,5)P_2 (Suh et al. 1988) and the unique phosphatases that convert the calcium-mobilizing molecule, InsP_3 (Majerus et al. 1986; Takazawa et al. 1989). Around the same time, it was also recognized that a conserved protein fold, the pleckstrin homology (PH) domain (Haslam et al. 1993; Mayer et al. 1993), is found in numerous peripheral membrane proteins, including many of the presumed PPIn-modifying enzymes. The function and target for this domain was initially puzzling (Musacchio et al. 1993) until it was recognized that selected PH domains showed specific binding to $InsP_3$ (Ferguson et al. 1995) and, in fact, to the PPIns lipids themselves (Lemmon et al. 1997). These studies led to the notion that PH domains are PPIns-recognition modules that recruit peripheral proteins to specific membrane locations through stereospecific PPIn interactions (Lemmon and Ferguson 2000). This view has since been refined, as some PH domains do not bind PPIns but instead interact with proteins (Balla 2005; Lemmon 2004), or even if they do recognize PPIns, their lipid binding does not serve as localization signals, but provides an allosteric regulatory input to their cognate interacting proteins (Campa et al. 2009).

PH domains are not the only domains that recognize PPIns. For instance, while searching for proteins that would mediate the effects of PI(3)P, a lipid that was initially linked to the regulation of endocytic trafficking (Schu et al. 1993), a conserved PI(3)P-binding segment was identified that was named the FYVE domain (Burd and Emr 1998; Gaullier et al. 1998; Misra and Hurley 1999). The name is an acronym that incorporates the first letters of the proteins in which it was first identified: **F**ab 1 (yeast orthologue of the human PI(3)P 5-kinase, PIKfyve), **Y**OTB (nematode protein, ZK632.12), **V**ac 1 (yeast vesicle transport protein), and the mammalian endosomal protein, **E**EA1. An additional PPIn-recognition domain was subsequently identified when it was found that Phox domains, initially described in the NADPH oxidase proteins $p40^{phox}$ and $p47^{phox}$ (Ponting 1996), showed specific binding to PI(3)P (Ago et al. 2001; Cheever et al. 2001; Ellson et al. 2001; Kanai et al. 2001; Song et al. 2001). Some PX domains can also bind other PPIns, including PI 3,4-bisphosphate ($PI(3,4)P_2$) (Kanai et al. 2001), or show limited PPIn-binding specificity or even no PPIns binding at all (Chandra et al. 2019).

Another domain that binds acidic phospholipids, including PS and the various PPIns species, is the C2 domains. C2 domains were initially described in protein kinase C (Nishizuka 1988) as structural segments that confer Ca^{2+} and PS regulation to these enzymes. C2 domains have since been identified in a great number of peripheral membrane proteins that possess Ca^{2+}-dependent binding to membranes enriched in anionic phospholipids (Corbalan-Garcia and Gomez-Fernandez 2014). One of the two C2 domains found in synaptotagmin-1, the C2B, binds $PI(4,5)P_2$ in addition to PS (Guillen et al. 2013), and the recognition of $PI(4,5)P_2$ by the C2B markedly increases the Ca^{2+} affinity of synaptotagmin-1 (Radhakrishnan et al. 2009). A similar $PI(4,5)P_2$-dependent binding of the Extended Synaptotagmins (E-Syts) to the PM is mediated by their numerous (3 to 5) C2 domains, and, in the case of ESyt-1, the binding is dynamically regulated by the cytosolic free Ca^{2+} levels (Giordano et al. 2013).

While PH-, FYVE-, PX-, and C2-domains (Figure 2) are the best recognized and most widely distributed PPIns-binding modules that are found in the proteome, additional domains that lend PPIn-specific recognition to proteins also exist. These include the ENTH, ANTH, CALM, and FERM domains that are found in proteins that play a role in clathrin-mediated endocytosis or provide a bridge between the PM and the actin cytoskeleton [reviewed in (Hammond and Balla 2015)] or the Tubby domain that is present in a family of proteins that play critical roles in signaling from primary cilia (Wang et al. 2018).

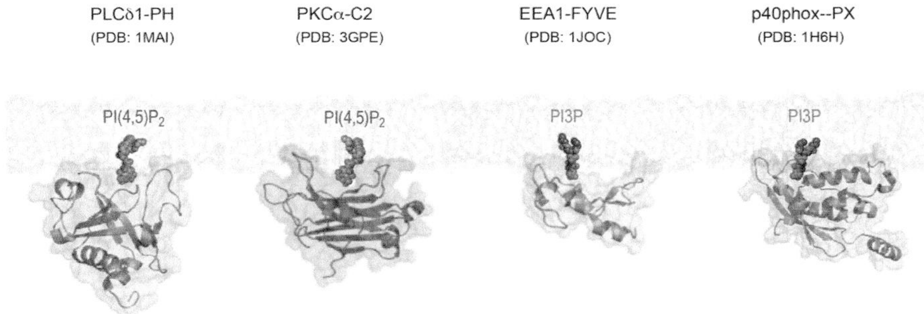

Figure 2. Structural comparison of the major protein domains exhibiting stereospecific PPIn-binding. Representative structures of the PH (**A**, PLCδ1 PH domain in complex with IP_3; PDB entry 1MAI), C2 (**B**, PKCα-C2 domain in complex with Ca^{2+} and $PI(4,5)P_2$; PDB entry 3GPE), FYVE (**C**, EEA1-FYVE domain bound to PI(3)P; extra cted from PDB entry 1JOC), and PX (**D**, p40phox PX domain in complex with PI(3)P; PDB entry 1H6H) domains are shown as cartoon representations of the secondary structure with transparent surfaces. For each domain, characteristic PPIn species have been oriented within the binding pockets, which are shown to be interfaced with a generic lipid bilayer. Please note the extent to which membrane-oriented protrusions surrounding the lipid-binding pockets must penetrate the cytosolic leaflet of the bilayer to coordinate the PPIn ligands. For further details, please refer to Section (Proteins Contain Selective Phosphoinositide-Recognition Modules) of the text. These structures were prepared using the PyMOL Molecular Graphics System, Version 2.0 Schrödinger, LLC.

Phosphoinositide-Binding Proteins Are Often Linked to Membrane Deformation

Most PPIn-interacting domains have been shown to possess hydrophobic loops or short amphipathic helices that can penetrate the lipid bilayer and directly deform membranes (Pang et al. 2014). Similarly, the FYVE domain of EEA1 (Kutateladze and Overduin 2001; Kutateladze et al. 2004), the PX domains of Bem1P, p40phox, and p47phox (Stahelin et al. 2003a; Stahelin et al. 2007) all exhibit penetration into the bilayer interface and membrane deforming properties.

More important for membrane deformation, however, are proteins that have specific membrane bending or curvature-recognizing domains. One of the largest families of proteins that contain PX domains is the sorting nexins (SNX), which play important roles in the recycling and retrieval of endocytic cargoes throughout the endosomal system (Cullen 2008). In addition to PX domains, SNXs can be classified into subgroups based on the presence of other protein modules (Hanley and Cooper 2020). One of those subgroups, called SNX-BAR proteins, is thought to directly sense and dynamically regulate membrane curvature using a unique protein fold called the BAR domain (van Weering et al. 2010). In general, BAR domains are intrinsically curved dimeric structures that are formed by an extended bundle of three α-helices. This domain was first identified as a conserved sequence found in the **B**in (Sakamuro et al. 1996), **A**mphiphysin (David et al. 1994), and yeast **R**vs (Sivadon et al. 1995) proteins, hence the BAR domain moniker. It was then demonstrated that purified amphiphysin and endophilin, another synaptic vesicle-enriched protein, can tubulate membranes *in vitro*, and the tubulating effect was mapped to the NH_2-terminal region that was conserved between the abovementioned three proteins (Farsad et al. 2001;

Takei et al. 1999). The first clues of how BAR domains sense or deform membranes came from the first BAR domain structures (Peter et al. 2004; Tarricone et al. 2001), which showed the α-helical bundles forming a curved banana-like shape, where the concave surface was lined with positively charged residues that interacted with the membrane through extensive electrostatic interactions (Zimmerberg and McLaughlin 2004). BAR domains have since been described in large numbers of proteins forming the most studied membrane-curving entities in eukaryotic cells (Simunovic et al. 2019) (Figure 3). BAR domains are now divided into three groups: (1) classical BAR and N-BAR, (2) F-BAR, and (3) I-BAR domains. The first two groups induce positive membrane curvature, whereas I-BARs promote negative curvature to form protrusions. It has been debated whether BAR domains actively deform membranes to generate curvature or simply recognize specific membrane curvatures and self-assemble to stabilize them. A more detailed description of the biophysics that contributes to the membrane deforming actions of BAR domains is beyond the scope of this Chapter and will certainly be covered in other Chapters within this collection as well as in other excellent recent reviews [e.g., Simunovic et al. (2019)]. It is important to note that while most BAR domain-containing proteins do not directly interact with PPIns—which makes the subclass of SNX-BAR proteins truly unique—but they do form direct protein-protein interactions with other proteins that possess selective PPIns-recognition domains. Good examples of these types of heteromeric complexes include the interaction of endophilins through their SH3 domain with the proline-rich sequence of dynamin-1 (Ringstad et al. 1997) or the assembly of amphiphysin together with dynamin-1 into ring-like structures on liposomes (Takei et al. 1999). In these cases, it is the resident dynamin that directly recognizes the PPIn lipids through its PH domain (see below).

Another family of proteins that have been associated with membrane deformation are the **e**ndosomal **s**orting **c**omplexes **r**equired for **t**ransport (ESCRT-0 to ESCRT-III), which were initially identified as critical effectors needed for sorting ubiquitylated membrane cargoes into intraluminal vesicles during maturation of multivesicular bodies (MVBs) (Hanson and Cashikar 2012; McCullough et al. 2013). Important roles for the ESCRT machinery have since been recognized in several other membrane-shaping processes, including exosome biogenesis (Colombo et al. 2013), outward budding of enveloped viruses including HIV (Carlton and Martin-Serrano 2007; Garrus et al. 2001), cytokinetic abscission (McCullough et al. 2013), and for the shaping of the sarcoplasmic reticulum (Lefebvre et al. 2016)—to mention a few of the best-characterized examples. In general, endosomal PI(3)P plays a critical role in the initiation and assembly of the ESCRT machinery. The yeast Vps27 and Hse1 proteins (Hrs and STAM in mammalian cells) form the ESCRT-0 complex, which is recruited to endosomes via the FYVE domain of Vps27/Hrs and initiates the translocation and assembly of the other members of the ESCRT complex (Katzmann et al. 2003). Moreover, the N-terminus of one of the ESCRT-II components in yeast, Vps36, contains a unique GLUE (**G**RAM-**l**ike, **u**biquitin binding in **E**AP45) domain that is essentially a split PH domain fold with a large intervening sequence (Teo et al. 2006). This domain binds PI(3)P in the yeast protein (Teo et al. 2006), but the GLUE domain of the mammalian homolog, EAP45, shows preferential binding to PI(3,4,5)P_3 and PI(3,4)P_2 (Alam et al. 2006; Hirano et al. 2006; Slagsvold et al. 2005).

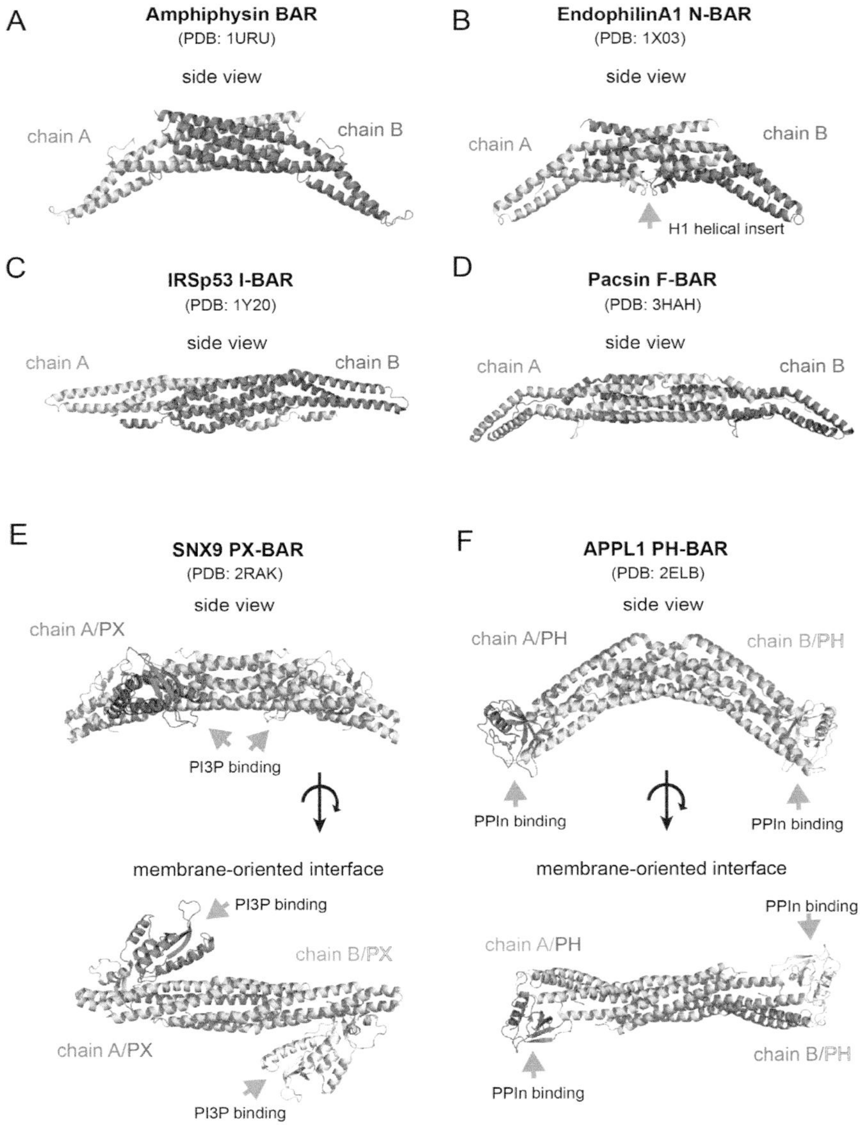

Figure 3. The structural diversity of the BAR domain superfamily. Representative structures of the classical BAR (**A**; Amphiphysin; PDB entry 1URU), N-BAR (**B**; EndophilinA1; PDB entry 1X03), I-BAR (**C**; IRSp53; PDB entry 1Y2O), and F-BAR (**D**; Pacsin1; PDB entry 3HAH) domains. Structures are shown as cartoon representations of the secondary structure, with individual chains of the obligate dimers colored green and magenta. Examples of the unique PX-BAR (**E**; SNX9; PDB entry 2RAK) and BAR-PH (**F**; APPL1; PDB entry 2ELB) domain variants that combine the canonical BAR domain with a flanking PPIn-binding module are also highlighted. In these multi-domain structures, the individual units of the assembled dimer are denoted by the green (BAR domain) and magenta (PPIn-binding domain) color combination for one chain together with the blue (BAR domain) and yellow (PPIn-binding domain) color combination for other chains. For further details, please refer to Section (Phosphoinositide-Binding Proteins Are Often Linked to Membrane Deformation) of the text. These structures were prepared using the PyMOL Molecular Graphics System, Version 2.0 Schrödinger, LLC.

Recent studies identified the Class II PI3K, PI3KC2a, and its lipid product, PI(3,4)P_2, as key regulators of cytokinetic abscission, with an essential element of this mechanism being PI(3,4)P_2-mediated membrane recruitment of the mammalian Vps36 component of the ESCRT-II complex (Gulluni et al. 2021). How different ESCRTs control the biophysics of the complex membrane remodeling events that occur during cytokinetic abscission has also been addressed using computational models based on quantitative imaging studies (Elia et al. 2012).

Phosphoinositides Regulate Signaling and Shaping of the Plasma Membrane

As already mentioned, PPIns in the PM have major roles as the precursors used for second messenger production [e.g., PI(4,5)P_2]. They also serve directly as signaling lipids [e.g., PI(3,4,5)P_3 or PI(3,4)P_2] as well as important regulators of the actin cytoskeleton [e.g., PI(4,5)P_2] or allosteric modulators of ion channels and transporters. With the exception of contributions to actin remodeling, most of these above-listed roles of PPIns are not associated directly with membrane deformation.

Perhaps the most prominent membrane-shaping processes are linked to localized changes in actin polymerization. Evidently, PPIn metabolism and turnover have been linked to the integrated regulation of actin dynamics in many complex membrane remodeling events, including chemotaxis (Swaney et al. 2010; Weiger and Parent 2012), endocytosis (Saarikangas et al. 2010; Zhang et al. 2012), and exocytosis (Trifaro et al. 2008; Wang and Thurmond 2009). PPIn lipids can directly bind to components of the actin cytoskeleton or its protein regulators or can indirectly control actin polymerization via influencing cytoplasmic Ca^{2+} changes or affecting the cycling of small GTP-binding proteins. Several actin capping and severing proteins bind PI(4,5)P_2 directly (Janmey 1994), while both PI(4)P and PI(4,5)P_2 can dissociate the capping protein gelsolin from the positive (+) end of actin filaments (Janmey 1994; Janmey and Stossel 1987) or inhibit the actin severing activities of cofilin and the actin depolymerizing factor, ADF (Ojala et al. 2001; Yonezawa et al. 1990; Zhao et al. 2010). Profilin and α-actinin, two additional proteins that promote actin assembly and crosslinking, are also regulated by interactions with PI(4,5)P_2 (Chaudhary et al. 1998; Goldschmidt-Clermont et al. 1990; Skare and Karlsson 2002). Alternatively, PI(4,5)P_2 levels can increase actin nucleation by the Arp2/3 complex via binding to N-WASP (**N**euronal **W**iskott-**A**ldrich **P**rotein) (Miki et al. 1996; Rohatgi et al. 2000). In addition, several actin-binding proteins contain FERM (**F**our-point one-**E**zrin-**R**adixin-**M**oesin) domains, which can recognize PI(4,5)P_2. These proteins, therefore, effectively serve as bridges between the PM and the actin cytoskeleton. (Christy et al. 1998; Hamada et al. 2000; Mani et al. 2011). Specific splice forms of PIP5Kγ are also associated with focal adhesions and can regulate focal adhesion dynamics via the local production of PI(4,5)P_2 (Ling et al. 2002).

PI(3,4,5)P_3 levels are also strongly linked to regulation of actin dynamics. The leading edge of migrating cells that move toward chemotactic cues is specifically marked by increased levels of PI(3,4,5)P_3 (Servant et al. 2000; Wang et al. 2002; Weiger and Parent 2012). The role of PI(3,4,5)P_3 in chemotaxis involves the recruitment of the ARNO/Cytohesin family of Arf GDP/GTP exchange factors

(GEFs) (Venkateswarlu et al. 1998). Activation of Arf6 is important for enriching these exchange factors in the PM (Cohen et al. 2007) and promoting the activity of PIP5Ks (Honda et al. 1999). There is also a strong connection between PI(3,4,5)P_3 production and activation of the Rho-family GTPases: Rho, Rac, and Cdc42. These unique Rho-family GTPases are responsible for regulating different forms of actin-mediated membrane deformations, including stress fibers (Rac), lamellopodia (Rho), or filopodia (Cdc42) (Heo et al. 2006). While PI(3,4,5)P_3 can recruit and activate several GEFs of the Rac protein, active Rac also activates the PI3K pathway (Campa et al. 2015).

Alternatively, independent of downstream PPIn production, increased PI(4)P has also been shown to activate actin polymerization. Genetic deletion of VAP proteins leads to massive accumulation of PI(4)P in the Golgi as well as endosomes, and the generation of actin comets from these endocytic structures is dependent on PI(4)P-mediated activation of the endosomal Arp2/3 activating **W**ASP **a**nd **S**CAR **h**omology (WASH) complex (Dong et al. 2016).

As cells actively undergo and integrate bulk endocytosis and exocytosis at the PM, the underlying processes require the coordinated shaping of membranes, ultimately resulting in localized changes to membrane fusion or fission. Not surprisingly, membrane-shaping proteins have major roles in both the fusion as well as fission processes, and there are several molecular steps at which interactions with PPIns species can facilitate the ability of a protein to deform the membrane or execute the fission or fusion step. These molecular events will be detailed below.

Endocytosis

During the formation of the clathrin-coated pits (CCPs) that serve as the seeds for clathrin-mediated endocytosis (CME), several proteins assemble in a sequential manner that need PPIns interaction. Direct binding of inositol-hexakisphosphate, which presumably mimics the headgroup of PI(4,5)P_2, was shown to bind the tetrameric clathrin adaptor, AP-2 (Voglmaier et al. 1992); an interaction that was subsequently validated in the solved structure (Collins et al. 2002). The N-terminal domain of Epsin1, another protein that interacts with clathrin and AP-2, shows sequence similarities with the N-terminal regions of the monomeric clathrin adaptors AP-180 (Mao et al. 2001) and its non-neuronal homolog, CALM (Ford et al. 2001), both of which bind PI(4,5)P_2. Lipid binding of the N-terminal domain of Epsin to PI(4,5)P_2 was also demonstrated to be essential for the endocytosis of EGF receptors (Itoh et al. 2001) by facilitating the ability of Epsin to penetrate membranes and ultimately bend them (Stahelin et al. 2003b). Accordingly, rapid elimination of PI(4,5)P_2 from the PM in intact cells caused the dissociation of several molecules, including Epsin or the tetrameric clathrin adaptor, AP-2, from the PM and prevented the assembly of the clathrin cage (Zoncu et al. 2007). Such rapidly inducible PI(4,5)P_2 depletion also inhibited the endocytosis of the transferrin and angiotensin receptors (Abe et al. 2008; Varnai et al. 2006).

Several other proteins associated with CME are known to assemble on the tubular membrane intermediates that emanate from the PM. This is most clearly demonstrated in cells in which the fission of the endocytic vesicles is impaired, such as after dynamin knockout or in response to over-expression of GTPase-deficient dynamin mutants. In dynamin-2 knockout MEF cells, several N- and F-BAR domain proteins were shown to accumulate on the extended tubules, including endophilin-A1, -A2, and -A3, as well as SNX-9 and -18 (Ferguson et al. 2009). It is not fully understood, however, what lipids are important in assembling these F-BAR proteins on the extended tubulations. Other F-BAR domain proteins, such as FCHo1/2, were found to play a role in the early phase CME, essentially acting as nucleators of the CCPs (Henne et al. 2010). These proteins directly bind and enrich PI(4,5)P_2 in CCPs (El Alaoui et al. 2022), as well as recruit other proteins, such as AP-2, intersectin, or Eps15. PACSINs (also called syndapins) belong to another F-BAR domain-containing family that plays a role in CME by establishing a link with the actin polymerization machinery (Dumont and Lehtonen 2022), and they bind PI(4,5)P_2 (Edeling et al. 2009). PACSINs also interact with dynamin, InPP phosphatases, OCRL, and Synaptojanin via the C-terminal SH3 domains (Billcliff et al. 2016; Dumont and Lehtonen 2022).

The fission of the endocytic vesicle from the PM during CME is mediated by the conserved GTPases, dynamin-1 and -2 (Hinshaw 2000). Protein-protein interactions mediated by the proline-rich domain of dynamin with SH3-domain-containing membrane shaping proteins, such as the endophilin A isoforms, play an important role in the assembly of dynamins on the neck of the endocytic vesicles (Gout et al. 1993; Rosendale et al. 2019; Simpson et al. 1999). At the same time, PH domain-mediated binding of PI(4,5)P_2 is critical for dynamin membrane localization and endocytosis, in general, as well as directly increasing the GTPase activity (Salim et al. 1996). Interestingly, enhanced self-assembly and the formation of the oligomeric dynamin ring also substantially enhance the PH domain-mediated membrane binding of dynamin (Klein et al. 1998).

The critical involvement of PPIns in endocytosis, including their role as regulators of synaptic vesicle recycling, has been inferred, in part, by the involvement of various PPIn kinases and phosphatases at numerous stages during the endocytic process. The inositol lipid phosphatase, Synaptojanin-1 (Synj1), was identified as an enriched synaptic protein (McPherson et al. 1996), and its genetic deletion was shown to result in the accumulation of clathrin-coated vesicles inside of nerve terminals (Cremona et al. 1999). Synj1, which is highly enriched in the brain, and the more widely expressed isoform, Synj2, play important roles in removing PI(4,5)P_2 at the late stages of CME, which allows for disassembly of the protein coat that is assembled on CCVs (Milosevic et al. 2011). It has been shown that Synj1-mediated PI(4,5)P_2 hydrolysis is also regulated by membrane curvature (Chang-Ileto et al. 2011). Other PPIn phosphatases were also described as key molecular players in CME. For example, the 5-phosphatase, OCRL, also associates with late-stage CCVs and contributes to the maturation of CCVs toward endocytic compartments that are marked by the APPL1 protein (Erdmann et al. 2007). Another PPIn 5-phosphatase, SHIP2, which acts on PI(3,4,5)P_3, is recruited to CCPs by intersectin and can regulate the rate of CCP maturation (Nakatsu et al. 2010).

In addition to the PPIns phosphatases, PI 4-kinase (Guo et al. 2003) and PIP 5-kinase (Wenk et al. 2001) enzymes were also shown to play critical roles in synaptic vesicle turnover. In particular, the small GTP-binding protein, Arf6, can stimulate

PIP5Kα (Honda et al. 1999) and PIP5Kγ activities, and the latter additionally promotes AP-2 and clathrin recruitment to synaptic membranes (Krauss et al. 2003). Another PPIns kinase, PI3KC2α, which generates $PI(3,4)P_2$, associates with CCPs and promotes the maturation of late-stage CCPs before their fission. The pool of $PI(3,4)P_2$ that is generated by PI3KC2α can recruit the BAR domain protein SNX9 to late-stage endocytic intermediates (Posor et al. 2013).

It is important to note that while PPIns and membrane-shaping proteins have been extensively studied in CME, their involvement in other forms of endocytosis, such as clathrin-independent endocytosis (Brown et al. 2001) or the internalization of caveolae (Larsson et al. 2023), as well as in pinocytosis (Schink et al. 2021) and phagocytosis (Levin et al. 2015), have all been demonstrated. Although most of the specific molecular players are different in these various forms of membrane internalization processes, the general principles are similar (Saheki and De Camilli 2012). A detailed discussion of these, however, is beyond the scope of this chapter.

Exocytosis

Exocytosis comes in different forms, depending on whether it is constitutive or regulated. In regulated exocytosis, the process may significantly differ based on the size of the vesicles to be released. For example, synaptic vesicles (SVs) are relatively small at about ~ 40 nm, while dense core neuroendocrine secretory vesicles (DCVs) are ~ 100–300 nm, and exocrine cell secretory vesicles can reach sizes up to 1–2 μm in diameter. It has been recognized long ago that exocytosis of DCVs requires $PI(4,5)P_2$ (Eberhard et al. 1990; Hay et al. 1995; Holz et al. 2000). In Ca^{2+}-activated exocytosis of neuroendocrine cells, the fusion of vesicles is preceded by an ATP-dependent priming step (Hay and Martin 1992), which requires unique PI-transfer proteins as well as PI4Ks and PIP5Ks (Hay et al. 1995; Wiedemann et al. 1996). Several subsequent studies showed that $PI(4,5)P_2$ was necessary for vesicle priming and was involved in controlling the size of the rapidly releasable vesicular pool (Aikawa and Martin 2005; Biton et al. 2023; Gong et al. 2005; Milosevic et al. 2005).

How the membrane fusion happens upon the increase in cytosolic Ca^{2+} remains the ultimate question that still needs to be understood. Critical proteins that regulate vesicle docking and fusion are the SNARE proteins (O'Connor et al. 1994). According to the SNARE hypothesis, the specific pairing of t-SNARES, which are anchored to the acceptor membranes, with the v-SNARES that are anchored to the vesicular membrane, provide the compartmental specificity to drive the heterotypic fusion between the discrete membranes. While SNARE proteins are unfolded when the vesicles are away from the membrane, they form a stable complex of four-helix bundles as the vesicles dock onto the target membrane. During synaptic transmission, three of the helices are provided on the PM by Syntaxin-1 and SNAP25 (N- and C-terminal helices), whereas one helix is provided by the v-SNARE, VAMP/Synaptobrevin, which is embedded in the synaptic vesicle (Rizo 2022). The energy to fuse the membranes is provided by forming the *cis*-SNARE complex during the zippering and fusion of the opposing membranes. SNARE proteins contain a characteristic 70-residue motif that repeats the seven residues that form the structural basis of the extended helical bundles. The unfolding of the SNARE complexes is mediated by an ATP-dependent step catalyzed by the NSF ATPase and its adaptor, SNAP [see (Rizo 2022; Sudhof and Rothman 2009) for details].

Although the SNARE complexes form the basic machinery needed for vesicle fusion, the fine-tuning, and synchronization of these molecular processes, especially in the case of the synchronized release of synaptic vesicles with sub-millisecond precision, requires additional proteins. These include the SM (Sec1/Munc18-like) proteins and the vesicle-associated Ca^{2+}-sensing protein, synaptotagmin-1 (Sudhof and Rothman 2009), among others. It is notable that several of the SNARE complex-associated proteins, including Syntaxin-1, CAPS, MUNC13, and synaptotagmin-1, all have PPIn recognizing motifs (Martin 2015). These protein-lipid interactions not only help direct and prime secretory vesicles to specific sites within the PM with unique lipid compositions, including localized enrichment in PI(4,5)P_2 (Aoyagi et al. 2005; Murray and Tamm 2011; Shin et al. 2018), but also lend Ca^{2+} sensitivity to the process.

While most studies focus on the role of PI(4,5)P_2 as the main PPIn that controls the exocytic process, recent studies have also pointed to defined roles for PI(4)P beyond its function as the precursor for PI(4,5)P_2 production. Specifically, in pituitary lactotrophs, prolactin secretion was inhibited by pharmacological inhibition of PI4KA without affecting the PI(4,5)P_2 levels in the PM or inhibiting the signaling downstream of voltage-gated Ca^{2+} channels (Kucka et al. 2021; Stojilkovic and Balla 2023). More studies are needed to learn whether PI(4)P and PI(4,5)P_2 are required for overlapping or distinct molecular events during endocrine secretion, as well as whether PI(4)P is also involved in other types of exocytic events.

Constitutive exocytosis requires a different set of proteins for vesicle docking and fusion. The multimeric protein complex named exocyst was identified in yeast as the major regulator of the actin-dependent polarized exocytosis of post-Golgi vesicles at the bud site (Novick et al. 1981; TerBush et al. 1996). Homologs of the exocyst components are also found in mammalian cells [reviewed in (Heider and Munson 2012). These multimeric complexes demarcate the sites of exocytosis, physically tether vesicles to the PM, and facilitate the assembly of the SNARE machinery that executes the final membrane fusion step. Many subunits of the exocyst complex interact with additional proteins, such as SNAREs and GTPases, which show binding to lipids. In particular, the N-terminal PH domain of Sec3 and a basic amino acid patch located in the Exo70 subunit were both shown to specifically bind to PI(4,5)P_2 (Baek et al. 2010; Liu et al. 2007) within the PM, and these same exocyst components are critical for identifying the site of vesicle docking and exocytosis (Boyd et al. 2004). A recent yeast study also showed the importance of the acyl-side chain composition of PPIns in the functional assembly of the exocyst, including specifically highlighting the importance of stearoyl side chains being present at the sn-1 position of PPIns for the regulation of the essential small GTP-binding protein, cdc42 (Laquel et al. 2022).

Phosphoinositides Shaping the Endomembrane Compartments

Membrane fission and fusion are equally important components of the pathways that govern the dynamics of vesicular trafficking between the various endomembrane compartments. The biological principles, and often even the molecular players, are very similar, or sometimes identical, to those introduced for the endocytic or exocytic events controlled at the PM. Therefore, these processes will only be discussed briefly below.

PPIns and Golgi Transport

The regulated exit of transport vesicles from the Golgi complex requires membrane remodeling processes similar to those seen at the PM. However, depending on the destination of these post-Golgi vesicles, the molecules that guide these events may differ. Throughout these processes, PI(4)P stands out as the primary PPIn species that regulates post-Golgi vesicular trafficking [see D'Angelo et al. (2012) and Waugh (2019) for detailed reviews], but most studies suggest that the role of PI(4)P is indirect and, as discussed above (see Section: Phosphoinositides and Subcellular Lipid Distribution), actually relates to changes in non-vesicular lipid transport. Given the central importance of these PI(4)P-dependent lipid-transport cycles, post-Golgi trafficking defects described in response to knock-downs of the various PI4Ks are commonly attributed to changes in the abundance of the counter-transported lipids, such as Cho or sphingolipids, within the Golgi compartment, rather than as a result of changes to the resident PPIn levels. The PI/PC transfer protein, Sec14, was one of the first proteins identified as critical for post-Golgi secretion in yeast (Bankaitis et al. 1989). Indeed, several studies have shown the impact of diverse lipids, including DAG, PA, lysophospholipids, sphingolipids, and Cho, on the budding process out of the Golgi (Stalder and Gershlick 2020) (von Blume and Hausser 2019), while relatively fewer studies have outlined roles for PI(4)P, or other PPIns, separate from effects on the activity of lipid transport proteins. Direct regulation of vesicular transport by PI(4)P-binding targets is mostly associated with clathrin-mediated cargo exit from the Golgi complex. For example, PI(4)P generated by PI4K2A was shown to attract the tetrameric clathrin adaptors AP1 (Wang et al. 2003) and AP3 (Craige et al. 2008; Salazar et al. 2005) but also to recruit the monomeric GGA adaptor proteins (Szentpetery et al. 2010; Wang et al. 2007). However, as established in early studies, clathrin-mediated budding from the Golgi is not the main transport route for exocytic cargoes or proteins destined to be delivered to the PM (Griffiths et al. 1985; Payne and Schekman 1985). Instead, clathrin was required to retain specific cargoes in the Golgi complex of yeast (Payne and Schekman 1989) or the Golgi-to-endosomal trafficking in mammalian cells (Doray et al. 2002; Mardones et al. 2007). A Golgi-localized protein with direct PI(4)P recognition, as well as a great impact on vesicular trafficking, is the oncogenic GOLPH3 protein, which links membranes of the Golgi complex to the actin cytoskeleton via binding to Myo-18B (Dippold et al. 2009; Kuna and Field 2019). GOLPH3 is also able to directly bend the membranes through the insertion of a hydrophobic loop into the bilayer (Rahajeng et al. 2019).

Remarkably, no clear pathway has emerged to define the molecular components that drive the creation, budding, translocation, and fusion of the post-Golgi vesicular carriers that are destined for the PM in mammalian cells. Instead, several candidate molecular assemblies have been shown to play minor roles in these processes, depending on the cargo and its destination (Stalder and Gershlick 2020). Many, if not all, of these assemblies are associated with PI4K enzymes or PI(4)P-recognizing proteins. However, one central regulator of Golgi functions, which controls many of these pathways, is the Arf family of small GTPases (D'Souza-Schorey and Chavrier 2006; Tan and Gleeson 2019). While Arf1 itself is not directly regulated by PPIns, Arf1 recruits PI4KB to membranes and stimulates its catalytic activity at the Golgi (Godi et al. 1999). Moreover, Arf1 binding is also critical for Golgi-specific binding of the PI(4)P-recognizing PH domains present in many of the essential lipid transfer proteins (Levine and Munro 2002). Notably, the PH domain of FAPP2 is also capable of curving and tubulating lipid membranes (Cao et al. 2009). The Arf interacting proteins, Arfaptin-1 and -2, are Golgi localized (Kanoh et al. 1997) and contain BAR domains that can deform membranes (see above). Intriguingly, both Arfaptins are localized to the Golgi complex through an interaction of their amphipathic helices (AH), which precedes the BAR domains, with PI(4)P. This interaction is regulated by protein kinase D (PKD)-dependent phosphorylation in the case of Arfaptin-1 (Cruz-Garcia et al. 2013). PKD also phosphorylates other PPIn-binding Golgi-localized proteins, including PI4KB, OSBP, and CERT, and thereby functions as a major regulator of both ER-Golgi lipid transport and in carrier formation at the TGN (Olayioye and Hausser 2012). The activity of Arf proteins, as defined by the transition between their GDP and GTP-bound forms, is controlled by numerous Arf GEF and Arf-GAP proteins (Donaldson and Jackson 2011). Several of these proteins were shown to be regulated by PPIn lipids, primarily PI(4)P and PI(4,5)P_2 (Donaldson and Jackson 2011; Kahn et al. 1996). Many of the Arf GEFs and GAPs contain at least one, and up to three (in GAPs), PH domains (Roy et al. 2016). Among the Arf GEFs, the PH domains of ARNO/Cytohesin were shown to be regulated by PI(4,5)P_2 and PI(3,4,5)P_3, although this regulation primarily occurs at the PM (Cohen et al., 2007). In the case of ARF-GAPs, lipid-binding to the PH domain acts as an allosteric regulator rather than a simple membrane localization signal (Jian et al. 2015; Roy et al. 2016).

Control of Endosomal Membrane Bending by PPIns

While PI(4,5)P_2 and PI(3,4,5)P_3 are the primary PPIns that contribute to PM-associated signaling and remodeling events, and PI(4)P dominates the regulatory functions associated with the Golgi/TGN compartment, the fate of the endosomal compartments is largely controlled by the 3-phosphorylated PPIn lipids, PI(3)P and PI(3,5)P_2 (Cullen and Carlton, 2012; Posor et al. 2022; Schink et al. 2013).

Early Endosomes

Early endosomes acquire PI(3)P, which can recruit the FYVE- and PX-domain-containing proteins that work closely with the Rab5 small GTPases (Hayakawa et al. 2006; Lawe et al. 2000). Early endosomes are also decorated with the Rab5-interacting proteins APPL1 and APPL2, which possess BAR-PH-PTB domains and mark an earlier stage of endosomal maturation that precedes the appearance of PI(3)P (Miaczynska et al. 2004; Zoncu et al. 2009). Both the isolated PH and PTB domains of APPL1 were shown to bind several of the anionic PPIns species as well as PS in lipid-strip binding assays (Chial et al. 2008), but it is not clear which PPIn, if any, contributes to the recruitment of APPL1 to early endosomes. The fact that APPL1 interacts with and recruits the PPIn 5-phosphatases, OCRL, and Inpp5B (Bohdanowicz et al. 2012) also raises the possibility that PI(4)P might be the PPIn lipid that binds to the PH domain of APPL1. Whether these APPL proteins are regulated by PPIns or not, the resident BAR domains are certainly involved in determining the curvature of early endosomes.

Sorting Endosomes

As early endosomes mature, their cargoes take different routes, with some returning to the PM and others progressing toward degradation via the late endosomal/lysosomal pathway. Such cargo-routing decisions are made at sorting endosomes, although the incredible diversity of the endosomal system, both in terms of morphologies and associated Rab proteins, makes it extremely hard to identify the exact stage(s) where endosomal fate determinations occur (Solinger and Spang 2022). Additionally, during endosome maturation, cargoes are retrieved from degradation and recycled back to the PM or the trans-Golgi network (TGN). This is accomplished by segregating specific cargoes into extended endosomal subdomains called the tubular endosomal network (TEN) (van Weering and Cullen 2014) (also called tubular recycling endosomes). Among the BAR domain-containing proteins, SNXs play an important role in deforming membranes at the TEN (van Weering et al. 2010). As discussed above (Section: Proteins Contain Selective Phosphoinositide-Recognition Modules), the PX domains present in many SNX proteins bind PI(3)P, although some may bind other PPIns. In addition to their PPIn lipid interactions, a variety of SNX proteins also function to define the cargo and route selection by direct interactions with other proteins. Discussion of these complexes and the molecular mechanisms they regulate are beyond the scope of this chapter, but further details can be found in several excellent reviews (Cullen and Korswagen 2012; Solinger and Spang 2022; van Weering and Cullen 2014).

Another protein family that is associated with the tubulated membrane structures in the TEN and plays an important role in the recycling of receptors from endosomes is the EHD1/RME-1 proteins. These proteins are unique among the epsin homology (EH) domain-containing proteins in that their EH domain is located at the C-terminus. They also contain a nucleotide (ATP)-binding G-domain that is sandwiched between two helical domains and provides interactions with membrane lipids (Naslavsky and Caplan 2011). Important features of the EHD/RME-1 proteins relevant to their membrane tubulating activities are reminiscent of those found in members of the dynamin-like protein superfamily. These include binding to nucleotides, the ability to tubulate liposomes *in vitro*, forming ring-like structures by homo-oligomerization around lipid tubules, and enhanced nucleotide hydrolysis in response to lipid binding (Daumke et al. 2007). Out of the four mammalian EHD proteins, EHD1 and EHD3 are localized to the TEN, while EHD2 and EHD4 are found at the PM and Rab5-positive early endosomes, respectively, and are important for cargo recycling to the

PM (Naslavsky and Caplan 2011). EHD1 was shown to assemble on PI(4)P- and PI(4,5)P_2-containing tubules and bind to PI(4)P via its EH domain (Jovic et al. 2009). Endosomal PI(4)P, produced by PI4K2A, has, indeed, been shown to play an important role in the cargo-sorting process (Zhu et al. 2022) and contributes to the tubulation of membranes in recycling endosomes, a process that requires the biogenesis of lysosome-related organelles complex 1 (BLOC-1) (Jani et al. 2022).

Multivesicular Bodies

Some cargoes are directed for degradation from the recycling endosomal compartment. Remarkably, the membranes that contain these cargoes are segregated away from the tubular intermediates that contain the cargoes that are destined for the PM or TGN. Membrane domains that contain cargoes for degradation bud inwardly into the lumen of the endosomes, giving rise to structures called multivesicular bodies (MVBs) (Katzmann et al. 2002; Piper and Katzmann 2007). Polyubiquitylation is one of the major mechanisms by which cargoes are marked for sorting into the degradative pathways (Staub and Rotin 2006). PI(3)P generated by Class III PI3K is, again, critical for the process that drives intraluminal vesicle formation by engaging with the earliest components of the ESCRT complex (Henne et al. 2011; Hurley 2008; Hurley and Emr 2006; Williams and Urbe 2007). The assembly of the ESCRT machinery and their regulation by PPIns has already been discussed above (see Section: Phosphoinositides Regulate Signaling and Shaping of the Plasma Membrane) and will not be elaborated on further here. What has to be discussed, though, is the enigmatic role of PI(3,5)P_2 and the synthetic kinase enzyme, PIKfyve, which generates this lipid from PI(3)P along the endocytic pathway (Rivero-Rios and Weisman 2022).

The importance of PIKfyve (Fab1 in yeast) for endo-lysosomal function was hinted at by early findings showing that the levels of PI(3,5)P_2 show a dramatic increase in yeast after hyperosmotic shock (Dove et al. 1997), a finding that has also been reproduced, although to a smaller extent, in mammalian cells (Sbrissa and Shisheva 2005). Yeast Fab1 mutants show defects in vacuole function (Yamamoto et al. 1995), which causes defects in sorting proteins into the MVB (Odorizzi et al. 1998). Inactivation of Fab1 using temperature-sensitive alleles or inhibition with selective small molecules causes enlargement of the vacuole in yeast (Yamamoto et al. 1995), while similar pharmacological interventions or transient knock-downs of PIKfyve cause pronounced swelling of endosomes in mammalian cells (Cai et al. 2013; Ikonomov et al. 2001; Jefferies et al. 2008). It is also important to note that the production of another minor PPIn, PI5P, also depends on the activity of PIKfyve, although the molecular targets of PI5P have proven to be even more enigmatic than those that associate with PI(3,5)P_2. Many studies have already documented the importance of PIKfyve/Fab1 in various aspects of mammalian and yeast physiology, as well as traced their dysregulation in human diseases (McCartney et al. 2014; Rivero-Rios and Weisman 2022; Shisheva 2012). Functionally speaking, PIKfyve has been associated with early endosomal sorting events and late endo-lysosomal trafficking pathways [see details in (Rivero-Rios and Weisman 2022)]. Yet, no specific molecular targets or mechanism(s) have been pinpointed to define the mode of PI(3,5)P_2 actions in these organelles. Potential regulators that may be sensitive to $PI(3,5)P_2$ include ion channels located in endo-lysosomes (Dong et al. 2010; Leray et al. 2022; Wang et al. 2012; Wilson et al. 2018), proton pumping V-ATPases (Banerjee et al. 2019), and the WD40 repeat-containing PROPPIN/WIPI proteins (Dove et al. 2004; Liang et al. 2019). Research in this area has been hampered by the lack of tools to detect the location of the minute amounts of $PI(3,5)P_2$ present in living cells or artificially manipulate membrane $PI(3,5)P_2$ levels and monitoring biological outcomes [see (Wills et al. 2018) for detailed coverage of imaging PPIn lipids in live cells]. The great interest and potential benefit from PIKfyve inhibition in preventing viral infection (Kang et al. 2020), combating cancer (Sharma et al. 2019), or neurodegeneration (Huang et al. 2021; Hung et al. 2023) certainly justifies efforts to understand better the molecular intricacies of $PI(3,5)P_2$ and PI5P functions.

Autophagy

Autophagy is another membrane remodeling process where PPIns have been identified as major initiation factors. The early phases of autophagy involve the formation of the isolation double membrane, also known as the phagophore, which is generated as the result of an orchestrated cascade of molecular events by the characteristic ATG proteins (Galluzzi et al. 2017; Yu et al. 2018). Critical elements of this cascade include the Class III PI3K complex I (PI3KC3/VPS34, PI3KR4/p150, BECN1/Beclin 1, Atg14) that locally produces PI(3)P, and the primary PI(3)P-dependent effectors, the WIPI proteins. PI(3)P has long been recognized as a key PPIn in the initiation of autophagy through recruitment of the Atg18-Atg2 complex (Obara and Ohsumi 2008); however, recent studies also implicated PI(4)P in these processes. Atg9, the only transmembrane protein that is part of the core autophagy machinery, is transported from the TGN to the expanding phagophore by the AP-4 adaptor protein, and this step is important for the lipidation of the Atg8 protein, LC3B (Mattera et al. 2017). ATG9 is a scramblase (Guardia et al. 2020; Maeda et al. 2020; Matoba et al. 2020) that plays an important role in the acquisition of lipids from donor membranes for the growing phagophore (Zhen and Stenmark 2023). Upon starvation, ATG9A-positive vesicles are formed from the TGN and deliver the lipid kinases, PI4K2A and PI4KB, to the expanding autophagic membranes. It is also notable that the resident PI(3)P also controls the recycling of Atg9 from endolysosomes back into the TGN via the PI(3)P-binding SNX4 proteins (Ravussin et al. 2021).

A characteristic hallmark of phagophore expansion is the lipidation, which classically involves phosphatidylethanolamine (PE)-conjugation of the ATG8 family proteins, which consist of LC3A, LC3B, LC3C, GABARAP, GABARAPL1, and GABARAPL2 (Yu et al. 2018). GABARAP was also found to recruit PI4K2A to autophagosomes (Wang et al. 2015), and PI4K2A downregulation can impair autophagic flux and inhibit autophagosome-lysosome fusion (Baba et al. 2019; Wang et al. 2015). Some of the roles of the PI(4)P that are made by PI4K2A in autophagosomes and Rab7-positive compartments involve localized conversion to $PI(4,5)P_2$. It was shown that $PI(4,5)P_2$ produced by PIP5KIγ can activate Rab7 GAP proteins of the TBC1D family and facilitate the liberation of PLEKHM1 (Baba et al. 2019), a protein important for tethering autophagosomes and lysosomes, to stabilize homotypic fusion by the vacuole protein sorting (HOPS) complex (see below)

(McEwan et al. 2015). Accordingly, genetic inactivation or depletion of PI4K2A or PIP5KIγ causes tubulation of Rab7-positive endosomes and inhibits lysosome-autophagosome fusion (Baba et al. 2019). Depletion of PI(4)P 5-kinase γ isoform 5 (PIP5K1Ci5) was also reported to impair the lysosomal degradation of E-cadherin (Schill et al. 2014) and the EGF receptor (Sun et al. 2013). An alternative source of PI(4,5)P_2, which involves the phosphorylation of PI5P by PI5P 4-kinases, was reported to be important for controlling autophagy flux (Lundquist et al. 2018). A more detailed discussion of the multiple roles of diverse PPIns during autophagy can be found in comprehensive recent reviews that specifically focus on this topic (Baba and Balla 2020; Palamiuc et al. 2020).

Late Endosomes and Lysosomes

The cargoes sorted into the intraluminal vesicles within MVBs are either released to the extracellular space as exosomes after the fusion of MVBs with the PM or degraded after the MVB fuses with lysosomes. The factors that are important for deciding which of these two fates the MVBs should enter are of great interest. During endosomal maturation, a Rab5-to-Rab7 switch effectively populates the surface of MVBs with Rab7 GTP-binding proteins (Nordmann et al. 2010; Poteryaev et al. 2010). PI(3)P is required to recruit and activate the Rab7 GEFs (Mon1-Czz1), which leads to the engagement of the HOPS complex (Cabrera et al., 2014) that is required for endosome-lysosome tethering. The actual fusion step is then mediated by the t-SNARE complex that includes the R-SNARE protein, VAMP7 (Luzio et al. 2010). PI(3)P and PI(4,5)P_2 were both shown to be enriched at the vertices of docked vacuoles in yeast (Fratti et al. 2004), and these lipids were also found to regulate vacuole fusion, although in the case of PI(4,5)P_2, the mechanism responsible for these functions remains unknown (Mayer et al. 2000). The t-SNARE Vam7 actually contains a PX domain and its interaction with the vacuole is mediated by PI(3)P (Cheever et al. 2001; Stroupe et al. 2006). The purified HOPS complex also showed binding to PPIns, although without any clear specificity (Stroupe et al. 2006).A critical process in the maintenance of lysosomal function is the retrieval of lysosomes from autophagolysosomes, a process called autophagic lysosome reformation (ALR; (Yu et al. 2010). PI(4)P and PI(4,5)P_2 have been implicated in regulating ALR. Specifically, PIP5KIα and PIP5KIβ, which are different from the PIP5KIγ isoform discussed above, were found to act on ALR initiation as well as promote the budding of the proto-lysosome membranes by recruitment of vesicle coat proteins (Rong et al. 2012). PI(4)P made by the Golgi-resident PI4KB plays important roles in lysosomal vesiculation and excision (Sridhar et al. 2013). The importance of these PPIn lipids in autophagosome formation and ALR is underscored by finding that the PI(4,5)P_2-targeting 5-phosphatases, OCRL (De Leo et al. 2016), Synaptojanin-1 (George et al. 2016), and Inpp5K (Eramo et al. 2021), all play important roles in these processes. The direct demonstration of PI(4,5)P_2 production within lysosomes or late endosomes using established PI(4,5)P_2 reporters has not been straightforward due to the minute amounts of this lipid in the endolysosomes relative to the PM. However, the presence of PI(4,5)P_2 in the Rab7 compartment has been recently demonstrated using the PLCδ_1-PH domain (Baba et al. 2019). Importantly, the exact molecular

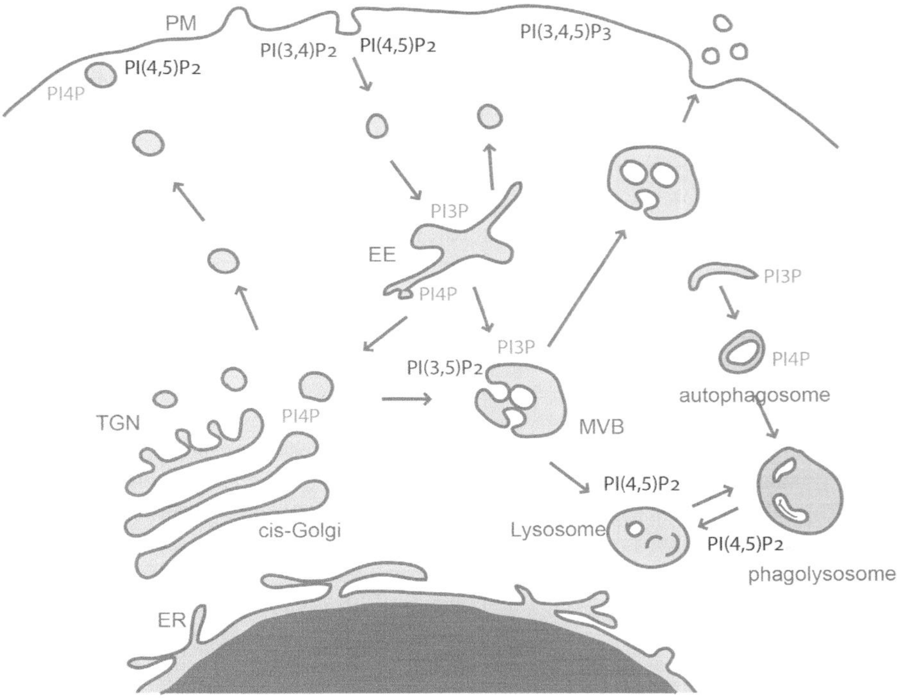

Figure 4. A simplified landscape of the subcellular PPIn distribution in a generic eukaryotic cell. Although the various PPIns are placed next to the organelle membranes where they have their major roles, no attempts were made to color the membranes according to their PPIn contents, as often more than one kind of PPIn species is found within specific membrane subdomains—even within the same organelle.

targets that respond to the local PI(4,5)P_2 or PI(4)P lipids that are produced during ALR are still awaiting identification.

Concluding Remarks

It is nearly impossible to find a subcellular function that is not controlled by PPIn lipids. Evidently, diverse combinations of PPIn kinases and phosphatases have been described as essential regulators for the integrated control of various membrane-associated processes. The involvement of these enzymes in human diseases has also been well documented, and the number of reports on this topic continues to increase rapidly with each passing day. Searching PubMed with the keywords "phosphoinositide and disease" retrieves only about 100 entries in the year 2000; however, as of 2020, this number has increased by over six-fold. The trend is even more striking when using the keyword "phosphoinositide", which yielded 100 entries in 1980 and increased to over 20,000 entries by 2020. The number of proteins that recognize and respond to PPIn lipids has also seen a similar expansion, often with stereospecific structural descriptions of their lipid-binding modes.

It has been hotly debated whether membrane domains with specific lipid-, and perhaps PPIn-composition pre-exist and determine what protein molecules are attracted to these specific membrane domains. Fatty acyl-side chain length and unsaturation of phospholipids as well as enrichment in sphingolipids and Cho, have been defining features of the lipid packing state of such membrane domains (Ouweneel et al. 2020). However, the existence of PPIn-enriched membrane domains has not been convincingly demonstrated with direct measurement methods (Ji et al. 2015; Pacheco et al. 2023; van Rheenen et al. 2005), even though it is well documented that PPIn lipids are uniquely enriched in the 1-stearoyl, 2-arachidonoyl side chains (Barneda et al. 2019). Having said that, membrane domains with specific lipid and protein compositions do exist, and their formation is mutually instructed by the lipid species present, including PPIns, and the relative binding properties of peripheral proteins.

While the subcellular distribution of PPIn lipids does show characteristic compartmentalization at the cellular scale, which often corresponds to organelle identity (Figure 4), it is increasingly apparent that PPIn lipids can control parallel signaling cascades even within the same organelle. The intricate compartmentalization of unique signaling nodes, often within spatially defined membrane subdomains with specific PPIn involvements, is becoming a recurring theme across diverse organelles. In fact, it seems likely that the spatial control of PPIn metabolism, in general, must ultimately contribute to the control of membrane sorting and thereby impact the selective enrichment of membrane-associated proteins within distinct destinations. The nascent membrane domains that will bud and eventually be pinched off from some compartments require the assembly of unique membrane-bending proteins. At the same time, local changes in the membrane lipid composition similarly facilitate the formation of the membrane geometries required for the remodeling processes. The local enrichment of unique lipid classes likely requires specialized lipid transport mechanisms as well as acute changes to the asymmetry between the inner and outer leaflets of the lipid bilayer. This explains the importance of the lipid flippase, floppase, and scramblase proteins that have now been implicated in almost every membrane remodeling process (Best et al. 2019). Several of these multi-transmembrane-panning proteins are regulated by interactions with PPIns (Hammond and Burke 2020), but the trans-bilayer movement of PPIns by these proteins has yet to be reported. PPIns not only influence the recruitment of proteins with membrane-deforming abilities, but they also dynamically regulate non-vesicular lipid transfer as well as influence the trafficking and activities of some of the flippase proteins (Dieudonne et al. 2022; Natarajan et al. 2009).

Identifying these different local PPIn pools, let alone imaging them, remains a major challenge in the field, as our current imaging tools would likely dissipate local PPIn enrichments or have no access to such localized pools if those are only formed transiently or are dedicated to proteins that interact with them tightly. Only the use of specific effectors responding to specific PPIns lipids, expressed at endogenous levels and being modified such that allows detection of their molecular states (i.e., bound or unbound to PPIns) would it be possible to demonstrate the existence of dedicated PPIn pools associated with specific signaling processes.

While this chapter lists several processes that are controlled by PPIn lipids and their targets, it cannot extensively cover all of these, as each deserves a review of its own. However, we hope that this Chapter will serve as a useful reference that helps to demonstrate that while the molecular machinery may differ, the regulatory principles that are associated with the PPIn lipids are much conserved across these otherwise diverse membrane remodeling events that occur throughout the cell. In understanding these complex molecular events at the membrane interface, the study of PPIn lipids has led to identifying important new targets for drug development that should provide new strategies to combat numerous human diseases.

Acknowledgment

This research was funded by the Intramural Research Program of the *Eunice Kennedy Shriver*, National Institute of Child Health and Human Development.

References

Abe, N., T. Inoue, T. Galvez et al. 2008. Dissecting the role of PtdIns(4,5)P2 in endocytosis and recycling of the transferrin receptor. J. Cell Sci. 121: 1488–1494.

Ago, T., R. Takeya, H. Hiroaki et al. 2001. The PX domain as a novel phosphoinositide-binding module. Biochem. Biophys. Res. Commun. 287: 733–738.

Aikawa, Y. and T.F. Martin. 2005. ADP-ribosylation factor 6 regulation of phosphatidylinositol-4,5-bisphosphate synthesis, endocytosis, and exocytosis. Methods Enzymol. 404: 422–431.

Alam, S.L., C. Langelier, F.G. Whitby et al. 2006. Structural basis for ubiquitin recognition by the human ESCRT-II EAP45 GLUE domain. Nat. Struct. Mol. Biol. 13: 1029–1030.

Antonny, B., J. Bigay and B. Mesmin. 2018. The Oxysterol-Binding Protein Cycle: Burning Off PI(4)P to Transport Cholesterol. Annu. Rev. Biochem. 87: 809–837.

Aoyagi, K., T. Sugaya, M. Umeda et al. 2005. The activation of exocytotic sites by the formation of phosphatidylinositol 4,5-bisphosphate microdomains at syntaxin clusters. J. Biol. Chem. 280: 17346–17352.

Baba, T. and T. Balla. 2020. Emerging roles of phosphatidylinositol 4-phosphate and phosphatidylinositol 4,5-bisphosphate as regulators of multiple steps in autophagy. J. Biochem. 168: 329–336.

Baba, T., D.J. Toth, N. Sengupta et al. 2019. Phosphatidylinositol 4,5-bisphosphate controls Rab7 and PLEKHM1 membrane cycling during autophagosome-lysosome fusion. EMBO J. 38: e100312.

Baek, K., A. Knodler, S.H. Lee et al. 2010. Structure-function study of the N-terminal domain of exocyst. subunit Sec3. J. Biol. Chem. 285: 10424–10433.

Balla, T. 2005. Inositol-lipid binding motifs: signal integrators through protein-lipid and protein-protein interactions. J. Cell Sci. 118: 2093–2104.

Balla, T. 2013. Phosphoinositides: tiny lipids with giant impact on cell regulation. Physiol Rev. 93: 1019–1137.

Banerjee, S., K. Clapp, M. Tarsio et al. 2019. Interaction of the late endo-lysosomal lipid PI(3,5)P2 with the Vph1 isoform of yeast V-ATPase increases its activity and cellular stress tolerance. J. Biol. Chem. 294: 9161–9171.

Banerji, S., M. Ngo, C.F. Lane et al. 2010. Oxysterol Binding Protein-dependent Activation of Sphingomyelin Synthesis in the Golgi Apparatus Requires Phosphatidylinositol 4-Kinase II{alpha}. Mol. Biol. Cell. 21: 4141–4150.

Bankaitis, V.A., D.E. Malehorn, S.D. Emr et al. 1989. The Saccharomyces cerevisiae SEC 14 gene encodes a cytosolic factor that is required for transport of secretory proteins from the yeast Golgi complex. J. Cell Biol. 108: 1271–1281.

Barneda, D., S. Cosulich, L. Stephens et al. 2019. How is the acyl chain composition of phosphoinositides created and does it matter? Biochem. Soc. Trans. 47: 1291–1305.

Beck, K.A. and J.H. Keen. 1991. Interaction of phosphoinositide cycle intermediates with the plasma membrane-associated clathrin assembly protein AP-2. J. Biol. Chem. 266: 4442–4447.

Berridge, M.J. 1984. Inositol trisphosphate and diacylglycerol as intracellular messengers. Biochem. J. 220: 345–360.

Best, J.T., F. Xu and T.R. Graham. 2019. Phospholipid flippases in membrane remodeling and transport carrier biogenesis. Curr. Opin. Cell Biol. 59: 8–15.

Billcliff, P.G. and M. Lowe. 2014. Inositol lipid phosphatases in membrane trafficking and human disease. Biochem. J. 461: 159–175.

Billcliff, P.G., C.J. Noakes, Z.B. Mehta et al. 2016. OCRL1 engages with the F-BAR protein pacsin 2 to promote biogenesis of membrane-trafficking intermediates. Mol. Biol. Cell. 27: 90–107.

Biton, T., N. Scher, S. Carmon et al. 2023. Fusion pore dynamics of large secretory vesicles define a distinct mechanism of exocytosis. J. Cell. Biol. 222.

Bohdanowicz, M., D.M. Balkin, P. De Camilli et al. 2012. Recruitment of OCRL and Inpp5B to phagosomes by Rab5 and APPL1 depletes phosphoinositides and attenuates Akt signaling. Mol. Biol. Cell. 23: 176–187.

Boyd, C., T. Hughes, M. Pypaert et al. 2004. Vesicles carry most exocyst subunits to exocytic sites marked by the remaining two subunits. Sec3p and Exo70p. J. Cell Biol. 167: 889–901.

Brown, F.D., A.L. Rozelle, H.L. Yin et al. 2001. Phosphatidylinositol 4,5-bisphosphate and Arf6-regulated membrane traffic. J. Cell Biol. 154: 1007–1017.

Burd, C.G. and S.D. Emr. 1998. Phosphatidylinositol(3)-phosphate signaling mediated by specific binding to RING FYVE domains. Mol. Cell. 2: 157–162.

Burke, J.E., J. Triscott, B.M. Emerling et al. 2022. Beyond PI3Ks: targeting phosphoinositide kinases in disease. Nat. Rev. Drug. Discov. 1–30.

Cabrera, M., S. Engelbrecht-Vandre and C. Ungermann. 2014. Function of the Mon1-Ccz1 complex on endosomes. Small GTPases. 5: 1–3.

Cai, X., Y. Xu, A.K. Cheung et al. 2013. PIKfyve, a class III PI kinase, is the target of the small molecular IL-12/IL-23 inhibitor apilimod and a player in Toll-like receptor signaling. Chem. Biol. 20: 912–921.

Campa, C.C., E. Ciraolo, A. Ghigo et al. 2015. Crossroads of PI3K and Rac pathways. Small GTPases. 6: 71–80.

Campa, F., H.Y. Yoon, V.L. Ha et al. 2009. A PH domain in the Arf GTPase-activating protein (GAP) ARAP1 binds phosphatidylinositol 3,4,5-trisphosphate and regulates Arf GAP activity independently of recruitment to the plasma membranes. J. Biol. Chem. 284: 28069–28083.

Cantley, L.C., K.R. Auger, C.L. Carpenter et al. 1991. Oncogenes and signal transduction. Cell. 64: 281–302.

Cao, X., U. Coskun, M. Rossle et al. 2009. Golgi protein FAPP2 tubulates membranes. Proc. Natl. Acad. Sci. U S. A. 106: 21121–21125.

Carlton, J.G. and J. Martin-Serrano. 2007. Parallels between cytokinesis and retroviral budding: a role for the ESCRT machinery. Science 316: 1908–1912.

Chandra, M., Y.K. Chin, C. Mas et al. 2019. Classification of the human phox homology (PX) domains based on their phosphoinositide binding specificities. Nat. Commun. 10: 1528.

Chang-Ileto, B., S.G. Frere, R.B. Chan et al. 2011. Synaptojanin 1-mediated PI(4,5)P2 hydrolysis is modulated by membrane curvature and facilitates membrane fission. Dev. Cell. 20: 206–218.

Chaudhary, A., J. Chen, Q.M. Gu et al. 1998. Probing the phosphoinositide 4,5-bisphosphate binding site of human profilin. Chem. Biol. 5: 273–281.

Cheever, M.L., T.K. Sato, T. de Beer et al. 2001. Phox domain interaction with PtdIns(3)P targets the Vam7 t-SNARE to vacuole membranes. Nat. Cell Biol. 3: 613–618.

Chial, H.J., R. Wu, C.V. Ustach et al. 2008. Membrane targeting by APPL1 and APPL2: dynamic scaffolds that oligomerize and bind phosphoinositides. Traffic. 9: 215–229.

Christy, A.H., A.C. Kim, S.M. Marfatia et al. 1998. The FERM domain: a unique module involved in the linkage of cytoplasmic proteins to the membrane. Trends Biochem. Sci. 23: 281–282.

Chung, J., F. Torta, K. Masai et al. 2015. Intracellular Transport. PI4P/phosphatidylserine countertransport at ORP5- and ORP8-mediated ER-plasma membrane contacts. Science 349: 428–432.
Cohen, L.A., A. Honda, P. Varnai et al. 2007. Active Arf6 recruits ARNO/Cytohesin GEFs to the PM by binding their PH domains. Mol. Biol. Cell. In press.
Collins, B.M., A.J. McCoy, H.M. Kent et al. 2002. Molecular architecture and functional model of the endocytic AP2 complex. Cell. 109: 523–535.
Colombo, M., C. Moita, G. van Niel et al. 2013. Analysis of ESCRT functions in exosome biogenesis, composition and secretion highlights the heterogeneity of extracellular vesicles. J. Cell Sci. 126: 5553–5565.
Corbalan-Garcia, S. and J.C. Gomez-Fernandez. 2014. Signaling through C2 domains: more than one lipid target. Biochim. Biophys. Acta. 1838: 1536–1547.
Craige, B., G. Salazar and V. Faundez. 2008. Phosphatidylinositol-4-Kinase Type II Alpha Contains an AP-3 Sorting Motif and a Kinase Domain that are both Required for Endosome Traffic. Mol. Biol. Cell.
Cremona, O., G. Di Paolo, M.R. Wenk et al. 1999. Essential role of phosphoinositide metabolism in synaptic vesicle recycling. Cell. 99: 179–188.
Cruz-Garcia, D., M. Ortega-Bellido, M. Scarpa et al. 2013. Recruitment of arfaptins to the trans-Golgi network by PI(4)P and their involvement in cargo export. EMBO J. 32: 1717–1729.
Cullen, P.J. 2008. Endosomal sorting and signalling: an emerging role for sorting nexins. Nat. Rev. Mol. Cell Biol. 9: 574–582.
Cullen, P.J. and J.G. Carlton. 2012. Phosphoinositides in the Mammalian Endo-lysosomal Network. Subcell Biochem. 59: 65–110.
Cullen, P.J. and H.C. Korswagen. 2012. Sorting nexins provide diversity for retromer-dependent trafficking events. Nat. Cell Biol. 14: 29–37.
D'Angelo, G., E. Polishchuk, G. Di Tullio et al. 2007. Glycosphingolipid synthesis requires FAPP2 transfer of glucosylceramide. Nature 449: 62–67.
D'Angelo, G., M. Vicinanza, C. Wilson et al. 2012. Phosphoinositides in Golgi complex function. Subcell Biochem. 59: 255–270.
D'Souza-Schorey, C. and P. Chavrier. 2006. ARF proteins: roles in membrane traffic and beyond. Nat. Rev. Mol. Cell Biol. 7: 347–358.
Daumke, O., R. Lundmark, Y. Vallis et al. 2007. Architectural and mechanistic insights into an EHD ATPase involved in membrane remodelling. Nature 449: 923–927.
David, C., M. Solimena and P. De Camilli. 1994. Autoimmunity in stiff-Man syndrome with breast cancer is targeted to the C-terminal region of human amphiphysin, a protein similar to the yeast proteins, Rvs167 and Rvs161. FEBS Lett. 351: 73–79.
De Leo, M.G., L. Staiano, M. Vicinanza et al. 2016. Autophagosome-lysosome fusion triggers a lysosomal response mediated by TLR9 and controlled by OCRL. Nat. Cell Biol. 18: 839–850.
de Saint-Jean, M., V. Delfosse, D. Douguet et al. 2011. Osh4p exchanges sterols for phosphatidylinositol 4-phosphate between lipid bilayers. J. Cell Biol. 195: 965–978.
Dickson, E.J. and B. Hille. 2019. Understanding phosphoinositides: rare, dynamic, and essential membrane phospholipids. Biochem. J. 476: 1–23.
Dieudonne, T., S.A. Herrera, M.J. Laursen et al. 2022. Autoinhibition and regulation by phosphoinositides of ATP8B1, a human lipid flippase associated with intrahepatic cholestatic disorders. Elife. 11.
Dippold, H.C., M.M. Ng, S.E. Farber-Katz et al. 2009. GOLPH3 bridges phosphatidylinositol-4-phosphate and actomyosin to stretch and shape the Golgi to promote budding. Cell. 139: 337–351.
Donaldson, J.G. and C.L. Jackson. 2011. ARF family G proteins and their regulators: roles in membrane transport, development and disease. Nat. Rev. Mol. Cell Biol. 12: 362–375.
Dong, R., Y. Saheki, S. Swarup et al. 2016. Endosome-ER Contacts Control Actin Nucleation and Retromer Function through VAP-Dependent Regulation of PI4P. Cell. 166: 408–423.
Dong, X.P., D. Shen, X. Wang et al. 2010. PI(3,5)P(2) controls membrane trafficking by direct activation of mucolipin Ca(2+) release channels in the endolysosome. Nat Commun. 1: 38.
Doray, B., K. Bruns, P. Ghosh et al. 2002. Interaction of the cation-dependent mannose 6-phosphate receptor with GGA proteins. J. Biol. Chem. 277: 18477–18482.

Dove, S.K., F.T. Cooke, M.R. Douglas et al. 1997. Osmotic stress activates phosphatidylinositol-3,5-bisphosphate synthesis. Nature 390: 187–192.

Dove, S.K., R.C. Piper, R.K. McEwen et al. 2004. Svp1p defines a family of phosphatidylinositol 3,5-bisphosphate effectors. Embo. J. 23: 1922–1933.

Dumont, V. and S. Lehtonen. 2022. PACSIN proteins *in vivo*: Roles in development and physiology. Acta Physiol (Oxf). 234: e13783.

Eberhard, D.A., C.L. Cooper, M.G. Low et al. 1990. Evidence that the inositol phospholipids are necessary for exocytosis. Biochem. J. 268: 15–25.

Edeling, M.A., S. Sanker, T. Shima et al. 2009. Structural requirements for PACSIN/Syndapin operation during zebrafish embryonic notochord development. PloS one. 4: e8150.

El Alaoui, F., I. Casuso, D. Sanchez-Fuentes et al. 2022. Structural organization and dynamics of FCHo2 docking on membranes. Elife. 11.

Elia, N., G. Fabrikant, M.M. Kozlov et al. 2012. Computational model of cytokinetic abscission driven by ESCRT-III polymerization and remodeling. Biophys. J. 102: 2309–2320.

Ellson, C.D., S. Gobert-Gosse, K.E. Anderson et al. 2001. PtdIns(3)P regulates the neutrophil oxidase complex by binding to the PX domain of p40phox. Nat. Cell Biol. 3: 679–682.

Eramo, M.J., R. Gurung, C.A. Mitchell et al. 2021. Bidirectional interconversion between PtdIns4P and PtdIns(4,5)P(2) is required for autophagic lysosome reformation and protection from skeletal muscle disease. Autophagy. 17: 1287–1289.

Erdmann, K.S., Y. Mao, H.J. McCrea et al. 2007. A role of the Lowe syndrome protein OCRL in early steps of the endocytic pathway. Dev. Cell. 13: 377–390.

Farsad, K., N. Ringstad, K. Takei et al. 2001. Generation of high curvature membranes mediated by direct endophilin bilayer interactions. J. Cell Biol. 155: 193–200.

Ferguson, K.M., M.A. Lemmon, J. Schlessinger et al. 1995. Structure of the high affinity complex of inositol trisphosphate with a phospholipase C plekstrin homology domain. Cell. 83: 1037–1046.

Ferguson, S.M., A. Raimondi, S. Paradise et al. 2009. Coordinated actions of actin and BAR proteins upstream of dynamin at endocytic clathrin-coated pits. Dev. Cell. 17: 811–822.

Ford, M.G., B.M. Pearse, M.K. Higgins et al. 2001. Simultaneous binding of PtdIns(4,5)P2 and clathrin by AP180 in the nucleation of clathrin lattices on membranes. Science 291: 1051–1055.

Fratti, R.A., Y. Jun, A.J. Merz et al. 2004. Interdependent assembly of specific regulatory lipids and membrane fusion proteins into the vertex ring domain of docked vacuoles. J. Cell Biol. 167: 1087–1098.

Galluzzi, L., E.H. Baehrecke, A. Ballabio et al. 2017. Molecular definitions of autophagy and related processes. EMBO J. 36: 1811–1836.

Garrus, J.E., U.K. von Schwedler, O.W. Pornillos et al. 2001. Tsg101 and the vacuolar protein sorting pathway are essential for HIV-1 budding. Cell. 107: 55–65.

Gaullier, J.M., A. Simonsen, A. D'Arrigo et al. 1998. FYVE fingers bind PtdIns (3) P. Nature. 394: 432–433.

George, A.A., S. Hayden, G.R. Stanton et al. 2016. Arf6 and the 5'phosphatase of synaptojanin 1 regulate autophagy in cone photoreceptors. Bioessays. 38 Suppl. 1: S119–135.

Ghanbarpour, A., D.P. Valverde, T.J. Melia et al. 2021. A model for a partnership of lipid transfer proteins and scramblases in membrane expansion and organelle biogenesis. Proc. Natl. Acad. Sci. U.S.A. 118.

Giordano, F., Y. Saheki, O. Idevall-Hagren et al. 2013. PI(4,5)P(2)-dependent and Ca(2+)-regulated ER-PM interactions mediated by the extended synaptotagmins. Cell. 153: 1494–1509.

Godi, A., A. Di Campi, A. Konstantakopoulos et al. 2004. FAPPs control Golgi-to-cell-surface membrane traffic by binding to ARF and PtdIns(4) P. Nat. Cell Biol. 6: 393–404.

Godi, A., P. Pertile, R. Meyers et al. 1999. ARF mediates recruitment of PtdIns-4-OH kinase-beta and stimulates synthesis of PtdIns(4,5)P2 on the Golgi complex. Nat. Cell Biol. 1: 280–287.

Goldschmidt-Clermont, P.J., L.M. Machesky, J.J. Baldassare et al. 1990. The actin binding protein profilin binds to PIP2 and inhibits its hydrolysis by phospholipase C. Science 247: 1575–1578.

Gong, L.W., G. Di Paolo, E. Diaz et al. 2005. Phosphatidylinositol phosphate kinase type I gamma regulates dynamics of large dense-core vesicle fusion. Proc. Natl. Acad. Sci. U.S.A. 102: 5204–5209.

Gout, I., R. Dhand, I.D. Hiles et al. 1993. The GTPase dynamin binds to and is activated by a subset of SH3 domains. Cell. 75: 25–36.

Griffiths, G., S. Pfeiffer, K. Simons et al. 1985. Exit of newly synthesized membrane proteins from the trans cisterna of the Golgi complex to the plasma membrane. J. Cell Biol. 101: 949–964.
Guardia, C.M., X.F. Tan, T. Lian et al. 2020. Structure of human ATG9A, the only transmembrane protein of the core autophagy machinery. Cell Rep. 31: 107837.
Guillen, J., C. Ferrer-Orta, M. Buxaderas et al. 2013. Structural insights into the Ca2+ and PI(4,5)P2 binding modes of the C2 domains of rabphilin 3A and synaptotagmin 1. Proc. Natl. Acad. Sci. U S A. 110: 20503–20508.
Guillen-Samander, A., Y. Wu, S.S. Pineda et al. 2022. A partnership between the lipid scramblase XK and the lipid transfer protein VPS13A at the plasma membrane. Proc. Natl. Acad. Sci. U S A. 119: e2205425119.
Gulluni, F., L. Prever, H. Li et al. 2021. PI(3,4)P2-mediated cytokinetic abscission prevents early senescence and cataract formation. Science 374: eabk0410.
Guo, J., M.R. Wenk, L. Pellegrini et al. 2003. Phosphatidylinositol 4-kinase type IIalpha is responsible for the phosphatidylinositol 4-kinase activity associated with synaptic vesicles. Proc. Natl. Acad. Sci. U.S.A. 100: 3995–4000.
Hamada, K., T. Shimizu, T. Matsui et al. 2000. Structural basis of the membrane targeting and unmasking mechanisms of the radixin FERM domain. EMBO J. 19: 4449–4462.
Hammond, G.R. and T. Balla. 2015. Polyphosphoinositide binding domains: Key to inositol lipid biology. Biochim. Biophys. Acta. 1851: 746–758.
Hammond, G.R.V. and J.E. Burke. 2020. Novel roles of phosphoinositides in signaling, lipid transport, and disease. Curr. Opin. Cell Biol. 63: 57–67.
Hanada, K., K. Kumagai, S. Yasuda et al. 2003. Molecular machinery for non-vesicular trafficking of ceramide. Nature 426: 803–809.
Hanley, S.E. and K.F. Cooper. 2020. Sorting Nexins in Protein Homeostasis. Cells. 10.
Hanson, P.I. and A. Cashikar. 2012. Multivesicular body morphogenesis. Annu. Rev. Cell. Dev. Biol. 28: 337–362.
Haslam, R.J., H.B. Koide and B.A. Hemmings. 1993. Pleckstrin domain homology. Nature 363: 309–310.
Hay, J.C., P.L. Fisette, G.H. Jenkins et al. 1995. ATP-dependent inositide phosphorylation required for Ca2+-activated secretion. Nature 374: 173–177.
Hay, J.C. and T.F. Martin. 1992. Resolution of regulated secretion into sequential MgATP-dependent and calcium-dependent stages mediated by distinct cytosolic proteins. J. Cell Biol. 119: 139–151.
Hayakawa, A., D. Leonard, S. Murphy et al. 2006. The WD40 and FYVE domain containing protein 2 defines a class of early endosomes necessary for endocytosis. Proc. Natl. Acad. Sci. U.S.A. 103: 11928–11933.
Heider, M.R. and M. Munson. 2012. Exorcising the exocyst complex. Traffic. 13: 898–907.
Henne, W.M., E. Boucrot, M. Meinecke et al. 2010. FCHo proteins are nucleators of clathrin-mediated endocytosis. Science 328: 1281–1284.
Henne, W.M., N.J. Buchkovich and S.D. Emr. 2011. The ESCRT pathway. Dev Cell. 21: 77–91.
Heo, W.D., T. Inoue, W.S. Park et al. 2006. PI(3,4,5)P3 and PI(4,5)P2 lipids target proteins with polybasic clusters to the plasma membrane. Science 314: 1458–1461.
Hilgemann, D.W. and R. Ball. 1996. Regulation of cardiac Na+,Ca2+ exchange and KATP potassium channels by PIP2. Science 273: 956–959.
Hinshaw, J.E. 2000. Dynamin and its role in membrane fission. Annu. Rev. Cell Dev. Biol. 16: 483–519.
Hirano, S., N. Suzuki, T. Slagsvold et al. 2006. Structural basis of ubiquitin recognition by mammalian Eap45 GLUE domain. Nat. Struct. Mol. Biol. 13: 1031–1032.
Holz, R.W., M.D. Hlubek, S.D. Sorensen et al. 2000. A pleckstrin homology domain specific for Ptdins-4,5-P2 and fused to green fluorescent protein identifies plasma membrane Ptdins-4,5-P2 as being important in exocytosis. J. Biol. Chem. 275: 17878–17885.
Honda, A., M. Nogami, T. Yokozeki et al. 1999. Phosphatidylinositol 4-phosphate 5-kinase alpha is a downstream effector of the small G protein ARF6 in membrane ruffle formation. Cell. 99: 521–532.
Hsu, N.Y., O. Ilnytska, G. Belov et al. 2010. Viral reorganization of the secretory pathway generates distinct organelles for RNA replication. Cell. 141: 799–811.
Huang, C.L., S. Feng and D.W. Hilgemann. 1998. Direct activation of inward rectifier potassium channels by PIP2 and its stabilization by Gbetagamma. Nature 391: 803–806.

Huang, P.T., S. Einav and C.R.M. Asquith. 2021. PIKfyve: a lipid kinase target for COVID-19, cancer and neurodegenerative disorders. Nat. Rev. Drug Discov. 20: 730.

Hung, S.T., G.R. Linares, W.H. Chang et al. 2023. PIKFYVE inhibition mitigates disease in models of diverse forms of ALS. Cell. 186: 786–802 e728.

Hurley, J.H. 2008. ESCRT complexes and the biogenesis of multivesicular bodies. Curr. Opin. Cell Biol. 20: 4–11.

Hurley, J.H. and S.D. Emr. 2006. The ESCRT complexes: structure and mechanism of a membrane-trafficking network. Annu. Rev. Biophys. Biomol. Struct. 35: 277–298.

Ikonomov, O.C., D. Sbrissa and A. Shisheva. 2001. Mammalian cell morphology and endocytic membrane homeostasis require enzymatically active phosphoinositide 5-kinase PIKfyve. J. Biol. Chem. 276: 26141–26147.

Itoh, T., S. Koshiba, T. Kigawa et al. 2001. Role of the ENTH domain in phosphatidylinositol-4,5-bisphosphate binding and endocytosis. Science 291: 1047–1051.

Jani, R.A., A. Di Cicco, T. Keren-Kaplan et al. 2022. PI4P and BLOC-1 remodel endosomal membranes into tubules. J. Cell Biol. 221.

Janmey, P.A. 1994. Phosphoinositides and calcium as regulators of cellular actin assembly and disassembly. Annu. Rev. Physiol. 56: 169–191.

Janmey, P.A. and T.P. Stossel. 1987. Modulation of gelsolin function by phosphatidylinositol 4,5-bisphosphate. Nature 325: 362–364.

Jefferies, H.B., F.T. Cooke, P. Jat et al. 2008. A selective PIKfyve inhibitor blocks PtdIns(3,5)P(2) production and disrupts endomembrane transport and retroviral budding. EMBO Rep. 9: 164–170.

Ji, C., Y. Zhang, P. Xu et al. 2015. Nanoscale landscape of phosphoinositides revealed by specific Pleckstrin Homology (PH) domains using single-molecule superresolution imaging in the plasma membrane. J. Biol. Chem. 290: 26978–26993.

Jian, X., W.K. Tang, P. Zhai et al. 2015. Molecular basis for cooperative binding of anionic phospholipids to the PH domain of the Arf GAP ASAP1. Structure 23: 1977–1988.

Jovic, M., F. Kieken, N. Naslavsky et al. 2009. Eps15 homology domain 1-associated tubules contain phosphatidylinositol-4-phosphate and phosphatidylinositol-(4,5)-bisphosphate and are required for efficient recycling. Mol. Biol. Cell. 20: 2731–2743.

Kahn, R.A., T. Teru and P.A. Randazzo. 1996. Effects of acid phospholipids on ARF activities: potential roles in membrane traffic. J. Lipid Mediat. Cell Signal. 14: 209–214.

Kanai, F., H. Liu, S.J. Field et al. 2001. The PX domains of p47phox and p40phox bind to lipid products of PI(3)K. Nat. Cell Biol. 3: 675–678.

Kang, Y.L., Y.Y. Chou, P.W. Rothlauf et al. 2020. Inhibition of PIKfyve kinase prevents infection by Zaire ebolavirus and SARS-CoV-2. Proc. Natl. Acad. Sci U.S.A. 117: 20803–20813.

Kanoh, H., B.T. Williger and J.H. Exton. 1997. Arfaptin 1, a putative cytosolic target protein of ADP-ribosylation factor, is recruited to Golgi membranes. J. Biol. Chem. 272: 5421–5429.

Katzmann, D.J., G. Odorizzi and S.D. Emr. 2002. Receptor downregulation and multivesicular-body sorting. Nat. Rev. Mol. Cell Biol. 3: 893–905.

Katzmann, D.J., C.J. Stefan, M. Babst et al. 2003. Vps27 recruits ESCRT machinery to endosomes during MVB sorting. J. Cell Biol. 162: 413–423.

Kim, Y.J., M.L. Hernandez and T. Balla. 2013. Inositol lipid regulation of lipid transfer in specialized membrane domains. Trends Cell. Biol. 23: 270–278.

Klein, D.E., A. Lee, D.W. Frank et al. 1998. The pleckstrin homology domains of dynamin isoforms require oligomerization for high affinity phosphoinositide binding. J. Biol. Chem. 273: 27725–27733.

Krauss, M., M. Kinuta, M.R. Wenk et al. 2003. ARF6 stimulates clathrin/AP-2 recruitment to synaptic membranes by activating phosphatidylinositol phosphate kinase type Igamma. J. Cell Biol. 162: 113–124.

Kucka, M., A.E. Gonzalez-Iglesias, M. Tomic et al. 2021. Calcium-prolactin secretion coupling in rat pituitary lactotrophs is controlled by PI4-Kinase Alpha. Front Endocrinol (Lausanne). 12: 790441.

Kumar, N., M. Leonzino, W. Hancock-Cerutti et al. 2018. VPS13A and VPS13C are lipid transport proteins differentially localized at ER contact sites. J. Cell Biol. 217: 3625–3639.

Kuna, R.S. and S.J. Field. 2019. GOLPH3: a Golgi phosphatidylinositol(4)phosphate effector that directs vesicle trafficking and drives cancer. J. Lipid Res. 60: 269–275.

Kutateladze, T. and M. Overduin. 2001. Structural mechanism of endosome docking by the FYVE domain. Science 291: 1793–1796.
Kutateladze, T.G., D.G.S. Capelluto, C.G. Ferguson et al. 2004. Multivalent mechanism of membrane insertion by the FYVE domain. J. Biol. Chem. 279: 3050–3057.
Laquel, P., E. Testet, K. Tuphile et al. 2022. Phosphoinositides containing stearic acid are required for interaction between Rho GTPases and the exocyst to control the late steps of polarized exocytosis. Traffic. 23: 120–136.
Larsson, E., B. Moren, K.A. McMahon et al. 2023. Dynamin2 functions as an accessory protein to reduce the rate of caveola internalization. J. Cell Biol. 222.
Lawe, D.C., V. Patki, R. Heller-Harrison et al. 2000. The FYVE domain of early endosome antigen 1 is required for both phosphatidylinositol 3-phosphate and Rab5 binding. Critical role of this dual interaction for endosomal localization. J. Biol. Chem. 275: 3699–3705.
Lefebvre, C., C. Largeau, X. Michelet et al. 2016. The ESCRT-II proteins are involved in shaping the sarcoplasmic reticulum in C. elegans. J. Cell Sci. 129: 1490–1499.
Lemmon, M.A. 2004. Pleckstrin homology domains: not just for phosphoinositides. Biochem. Soc. Trans. 32: 707–711.
Lemmon, M.A., M. Falasca, K.M. Ferguson et al. 1997. Regulatory recruitment of signalling molecules to the cell membrane by plekstrin-homology domains. Trends Cell Biol. 7: 237–242.
Lemmon, M.A. and K.M. Ferguson. 2000. Signal-dependent membrane targeting by pleckstrin homology (PH) domains. Biochem. J. 350: 1–18.
Leonzino, M., K.M. Reinisch and P. De Camilli. 2021. Insights into VPS13 properties and function reveal a new mechanism of eukaryotic lipid transport. Biochim Biophys Acta Mol. Cell Biol. Lipids. 1866: 159003.
Leray, X., J.K. Hilton, K. Nwangwu et al. 2022. Tonic inhibition of the chloride/proton antiporter ClC-7 by PI(3,5)P2 is crucial for lysosomal pH maintenance. Elife. 11.
Levin, R., S. Grinstein and D. Schlam. 2015. Phosphoinositides in phagocytosis and macropinocytosis. Biochim. Biophys. Acta. 1851: 805–823.
Levine, T.P. and S. Munro. 1998. The pleckstrin-homology domain of oxysterol-binding protein recognizes a determinant specific to Golgi membranes. Curr. Biol. 8: 729–739.
Levine, T.P. and S. Munro. 2002. Targeting of Golgi-specific pleckstrin homology domains involves both PtdIns 4-kinase-dependent and -independent components. Curr. Biol. 12: 695–704.
Li, Y.E., Y. Wang, X. Du et al. 2021. TMEM41B and VMP1 are scramblases and regulate the distribution of cholesterol and phosphatidylserine. J. Cell Biol. 220.
Liang, R., J. Ren, Y. Zhang et al. 2019. Structural Conservation of the Two Phosphoinositide-Binding Sites in WIPI Proteins. J. Mol. Biol. 431: 1494–1505.
Ling, K., R.L. Doughman, A.J. Firestone et al. 2002. Type I gamma phosphatidylinositol phosphate kinase targets and regulates focal adhesions. Nature 420: 89–93.
Liu, J., X. Zuo, P. Yue et al. 2007. Phosphatidylinositol 4,5-bisphosphate mediates the targeting of the exocyst to the plasma membrane for exocytosis in mammalian cells. Mol. Biol. Cell. 18: 4483–4492.
Lundquist, M.R., M.D. Goncalves, R.M. Loughran et al. 2018. Phosphatidylinositol-5-Phosphate 4-Kinases regulate cellular lipid metabolism by facilitating autophagy. Mol. Cell. 70: 531–544 e539.
Luzio, J.P., S.R. Gray and N.A. Bright. 2010. Endosome-lysosome fusion. Biochem. Soc. Trans. 38: 1413–1416.
Maeda, S., H. Yamamoto, L.N. Kinch et al. 2020. Structure, lipid scrambling activity and role in autophagosome formation of ATG9A. Nat. Struct. Mol. Biol. 27: 1194–1201.
Majerus, P.W., T.M. Connolly, H. Deckmyn et al. 1986. The metabolism of phosphoinositide-derived messenger molecules. Science 234: 1519–1526.
Mani, T., R.F. Hennigan, L.A. Foster et al. 2011. FERM domain phosphoinositide binding targets merlin to the membrane and is essential for its growth-suppressive function. Mol. Cell Biol. 31: 1983–1996.
Mao, Y., J. Chen, J.A. Maynard et al. 2001. A novel all helix fold of the AP180 amino-terminal domain for phosphoinositide binding and clathrin assembly in synaptic vesicle endocytosis. Cell. 104: 433–440.
Mardones, G.A., P.V. Burgos, D.A. Brooks et al. 2007. The trans-Golgi network accessory protein p56 promotes long-range movement of GGA/clathrin-containing transport carriers and lysosomal enzyme sorting. Mol. Biol. Cell. 18: 3486–3501.

Martin, T.F. 2015. PI(4,5)P(2)-binding effector proteins for vesicle exocytosis. Biochim Biophys Acta. 1851: 785–793.

Matoba, K., T. Kotani, A. Tsutsumi et al. 2020. Atg9 is a lipid scramblase that mediates autophagosomal membrane expansion. Nat. Struct. Mol. Biol. 27: 1185–1193.

Mattera, R., S.Y. Park, R. De Pace et al. 2017. AP-4 mediates export of ATG9A from the trans-Golgi network to promote autophagosome formation. Proc. Natl. Acad. Sci. U.S.A. 114: E10697–E10706.

Mayer, A., D. Scheglmann, S. Dove et al. 2000. Phosphatidylinositol 4,5-bisphosphate regulates two steps of homotypic vacuole fusion. Mol. Biol. Cell. 11: 807–817.

Mayer, B.J., R. Ren, K.L. Clark et al. 1993. A putative modular domain present in diverse signaling proteins. Cell. 73: 629–630.

McCartney, A.J., Y. Zhang and L.S. Weisman. 2014. Phosphatidylinositol 3,5-bisphosphate: low abundance, high significance. Bioessays. 36: 52–64.

McCullough, J., L.A. Colf and W.I. Sundquist. 2013. Membrane fission reactions of the mammalian ESCRT pathway. Annu Rev Biochem. 82: 663–692.

McEwan, D.G., D. Popovic, A. Gubas et al. 2015. PLEKHM1 regulates autophagosome-lysosome fusion through HOPS complex and LC3/GABARAP proteins. Mol Cell. 57: 39–54.

McPherson, P.S., E.P. Garcia, V.I. Slepnev et al. 1996. A presynaptic inositol-5-phosphatase. Nature 379: 353–357.

Mesmin, B., J. Bigay, J. Moser von Filseck et al. 2013. A four-step cycle driven by PI(4)P hydrolysis directs sterol/PI(4)P exchange by the ER-Golgi tether OSBP. Cell. 155: 830–843.

Miaczynska, M., S. Christoforidis, A. Giner et al. 2004. APPL proteins link Rab5 to nuclear signal transduction via an endosomal compartment. Cell. 116: 445–456.

Miki, H., K. Miura and T. Takenawa. 1996. N-WASP, a novel actin-depolymerizing protein, regulates the cortical cytoskeletal rearrangement in a PIP2-dependent manner downstream of tyrosine kinases. EMBO J. 15: 5326–5335.

Milosevic, I., S. Giovedi, X. Lou et al. 2011. Recruitment of endophilin to clathrin-coated pit necks is required for efficient vesicle uncoating after fission. Neuron 72: 587–601.

Milosevic, I., J.B. Sorensen, T. Lang et al. 2005. Plasmalemmal phosphatidylinositol-4,5-bisphosphate level regulates the releasable vesicle pool size in chromaffin cells. J. Neurosci. 25: 2557–2565.

Misra, S. and J.H. Hurley. 1999. Crystal structure of a phosphatidylinositol 3-phosphate-specific membrane-targeting motif, the FYVE domain of Vps27p. Cell. 97: 657–666.

Moser von Filseck, J., A. Copic, V. Delfosse et al. 2015. INTRACELLULAR TRANSPORT. Phosphatidylserine transport by ORP/Osh proteins is driven by phosphatidylinositol 4-phosphate. Science 349: 432–436.

Murray, D.H. and L.K. Tamm. 2011. Molecular mechanism of cholesterol- and polyphosphoinositide-mediated syntaxin clustering. Biochemistry 50: 9014–9022.

Musacchio, A., T. Gibson, P. Rice et al. 1993. The PH domain: a common piece in the structural patchwork of signalling proteins. Trends Biochem. Sci. 18: 343–348.

Nakatsu, F., R.M. Perera, L. Lucast et al. 2010. The inositol 5-phosphatase SHIP2 regulates endocytic clathrin-coated pit dynamics. J. Cell Biol. 190: 307–315.

Naslavsky, N. and S. Caplan. 2011. EHD proteins: key conductors of endocytic transport. Trends Cell Biol. 21: 122–131.

Natarajan, P., K. Liu, D.V. Patil et al. 2009. Regulation of a Golgi flippase by phosphoinositides and an ArfGEF. Nat Cell Biol. 11: 1421–1426.

Nishizuka, Y. 1988. The molecular heterogeneity of protein kinase C and its implications for cellular regulation. Nature 34: 661–665.

Nordmann, M., M. Cabrera, A. Perz et al. 2010. The Mon1-Ccz1 complex is the GEF of the late endosomal Rab7 homolog Ypt7. Curr Biol. 20: 1654–1659.

Novick, P., S. Ferro and R. Schekman. 1981. Order of events in the yeast secretory pathway. Cell. 25: 461–469.

O'Connor, V., G.J. Augustine and H. Betz. 1994. Synaptic vesicle exocytosis: molecules and models. Cell. 76: 785–787.

Obara, K. and Y. Ohsumi. 2008. Dynamics and function of PtdIns(3)P in autophagy. Autophagy. 4: 952–954.

Odorizzi, G., M. Babst and S.D. Emr. 1998. Fab1p PtdIns(3)P 5-kinase function essential for protein sorting in the multivesicular body. Cell. 95: 847–858.

Ojala, P.J., V. Paavilainen and P. Lappalainen. 2001. Identification of yeast cofilin residues specific for actin monomer and PIP2 binding. Biochemistry 40: 15562–15569.

Olayioye, M.A. and A. Hausser. 2012. Integration of non-vesicular and vesicular transport processes at the Golgi complex by the PKD-CERT network. Biochim Biophys Acta. 1821: 1096–1103.

Ouweneel, A.B., M.J. Thomas and M.G. Sorci-Thomas. 2020. The ins and outs of lipid rafts: functions in intracellular cholesterol homeostasis, microparticles, and cell membranes: Thematic Review Series: Biology of Lipid Rafts. J. Lipid Res. 61: 676–686.

Pacheco, J., A.C. Cassidy, J.P. Zewe et al. 2023. PI(4,5)P2 diffuses freely in the plasma membrane even within high-density effector protein complexes. J. Cell Biol. 222.

Palamiuc, L., A. Ravi and B.M. Emerling. 2020. Phosphoinositides in autophagy: current roles and future insights. FEBS J. 287: 222–238.

Pang, X., J. Fan, Y. Zhang et al. 2014. A PH domain in ACAP1 possesses key features of the BAR domain in promoting membrane curvature. Dev Cell. 31: 73–86.

Payne, G.S. and R. Schekman. 1985. A test of clathrin function in protein secretion and cell growth. Science 230: 1009–1014.

Payne, G.S. and R. Schekman. 1989. Clathrin: A role in the intracellular retention of a Golgi membrane protein. Science 245: 1358–1365.

Peter, B.J., H.M. Kent, I.G. Mills et al. 2004. BAR domains as sensors of membrane curvature: the amphiphysin BAR structure. Science 303: 495–499.

Piper, R.C. and D.J. Katzmann. 2007. Biogenesis and function of multivesicular bodies. Annu. Rev. Cell Dev. Biol. 23: 519–547.

Ponting, C.P. 1996. Novel domains in NADPH oxydase subunits, sorting nexins, and PtdIns 3-kinases: binding partners of SH3 domains? Prot. Sci. 5: 2353–2357.

Posor, Y., M. Eichhorn-Gruenig, D. Puchkov et al. 2013. Spatiotemporal control of endocytosis by phosphatidylinositol-3,4-bisphosphate. Nature 499: 233–237.

Posor, Y., W. Jang and V. Haucke. 2022. Phosphoinositides as membrane organizers. Nat. Rev. Mol. Cell Biol. 23: 797–816.

Poteryaev, D., S. Datta, K. Ackema et al. 2010. Identification of the switch in early-to-late endosome transition. Cell. 141: 497–508.

Prinz, W.A., A. Toulmay and T. Balla. 2020. The functional universe of membrane contact sites. Nat. Rev. Mol. Cell Biol. 21: 7–24.

Radhakrishnan, A., A. Stein, R. Jahn et al. 2009. The Ca2+ affinity of synaptotagmin 1 is markedly increased by a specific interaction of its C2B domain with phosphatidylinositol 4,5-bisphosphate. J. Biol. Chem. 284: 25749–25760.

Radulovic, M., E.M. Wenzel, S. Gilani et al. 2022. Cholesterol transfer via endoplasmic reticulum contacts mediates lysosome damage repair. EMBO J. 41: e112677.

Rahajeng, J., R.S. Kuna, S.L. Makowski et al. 2019. Efficient Golgi Forward Trafficking Requires GOLPH3-Driven, PI4P-Dependent Membrane Curvature. Dev Cell. 50: 573–585 e575.

Ravussin, A., A. Brech, S.A. Tooze et al. 2021. The phosphatidylinositol 3-phosphate-binding protein SNX4 controls ATG9A recycling and autophagy. J. Cell Sci. 134.

Reiss, S., I. Rebhan, P. Backes et al. 2011. Recruitment and activation of a lipid kinase by hepatitis C virus NS5A is essential for integrity of the membranous replication compartment. Cell Host Microbe. 9: 32–45.

Ringstad, N., Y. Nemoto and P. De Camilli. 1997. The SH3p4/Sh3p8/SH3p13 protein family: binding partners for synaptojanin and dynamin via a Grb2-like Src homology 3 domain. Proc. Natl. Acad. Sci. U.S.A. 94: 8569–8574.

Rivero-Rios, P. and L.S. Weisman. 2022. Roles of PIKfyve in multiple cellular pathways. Curr. Opin. Cell Biol. 76: 102086.

Rizo, J. 2022. Molecular Mechanisms Underlying Neurotransmitter Release. Annu. Rev. Biophys. 51: 377–408.

Rohatgi, R., H.Y. Ho and M.W. Kirschner. 2000. Mechanism of N-WASP activation by CDC42 and phosphatidylinositol 4, 5-bisphosphate. J. Cell Biol. 150: 1299–1310.

Rong, Y., M. Liu, L. Ma et al. 2012. Clathrin and phosphatidylinositol-4,5-bisphosphate regulate autophagic lysosome reformation. Nat. Cell Biol. 14: 924–934.

Rosendale, M., T.N.N. Van, D. Grillo-Bosch et al. 2019. Functional recruitment of dynamin requires multimeric interactions for efficient endocytosis. Nat. Commun. 10: 4462.

Roy, N.S., M.E. Yohe, P.A. Randazzo et al. 2016. Allosteric properties of PH domains in Arf regulatory proteins. Cellular Logistics. 6: e1181700.

Ruderman, N.B., R. Kapeller, M.F. White et al. 1990. Activation of phosphatidylinositol 3-kinase by insulin. Proc. Natl. Acad. Sci. U.S.A. 87: 1411–1415.

Saarikangas, J., H. Zhao and P. Lappalainen. 2010. Regulation of the actin cytoskeleton-plasma membrane interplay by phosphoinositides. Physiol. Rev. 90: 259–289.

Saheki, Y. and P. De Camilli. 2012. Synaptic vesicle endocytosis. Cold Spring Harb. Perspect. Biol. 4: a005645.

Sakamuro, D., K.J. Elliott, R. Wechsler-Reya et al. 1996. BIN1 is a novel MYC-interacting protein with features of a tumour suppressor. Nat. Genet. 14: 69–77.

Salazar, G., B. Craige, B.H. Wainer et al. 2005. Phosphatidylinositol-4-kinase type II alpha is a component of adaptor protein-3-derived vesicles. Mol Biol Cell. 16: 3692–3704.

Salim, K., M.J. Bottomley, E. Querfurth et al. 1996. Distinct specificity in the recognition of phosphoinositides by the plekstrin homology domains of dynamin and Bruton's tyrosine kinase. EMBO J. 15: 6241–6250.

Sasaki, T., S. Takasuga, J. Sasaki et al. 2009. Mammalian phosphoinositide kinases and phosphatases. Prog. Lipid Res. 48: 307–343.

Sbrissa, D. and A. Shisheva. 2005. Acquisition of unprecedented phosphatidylinositol 3,5-bisphosphate rise in hyperosmotically stressed 3T3-L1 adipocytes, mediated by ArPIKfyve-PIKfyve pathway. J. Biol Chem. 280: 7883–7889.

Schill, N.J., A.C. Hedman, S. Choi et al. 2014. Isoform 5 of PIPKIgamma regulates the endosomal trafficking and degradation of E-cadherin. J. Cell Sci. 127: 2189–2203.

Schink, K.O., C. Raiborg and H. Stenmark. 2013. Phosphatidylinositol 3-phosphate, a lipid that regulates membrane dynamics, protein sorting and cell signalling. Bioessays. 35: 900–912.

Schink, K.O., K.W. Tan, H. Spangenberg et al. 2021. The phosphoinositide coincidence detector Phafin2 promotes macropinocytosis by coordinating actin organisation at forming macropinosomes. Nat. Commun. 12: 6577.

Schu, P.V., K. Takegawa, M.J. Fry et al. 1993. Phosphatidylinositol 3-kinase encoded by yeast VPS34 gene essential for protein sorting. Science 260: 88–91.

Servant, G., O.D. Weiner, P. Herzmark et al. 2000. Polarization of chemoattractant receptor signaling during neutrophil chemotaxis. Science 287: 1037–1040.

Sharma, G., C.M. Guardia, A. Roy et al. 2019. A family of PIKFYVE inhibitors with therapeutic potential against autophagy-dependent cancer cells disrupt multiple events in lysosome homeostasis. Autophagy. 15: 1694–1718.

Shin, W., L. Ge, G. Arpino et al. 2018. Visualization of Membrane Pore in Live Cells Reveals a Dynamic-Pore Theory Governing Fusion and Endocytosis. Cell. 173: 934–945 e912.

Shisheva, A. 2012. PIKfyve and its Lipid products in health and in sickness. Curr. Top. Microbiol. Immunol. 362: 127–162.

Simonsen, A., R. Lippe, S. Christoforidis et al. 1998. EEA1 links PI(3)K function to Rab5 regulation of endosome fusion. Nature 394: 494–498.

Simpson, F., N.K. Hussain, B. Qualmann et al. 1999. SH3-domain-containing proteins function at distinct steps in clathrin-coated vesicle formation. Nat. Cell Biol. 1: 119–124.

Simunovic, M., E. Evergren, A. Callan-Jones et al. 2019. Curving Cells Inside and Out: Roles of BAR Domain Proteins in Membrane Shaping and Its Cellular Implications. Annu. Rev. Cell Dev. Biol. 35: 111–129.

Sivadon, P., F. Bauer, M. Aigle et al. 1995. Actin cytoskeleton and budding pattern are altered in the yeast rvs161 mutant: the Rvs161 protein shares common domains with the brain protein amphiphysin. Mol. Gen. Genet. 246: 485–495.

Skare, P. and R. Karlsson. 2002. Evidence for two interaction regions for phosphatidylinositol(4,5)-bisphosphate on mammalian profilin I. FEBS Lett. 522: 119–124.

Slagsvold, T., R. Aasland, S. Hirano et al. 2005. Eap45 in mammalian ESCRT-II binds ubiquitin via a phosphoinositide-interacting GLUE domain. J. Biol. Chem. 280: 19600–19606.

Sohn, M., P. Ivanova, H.A. Brown et al. 2016. Lenz-Majewski mutations in PTDSS1 affect phosphatidylinositol 4-phosphate metabolism at ER-PM and ER-Golgi junctions. Proc. Natl. Acad. Sci. U.S.A. 113: 4314–4319.

Solinger, J.A. and A. Spang. 2022. Sorting of cargo in the tubular endosomal network. Bioessays. 44: e2200158.

Song, X., W. Xu, A. Zhang et al. 2001. Phox homology domains specifically bind phosphatidylinositol phosphates. Biochemistry 40: 8940–8944.

Sridhar, S., B. Patel, D. Aphkhazava et al. 2013. The lipid kinase PI4KIIIbeta preserves lysosomal identity. EMBO J. 32: 324–339.

Stahelin, R.V., A. Burian, K.S. Bruzik et al. 2003a. Membrane binding mechanisms of the PX domains of NADPH oxidase p40phox and p47phox. J. Biol. Chem. 278: 14469–14479.

Stahelin, R.V., D. Karathanassis, D. Murray et al. 2007. Structural and membrane binding analysis of the Phox homology domain of Bem1p: basis of phosphatidylinositol 4-phosphate specificity. J. Biol. Chem. 282: 25737–25747.

Stahelin, R.V., F. Long, B.J. Peter et al. 2003b. Contrasting membrane interaction mechanisms of AP180 N-terminal homology (ANTH) and epsin N-terminal homology (ENTH) domains. J. Biol. Chem. 278: 28993–28999.

Stalder, D. and D.C. Gershlick. 2020. Direct trafficking pathways from the Golgi apparatus to the plasma membrane. Semin. Cell Dev. Biol. 107: 112–125.

Staub, O. and D. Rotin. 2006. Role of ubiquitylation in cellular membrane transport. Physiol. Rev. 86: 669–707.

Stojilkovic, S.S. and T. Balla. 2023. PI(4,5)P2-dependent and -independent roles of PI4P in the control of hormone secretion by pituitary cells. Front. Endocrinol. (Lausanne). 14: 1118744.

Stroupe, C., K.M. Collins, R.A. Fratti et al. 2006. Purification of active HOPS complex reveals its affinities for phosphoinositides and the SNARE Vam7p. EMBO J. 25: 1579–1589.

Sudhof, T.C. and J.E. Rothman. 2009. Membrane fusion: grappling with SNARE and SM proteins. Science 323: 474–477.

Suh, P.G., S.H. Ryu, K.H. Moon et al. 1988. Cloning and sequence of multiple forms of phospholipase-C. Cell. 54: 161–169.

Sun, Y., A.C. Hedman, X. Tan et al. 2013. Endosomal type Igamma PIP 5-kinase controls EGF receptor lysosomal sorting. Dev. Cell. 25: 144–155.

Swaney, K.F., C.H. Huang and P.N. Devreotes. 2010. Eukaryotic chemotaxis: a network of signaling pathways controls motility, directional sensing, and polarity. Annu. Rev. Biophys. 39: 265–289.

Szentpetery, Z., P. Varnai and T. Balla. 2010. Acute manipulation of Golgi phosphoinositides to assess their importance in cellular trafficking and signaling. Proc. Natl. Acad. Sci. U.S.A. 107: 8225–8230.

Tai, A.W., Y. Benita, L.F. Peng et al. 2009. A functional genomic screen identifies cellular cofactors of hepatitis C virus replication. Cell Host. Microbe. 5: 298–307.

Takazawa, K., H. Passareiro, J.E. Dumont et al. 1989. Purification of bovine brain inositol 1,4,5-trisphosphate 3-kinase. Biochem. J. 261: 483–488.

Takei, K., V.I. Slepnev, V. Haucke et al. 1999. Functional partnership between amphiphysin and dynamin in clathrin-mediated endocytosis. Nat. Cell Biol. 1: 33–39.

Tan, J.X. and T. Finkel. 2022. A phosphoinositide signalling pathway mediates rapid lysosomal repair. Nature 609: 815–821.

Tan, J.Z.A. and P.A. Gleeson. 2019. Cargo Sorting at the trans-Golgi Network for Shunting into Specific Transport Routes: Role of Arf Small G Proteins and Adaptor Complexes. Cells. 8.

Tarricone, C., B. Xiao, N. Justin et al. 2001. The structural basis of Arfaptin-mediated cross-talk between Rac and Arf signalling pathways. Nature 411: 215–219.

Teo, H., D.J. Gill, J. Sun et al 2006. ESCRT-I core and ESCRT-II GLUE domain structures reveal role for GLUE in linking to ESCRT-I and membranes. Cell. 125: 99–111.

TerBush, D.R., T. Maurice, D. Roth et al. 1996. The Exocyst is a multiprotein complex required for exocytosis in Saccharomyces cerevisiae. EMBO J. 15: 6483–6494.

Toth, B., A. Balla, H. Ma et al. 2006. Phosphatidylinositol 4-kinase IIIbeta regulates the transport of ceramide between the endoplasmic reticulum and Golgi. J. Biol. Chem. 281: 36369–36377.

Toulmay, A. and W.A. Prinz. 2012. A conserved membrane-binding domain targets proteins to organelle contact sites. J. Cell Sci. 125: 49–58.
Trifaro, J.M., S. Gasman and L.M. Gutierrez. 2008. Cytoskeletal control of vesicle transport and exocytosis in chromaffin cells. Acta Physiol. (Oxf). 192: 165–172.
Vaillancourt, F.H., L. Pilote, M. Cartier et al. 2009. Identification of a lipid kinase as a host factor involved in hepatitis C virus RNA replication. Virology. 387: 5–10.
van Rheenen, J., E.M. Achame, H. Janssen et al. 2005. PIP2 signaling in lipid domains: a critical re-evaluation. Embo J. 24: 1664–1673.
van Weering, J.R and P.J. Cullen. 2014. Membrane-associated cargo recycling by tubule-based endosomal sorting. Semin Cell Dev Biol. 31: 40–47.
van Weering, J.R., P. Verkade and P.J. Cullen. 2010. SNX-BAR proteins in phosphoinositide-mediated, tubular-based endosomal sorting. Semin. Cell Dev. Biol. 21: 371–380.
Vance, J.E. 2015. Phospholipid synthesis and transport in mammalian cells. Traffic. 16: 1–18.
Varnai, P., B. Thyagarajan, T. Rohacs et al. 2006. Rapidly inducible changes in phosphatidylinositol 4,5-bisphosphate levels influence multiple regulatory functions of the lipid in intact living cells. J. Cell Biol. 175: 377–382.
Venkateswarlu, K., P.B. Oatey, J.M. Tavare et al. 1998. Insulin-dependent translocation of ARNO to the plasma membrane of adipocytes requires phosphatidylinositol 3-kinase. Curr. Biol. 8: 463–466.
Voglmaier, S.M., J.H. Keen, J.E. Murphy et al. 1992. Inositol hexakisphosphate receptor identified as the clathrin assembly protein AP-2. Biochem. Biophys. Res. Commun. 187: 158–163.
von Blume, J. and A. Hausser. 2019. Lipid-dependent coupling of secretory cargo sorting and trafficking at the trans-Golgi network. FEBS Lett. 593: 2412–2427.
Wang, F., P. Herzmark, O.D. Weiner et al. 2002. Lipid products of PI(3)Ks maintain persistent cell polarity and directed motility in neutrophils. Nat. Cell Biol. 4: 513–518.
Wang, H., H.Q. Sun, X. Zhu et al. 2015. GABARAPs regulate PI4P-dependent autophagosome:lysosome fusion. Proc. Natl. Acad. Sci. U.S.A. 112: 7015–7020.
Wang, J., H.Q. Sun, E. Macia et al. 2007. PI4P Promotes the Recruitment of the GGA Adaptor Proteins to the Trans-Golgi Network and Regulates Their Recognition of the Ubiquitin Sorting Signal. Mol. Biol. Cell.
Wang, M., Z. Xu and Y. Kong. 2018. The tubby-like proteins kingdom in animals and plants. Gene. 642: 16–25.
Wang, X., X. Zhang, X.P. Dong et al. 2012. TPC proteins are phosphoinositide- activated sodium-selective ion channels in endosomes and lysosomes. Cell. 151: 372–383.
Wang, Y.J., J. Wang, H.Q. Sun et al. 2003. Phosphatidylinositol 4 phosphate regulates targeting of clathrin adaptor AP-1 complexes to the Golgi. Cell. 114: 299–310.
Wang, Z. and D.C. Thurmond. 2009. Mechanisms of biphasic insulin-granule exocytosis—roles of the cytoskeleton, small GTPases and SNARE proteins. J. Cell Sci. 122: 893–903.
Waugh, M.G. 2019. The Great Escape: how phosphatidylinositol 4-kinases and PI4P promote vesicle exit from the Golgi (and drive cancer). Biochem. J. 476: 2321–2346.
Weiger, M.C. and C.A. Parent. 2012. Phosphoinositides in chemotaxis. Subcell Biochem. 59: 217–254.
Wenk, M.R., L. Pellegrini, V.A. Klenchin et al. 2001. PIP kinase Igamma is the major PI(4,5)P(2) synthesizing enzyme at the synapse. Neuron. 31: 79–88.
Wiedemann, C., T. Schäfer and M.M. Burger. 1996. Chromaffin granule-associated phosphatidylinositol 4-kinase activity is required for stimulated secretion. EMBO J. 15: 2094–2101.
Williams, R.L. and S. Urbe. 2007. The emerging shape of the ESCRT machinery. Nat. Rev. Mol. Cell Biol. 8: 355–368.
Wills, R.C., B.D. Goulden and G.R.V. Hammond. 2018. Genetically encoded lipid biosensors. Mol. Biol. Cell. 29: 1526–1532.
Wilson, Z.N., A.L. Scott, R.D. Dowell et al. 2018. PI(3,5)P(2) controls vacuole potassium transport to support cellular osmoregulation. Mol. Biol. Cell. 29: 1718–1731.
Yamamoto, A., D.B. DeWald, I.V. Boronenkov et al. 1995. Novel PI(4)P 5-kinase homologue, Fab1p, essential for normal vacuole function and morphology in yeast. Mol. Biol. Cell. 6: 525–539.
Yonezawa, N., E. Nishida, K. Iida et al. 1990. Inhibition of the interactions of cofilin, destrin, and deoxyribonuclease I with actin by phosphoinositides. J. Biol. Chem. 265: 8382–8386.

Yu, L., Y. Chen and S.A. Tooze. 2018. Autophagy pathway: Cellular and molecular mechanisms. Autophagy. 14: 207–215.
Yu, L., C.K. McPhee, L. Zheng et al. 2010. Termination of autophagy and reformation of lysosomes regulated by mTOR. Nature 465: 942–946.
Zhang, L., Y.S. Mao, P.A. Janmey et al. 2012. Phosphatidylinositol 4, 5 bisphosphate and the actin cytoskeleton. Subcell Biochem. 59: 177–215.
Zhao, H., M. Hakala and P. Lappalainen. 2010. ADF/cofilin binds phosphoinositides in a multivalent manner to act as a PIP(2)-density sensor. Biophys. J. 98: 2327–2336.
Zhen, Y. and H. Stenmark. 2023. Autophagosome Biogenesis. Cells. 12.
Zhu, Y., S. Li, A. Jaume et al. 2022. Type II phosphatidylinositol 4-kinases function sequentially in cargo delivery from early endosomes to melanosomes. J. Cell Biol. 221.
Zimmerberg, J. and S. McLaughlin. 2004. Membrane curvature: how BAR domains bend bilayers. Curr. Biol. 14: R250–252.
Zoncu, R., R.M. Perera, D.M. Balkin et al. 2009. A phosphoinositide switch controls the maturation and signaling properties of APPL endosomes. Cell. 136: 1110–1121.
Zoncu, R., R.M. Perera, R. Sebastian et al. 2007. Loss of endocytic clathrin-coated pits upon acute depletion of phosphatidylinositol 4,5-bisphosphate. Proc. Natl. Acad. Sci. U.S.A. 104: 3793–3798.

12

Shaping Membranes in Cell-Cell Fusion[#]

Kamran Melikov and *Leonid V. Chernomordik**

Introduction

Protein-mediated changes in the membranes' shapes play critical roles in membrane fusion processes, as diverse as synaptic transmission, protein trafficking, viral infection, and the formation of multinucleated cells. These cellular processes dramatically vary in sizes of the fusing structures (the estimated surface *areas of placental syncytiotrophoblast* and synaptic vesicle are ~ 10 m^2 and ~ 5×10^{-15} m^2, respectively), physiological functions, and proteins involved (often referred to as fusion proteins or fusogens). Despite this diversity, many barriers fusing membrane bilayers must overcome related to the general properties of the membranes and determine some of the shared key properties of disparate fusion processes, including cell-cell fusion, which is the focus of this chapter. To establish immediate bilayer contact required for the fusion, proteins at the surfaces of the apposed membranes must move apart to bare lipid bilayers. The further approach of the bilayers should overcome strong short-range repulsion between polar surfaces in water at nanometer separations. Furthermore, the powerful hydrophobic effect stabilizes a continuous double layer of lipid molecules at the basis of all biological membranes. The continuity and integrity of the membranes are critically important for the barrier function of the membranes but must be transiently disrupted in any fusion event. Because the energy costs of bringing lipid monolayers into immediate contact and breaking monolayer continuity increase with the areas of the membrane contact and monolayer disruption, early fusion intermediates are expected to develop in strongly bent membranes to minimize these areas. Thus, local bending deformations of lipid

Section on Membrane Biology, the *Eunice Kennedy Shriver* National Institute of Child Health and Human Development, National Institutes of Health, Bethesda, MD 20892, USA.

* Corresponding author: chernoml@mail.nih.gov

[#] This chapter is dedicated to the memory of the late Professor Yuri A. Chizmadzhev, our beloved and most supportive mentor and outstanding scientist.

bilayers and lipid monolayers play an essential role in establishing pre-fusion tight contact between the protein-depleted membrane patches and in the subsequent restructuring of these membrane bilayers.

Bringing Membranes Into Pre-Fusion Contact

Cell membrane shapes and protein machines that dynamically control these shapes play an important role in establishing the properties of the pre-fusion junction of the cells. The architecture and dynamics of the fusion sites have been extensively studied in myoblast fusion in *Drosophila* embryos, and the results are relevant for the myoblast and other cell-cell fusions in mammals (Lee and Chen 2019; Kim and Chen 2019). The fusion during myotube formation in *Drosophila* embryos is asymmetric, with the fusion of competent myoblasts with the muscle founder cells. Fusion-competent myoblasts generate finger-like actin-rich protrusions extending into muscle founder cells. These protrusions have an average diameter of 250 nm and length of up to 2 μm and are often called podosome-like structures with the whole fusion site referred to as a fusion synapse (Lee and Chen 2019; Kim and Chen 2019). The formation of these protrusions depends on the formation of branched actin filaments regulated through the Arp2/3 complex and actin nucleation-promoting factors WASP and SCAR. The muscle founder cell also actively participates in the formation of the fusion synapse by forming a dense actomyosin cortex under the podosome-like structure. The main proteins involved in the process on the side of the muscle founder cell are non-muscle myosin II and spectrin. Fusion synapse formation is important for efficient fusion but is insufficient to induce fusion in the absence of active protein fusogens (Shilagardi et al. 2013). It has been suggested that the formation of actin-enriched protrusions in fusion-competent myoblast and the stiffening of the cell cortex in the founder cells bring PMs of the apposed cells into tight contact and primes the membranes for fusion.

Recently, Fletcher and coauthors explored the dependence of cell-cell fusion induced by the fusion-associated small transmembrane (FAST) proteins, syncytium-inducing protein fusogens of some non-enveloped reoviruses, on actin cytoskeleton of the infected cells (Chan et al. 2021; Chan et al. 2020). Although syncytium formation is not obligatory for the life cycle of these reoviruses, it provides some competitive advantage. FAST proteins have a single transmembrane domain, a small ectodomain, and a small cytosolic domain but drive fusion all the way from hemifusion stalk to pore opening, although they rely on cellular adhesion proteins to provide cell-cell contacts (Duncan 2019). It was discovered that cell-cell fusion induced by two evolutionarily divergent FAST proteins (p14 from reptilian orthoreovirus and p22 from aquareovirus) depends on N-WASP-mediated activation of branched actin assembly. p14 triggers branched actin assembly through interactions of its cytosolic tail with adaptor protein Grb2. In contrast, the cytosolic tail of p22 interacts with another adapter protein, Intersectin-1, to activate Cdc42 GTPase, which then activates N-WASP mediated assembly of branched actin. It was suggested that FAST protein-mediated branched actin assembly induces membrane protrusions similar to podosome-like protrusions observed in myoblasts. Since FAST proteins have small ectodomains, such actin-rich protrusion may bring two membranes in close contact—a job performed by structural rearrangements of large and complex ectodomains in case of fusion proteins of enveloped viruses and structurally similar developmental fusogens. Additionally, curvatures generated at the protrusion tip may further promote the formation of early fusion intermediates. In analogy to the work on EFF-1 mediated fusion of *Drosophila* S2R+ cells (Shilagardi et al. 2013), showing that actin-propelled protrusions are insufficient for fusion, formation of the protrusions by FAST protein lacking fusogenic ectodomain does not lead to fusion (Chan et al. 2020)). Finding that co-expression of the mutant lacking ectodomain and the mutant lacking cytosolic domain, which is unable to induce branched actin assembly, also does not result in cell-cell fusion suggests that actin-rich protrusions may play a role in concentrating fusion proteins in the zone of a close membrane contact at the tip of protrusion.

Membrane Fusion and Proteins That Mediate it

It appears that at a stage of lipid rearrangements in membrane bilayer fusion, diverse biological fusion processes converge to the conserved pathway intrinsic for protein-free lipid bilayers (reviewed in Brukman et al. 2019). Fusion starts with hemifusion, a merger of the proximal monolayers of the opposing membranes (in the case of cell-cell fusion, a merger of outer monolayers of two plasma membranes, PMs) without the merger of distal monolayers of the fusing bilayers (inner monolayers of the PMs). A subsequent expansion of a nascent hemifusion connection, also called hemifusion stalk, brings the distal lipid monolayers of the membrane bilayers together into a hemifusion diaphragm (HD). Finally, the opening and expanding of a fusion pore within the HD merges two membrane-enclosed volumes into one and completes the fusion process. Formation of both hemifusion and fusion pore intermediates proceeds by strongly bending lipid monolayers: proximal monolayers of the membranes during hemifusion and distal monolayers in pore formation, respectively. Therefore, the energy cost of the formation of each of these fusion intermediates depends on the elastic properties of the membranes and their propensity to bend in the required direction. While net curvature of the proximal monolayers, which form a convex hydrophilic surface of the stalk, is, by convention, considered as negative (smaller cross-section area in the plane of polar headgroups of the lipids than in the plane of their hydrophobic acyl chains), the net curvature of the distal monolayers, which form a concave rim of the fusion pore, is positive.

The elastic energy of membrane bending into fusion intermediates depends upon the difference between the net curvature of the lipid monolayers in these intermediates and the spontaneous curvature of the monolayers in their unstressed state (reviewed in Chernomordik and Kozlov 2003). Truncated cone-shaped lipids like diacylglycerol, unsaturated phosphatidylethanolamine, and cis-unsaturated fatty acids, with the cross-sectional area of polar heads smaller than the cross-sectional area of the hydrophobic domains, favor negative curvature and are conventionally described as lipids of negative spontaneous curvature. In contrast, inverted cone-shaped lipids like lysophosphatidylcholine (LPC) are described as lipids of positive spontaneous curvature. Finally, phosphatidylcholine (PC), the most abundant lipid in mammalian membranes, has an almost cylindrical shape and small spontaneous curvature and, thus, favors a flat monolayer. Biological membranes feature a complex mix of lipids as well as transmembrane and peripheral proteins, which can contribute to the elastic properties of the membranes (Kozlov and Taraska 2023). However, the overall membrane composition is usually balanced to support monolayers of the required curvatures (close to flat in the PM case). Single-component protein-free lipid bilayers formed from PC do not fuse or even hemifuse spontaneously, as expected, considering that bending the monolayers out of their spontaneous curvature prompts resistance by an elastic force. Thus, protein fusogens must overcome the elastic resistance of the membrane bilayer to prime the membranes for fusion and drive the formation of fusion intermediates.

Even for the best characterized biological fusion processes mediated by viral and intracellular protein fusogens, where we know both the identity and the structures of the proteins involved (for instance, HIV Env and Dengue virus E protein; SNARE complex proteins and their regulators), the specific coupling between protein restructuring and membrane remodeling remains to be clarified. Still, less is known about the mechanisms of cell-cell fusion processes, which is the focus of this review. Below, we briefly describe several examples of cell-cell fusion processes and then concentrate on the emerging mechanistic insights on the cell-cell fusion stage in the development and regeneration of skeletal muscle that emphasize the importance of membrane shaping in this fusion process.

Cell-Cell Fusion

Fusion-generated multinucleated cells carry out many critical physiological functions (reviewed in Brukman et al. 2019). Sperm-egg fusion is essential for the propagation of sexually reproducing organisms. The syncytiotrophoblast, a specialized epithelium derived from billions of mononuclear cytotrophoblasts, is the placental barrier between maternal and fetal blood. The fusion of skeletal muscle precursor cells, myoblasts, into multinucleated myotubes forms and maintains mature skeletal muscle. Osteoclasts, multinucleated cells responsible for the resorption of old and damaged bones in the lifelong remodeling of our skeletons, are derived through the fusion of mononucleated precursors. These and other cell-cell fusion processes and multinucleated cells generated by fusion have strikingly different properties and serve strikingly different functions. However, different cell-cell fusions also share some mechanistic motifs. In all cases, the relatively fast (seconds to minutes) actual cell fusion event is preceded by complex, unsynchronized, and relatively slow (days) differentiation processes that prepare the cells for fusion. These pre-fusion processes ensure that appropriate cells, and only appropriate cells, express the complete set of proteins involved in fusion and establish tight contacts at the right time. Slow and asynchronous development of fusion competence complicates the identification of the proteins involved in the actual fusion event rather than the pre-fusion stages of the formation of multinucleated cells.

Another more specific mechanistic feature conserved for all cell-cell fusion processes is a non-apoptotic exposure of phosphatidylserine (PS) at the surface of fusion-committed cells (reviewed in Whitlock and Chernomordik 2021). PS, an anionic lipid usually present only in the inner monolayer of the PM bilayer of healthy cells, is delivered to the surface of differentiating mononucleated trophoblasts, myoblasts, and osteoclast precursors by lipid scramblases of the TMEM16 family. Importantly, PS exposure not only accompanies the development of the cells' fusion competency but is also required for fusion. Blocking the externalization of PS suppresses fusion. The functional importance of PS exposure is further confirmed by many studies demonstrating the involvement of PS-binding proteins, Annexins, and PS receptors at the surface of the cells in the diverse cell-cell fusions. Finally, blocking exposed PS with PS-binding protein not involved in fusion (for instance, lactadherin) inhibits cell fusion (reviewed in Whitlock and Chernomordik 2021).

In addition to conserved dependence on PS exposure, diverse cell-cell fusions are similarly inhibited by an inverted cone lipid LPC. Myoblast fusion, osteoclast fusion, fusion modeling formation of multinucleated epithelial cells in *Caenorhabditis elegans (*reviewed in Brukman et al. 2019), and fusion mediated by trophoblast fusogen syncytin 1 (K. Melikov, unpublished data, Figure 1A), and are all reversibly suppressed by LPC at a stage upstream of the lipid mixing between the PMs of differentiating cells. Since lipid mixing manifests hemifusion and LPC generates positive, spontaneous curvature, this conserved inhibition suggests that membrane merger in different developmental cell-cell fusions, like fusions mediated by viral and intracellular fusogens, starts with bending the outer monolayers of the PMs into hemifusion stalk, which has net negative curvature (reviewed in Brukman et al. 2019).

In addition to serving as a probe for evaluating the net curvature of fusion intermediates, LPC application has been successfully utilized to identify the proteins acting at the fusion stage. While the hemifusion inhibitor LPC is not conducive to the concave geometry of the hemifusion stalk, we found no notable effects of LPC on pre-fusion stages of differentiation of myoblasts and osteoclast precursors

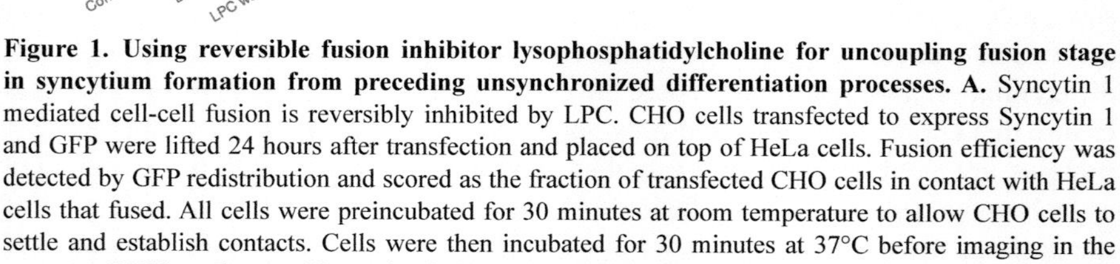

Figure 1. Using reversible fusion inhibitor lysophosphatidylcholine for uncoupling fusion stage in syncytium formation from preceding unsynchronized differentiation processes. A. Syncytin 1 mediated cell-cell fusion is reversibly inhibited by LPC. CHO cells transfected to express Syncytin 1 and GFP were lifted 24 hours after transfection and placed on top of HeLa cells. Fusion efficiency was detected by GFP redistribution and scored as the fraction of transfected CHO cells in contact with HeLa cells that fused. All cells were preincubated for 30 minutes at room temperature to allow CHO cells to settle and establish contacts. Cells were then incubated for 30 minutes at 37°C before imaging in the presence (LPC) or absence (Control) of 300 μM lauroyl LPC. For the LPC wash condition, cells were washed after 30 minutes of incubation in the presence of LPC and then incubated in an LPC-free medium for an additional 30 minutes at 37°C. **B.** Fusion of myoblasts and osteoclast precursors can be efficiently and reversibly blocked by incubation with LPC for up to 16 hours, allowing synchronization of fusion reaction upon LPC removal by washing the cells with LPC-free medium. Insert illustrates inhibition of the hemifusion stalk formation by inverted-cone-shaped lipid LPC that does not fit into net negative curvature of lipid monolayer in the hemifusion connection.

(Leikina et al. 2013; Verma et al. 2014). Keeping differentiating cells in the presence of LPC allowed us to accumulate ready-to-fuse cells upstream of hemifusion. After lauroyl LPC was removed by washing the cells in the LPC-free medium, cells undergo synchronized fusion relatively rapidly (within 30–90 minutes in comparison to several days required for the differentiation), affording us the ability to assess the function of proteins specifically in the membrane fusion stage decoupled from upstream differentiation (Figure 1B).

As mentioned above, lipid bilayers of the biological membranes are not expected to fuse spontaneously, with the possible exception of bilayers of small vesicles subjected to strong elastic stresses that may be released by fusion. This notion is supported by the lack of lipid mixing between cell membranes brought by the fusion-incompetent form of influenza virus hemagglutinin into extended and very close contacts. The relatively constant intermembrane distance of ~ 13 nm close to the height of the HA ectodomain (Leikina et al. 2004) is considerably smaller than several tens of nanometers interbilayer distances in contacts generated by conventional cell-cell adhesion machinery (Vanslembrouck et al. 2022). Finding that even these close contacts do not facilitate hemifusion in the absence of active protein fusogens emphasizes the importance of these proteins.

Biological membrane fusion is thought to be mediated by the proteins that bring membrane bilayers into a local tight contact and generate elastic stresses that can be released by hemifusion and fusion. Some proteins implicated in several cell-cell fusion processes have been identified with different degrees of certainty (Brukman et al. 2019). In many cases, further work is needed to clarify whether these proteins are bona fide fusogens and directly mediate fusion or rather regulators of the activity of the fusogens, as is the case with La protein found to regulate osteoclast fusion but not induce fusion on its own (Whitlock et al. 2023). Some of the better-established fusogens, for instance, syncytin 1 (involved in trophoblast fusion, osteoclast fusion, and cancer cell fusion), EFF1 (fusion of epithelial cells in *Caenorhabditis elegance*) and HAP2 (gamete fusion in *Chlamydomonas* and *Plasmodium*), are thought to drive membrane bending by restructuring from the metastable pre-fusion form to a low energy post-fusion conformation (Brukman et al. 2019). Some proteins (for instance, annexins involved in myoblast fusion and osteoclast fusion) can generate curvature by forming, likely together with other partner proteins, rigid negative curvature scaffolds at the surface of the membranes. Protein scaffolds of the negative curvature can drive the transition from early to late fusion intermediates (Kozlov and Chernomordik 2002). Syncytin 1 and HAP2 can also generate positive monolayer curvature by shallowly inserting their amphiphilic regions (referred to as fusion peptide and fusion loop, respectively) into the proximal monolayer of the opposite membrane.

Fusion machines mediating different fusion processes vary in the number of proteins involved and the distribution of the functions between different proteins. In many fusion processes, the same protein or protein complex reshapes the membranes into a hemifusion stalk structure and then into a fusion pore opening. This is the case for most fusion reactions induced by the better characterized viral glycoproteins belonging to class I (for instance, influenza virus hemagglutinin) and class II (for instance, Dengue virus E protein). The first identified developmental fusion protein, EFF-1, is also solely responsible for driving the fusion of epithelial cells in *Caenorhabditis elegans* (Brukman et al. 2019). It is unsurprising given the close structural similarity of EFF-1 to class II fusion glycoproteins. While SNARE-dependent intracellular fusion involves several proteins, including SNAREs, synaptotagmin, and complexin, the entire fusion pathway, from pre-fusion contact to fusion pore opening, is driven by the same protein complex (Kweon et al. 2017).

In contrast to the processes where the membrane shape transformations in hemifusion and fusion pore are mediated by the same protein or protein complex, myoblast fusion during the development and repair of skeletal muscles in mammals depends on two essential muscle-specific proteins—Myomaker and Myomerger that do not form a complex and are responsible for different stages of the fusion pathway (Leikina et al. 2018). Both proteins are required for the efficient formation of multinucleated myotubes (Millay et al. 2013; Bi et al. 2017; Zhang et al. 2017),

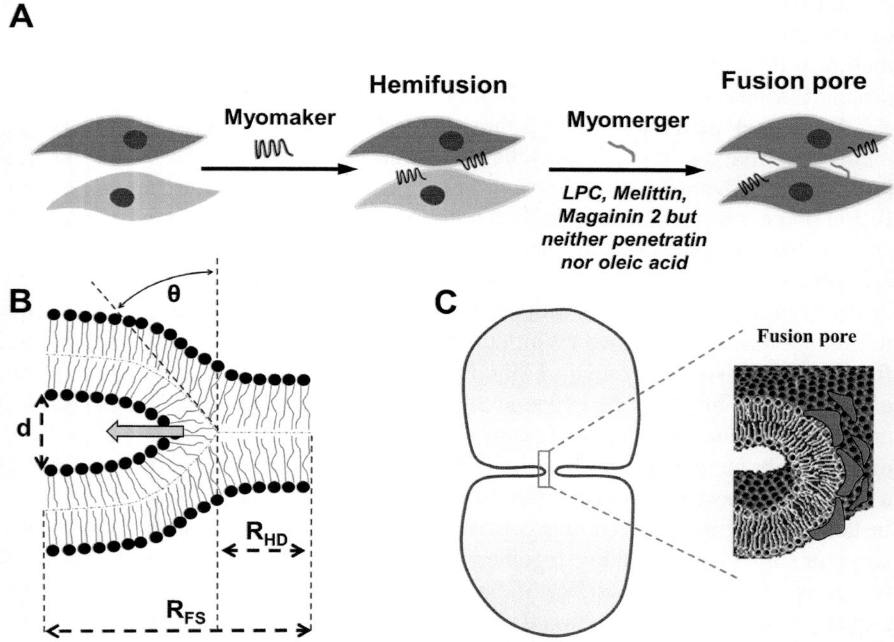

Figure 2. Muscle-specific proteins at the surface of myoblasts work one after another to open a fusion pore, and subsequent pore expansion depends on intracellular curvature-generating proteins. **A**. Schematic of the HD model. R_{HD} is the radius of HD, R_{FS} is the radius of the fusion site, d is the intermembrane distance, and θ is a junction angle. The arrow indicates elastic forces acting to increase the radius of HD (shown by arcs of different shades) upon insertion of molecules with positive, spontaneous curvature into proximal monolayers. Adapted from Golani et al. (2021). **B**. Myomaker and Myomerger act on different stages of myoblast fusion reaction. Myomaker induces myoblast hemifusion, and Myomerger induces positive curvature in contacting monolayers to promote a transition from the hemifusion-to-fusion pore opening. Adapted from Leikina et al. (2018). **C**. The membrane bilayer at the rim of the fusion pore expanding within the tight cell-cell contact has a strong positive curvature. Intracellular membrane-shaping proteins supporting this curvature (shown as dark shapes accumulating at the inner rim of the pore) lower the elastic energy of bending and promote pore expansion. Adapted from Richard et al. (2011).

and both act at the stage of actual fusion event as demonstrated by the fusion synchronization approach (Gamage et al. 2017; Leikina et al. 2018), Myomaker, a multipass membrane protein with seven transmembrane proteins, is involved in hemifusion and induces hemifusion even in Myomerger deficient cells (Figure 2A). On the other hand, Myomerger, a small (~ 9.5 kDa) single-pass transmembrane protein with most of the protein forming extracellular domain, also referred to in the literature as Myomixer and Minion (Millay et al. 2013; Bi et al. 2017; Zhang et al. 2017), promotes the opening of the fusion pore in cells arrested at the hemifusion stage of the process (Leikina et al. 2018). Importantly, Myomerger promotes pore formation in the heterologous fusion system in the absence of Myomaker, confirming that Myomaker and Myomerger work independently to promote different stages of the fusion process (Leikina et al. 2018). The mechanisms by which Myomaker directly or indirectly promotes hemifusion remain perplexing, especially considering that the protein structure is strikingly different from the structure of all well-characterized fusogens (Brukman et al. 2019). In contrast, there was significant progress in understanding the mechanism of Myomerger action (Gamage et al. 2022; Golani et al. 2021). Most of the work on this mechanism has focused on the extracellular domain of Myomerger, referred to as Myomerger ectodomain. Extracellular application of the synthetic or recombinant peptide corresponding to the Myomerger ectodomain rescues syncytium formation in Myomerger deficient myoblasts arrested at the hemifusion stage, suggesting that the transmembrane domain of this protein is dispensable for its function (Leikina et al. 2018).

Myomerger ectodomain was found to directly induce leakage of liposomes, suggesting that this protein supports the formation of pores in the lipid bilayer. The hypothesis that Myomerger alters cell membrane properties that are important for pore formation was further supported by the finding that PMs of differentiating Myomerger-expressing cells are easier to permeabilize by the tension generated by the application of the hypotonic osmotic shock. A decrease in the critical tension required for permeabilization of PM indicated that Myomerger increases the propensity for the membrane lipid bilayer to form a pore. Since the rim of a pore in the lipid bilayer features convex (positive) curvature, these data suggested that the Myomerger ectodomain generates positive, spontaneous curvature in the lipid monolayer with which this protein interacts.

In a follow-up study, we directly explored the effects of the Myomerger ectodomain on the spontaneous curvature using a novel approach to evaluate the sign of the spontaneous curvature induced by membrane-interacting peptides (Golani et al. 2021). The technique measures changes in the fluorescence resonance energy transfer (FRET) between donor and acceptor dyes attached to the lipids upon adding membrane-inserting molecules. Two sets of FRET pairs are compared—one with fluorescent dyes attached to the lipid head group and another with dyes attached in the lipid tail region. The qualitative assumption of the approach is that positive spontaneous curvature-inducing peptides will induce more pronounced separation and, therefore, a more significant decrease in FRET for probes located in the headgroup region than for FRET pair located in the lipid tail region. The opposite effect is expected for the peptides inducing negative spontaneous curvature. A detailed quantitative analysis of the method is still being developed. Therefore, in

(Golani et al. 2021), the effects of the Myomerger ectodomain insertion are compared with those of inverted-cone-shaped LPC and cylinder-shaped PC insertions. As expected, LPC has a more significant effect on FRET between the probes in the lipid headgroups than on FRET between the probes in lipid acyl chains, compared to the dioleoyl PC. In the case of Myomerger ectodomain, the difference in FRET for headgroup attached pair of probes vs. FRET for tail-attached probes is even more significant than for LPC, confirming the hypothesis that Myomerger induces a strong positive, spontaneous curvature of lipid monolayer (Leikina et al. 2018).

Notably, while induction of a positive, spontaneous curvature by the Myomerger ectodomain in the outer monolayers of PM can be expected to promote the formation of pores in PM and membrane leakage directly, the HD in cell-cell fusion is formed from the inner monolayers of fusing PMs and Myomerger in the outer monolayers can influence fusion pore formation in the HD only indirectly (Figure 2B). Indeed, the detailed theoretical analysis demonstrates that peptides inducing positive, spontaneous curvature in proximal monolayers can indirectly promote a transition from hemifusion to the opening of a nascent fusion pore in an expanding HD (Golani et al. 2021). The model considered a fusion site with a fixed overall radius and intermembrane distance. In contrast, the radius of the HD and the junction angle, which characterizes the structure of the rim of the HD, can change. In this model, the change in the spontaneous curvature of the outer monolayer to more positive values leads to a change in the junction angle and HD radius. This, in turn, results in additional elastic stresses around the junction region and additional lateral tension in the HD region. In effect, the rim of the HD mechanically couples proximal and distal monolayers within the fusion site (Golani et al. 2021). Using values of the elastic and geometric parameters of the model that are within experimentally observed and/or generally assumed ranges indicates that an increase of the spontaneous curvature of the outer monolayers equivalent to the addition of 10-mole percent of LPC can lead to up to 50-fold acceleration of fusion pore formation. Although the exact amplitude of the effects on the fusion rate cannot be obtained from the model due to insufficient information about important elastic moduli and geometric parameters of the fusion sites, it provides an insightful qualitative description of the coupling mechanism between the inner and outer monolayer during fusion. Interestingly, in larger fusion sites, fusion pores open less efficiently and are predicted to depend strongly on the increase in the positive spontaneous curvature. One may hypothesize that fusion sites formed when one of the fusing membranes is highly curved, such as viral and intracellular fusion, will be less reliant on additional curvature stress introduced by fusion protein, explaining why a single fusion complex can drive fusion all the way in those cases.

The hypothesis that Myomerger drives fusion pore opening by inducing positive, spontaneous curvature in the outer monolayers of the fusing cells has extensive experimental support (Golani et al. 2021; Leikina et al. 2018). First, the promotion of hemifusion-to-fusion transition in Myomerger deficient cells by applying recombinant Myomerger ectodomain can be replicated by the extracellular addition of LPC and either of two positive curvature-inducing peptides, magainin-2, and melittin, but not by negative curvature-inducing peptide penetratin and lipid oleic acid. Similarly, LPC and Myomerger promote a transition from hemifusion to fusion pore opening in a heterologous system of fusion between influenza hemagglutinin-expressing cells and red blood cells. This data experimentally demonstrates the mechanical coupling between proximal and distal monolayers around the HD. Second, applying the Myomerger ectodomain at the start of the fusion reaction inhibits both hemifusion and complete fusion, as expected for the positive, spontaneous curvature-generating peptide. Notably, concentrations of both LPC and Myomerger that inhibit hemifusion are higher than those that promote hemifusion-to-fusion pore opening transition, further supporting the idea that both act through the same mechanism, thus generating positive, spontaneous curvature. The observation that a significantly higher concentration of the Myomerger is needed to block hemifusion can explain why the timing of Myomerger surface expression is not tightly regulated in fusing myoblasts.

Myomerger ectodomain contains two alpha-helical domains followed by the unstructured tail (Leikina et al. 2018; Shi et al. 2017). The membrane-proximal alpha helix (helix-1) is amphipathic, while the other alpha helix (helix-2) is hydrophobic. Both alpha helices are necessary for the full fusion activity of the Myomerger, while the unstructured tail is dispensable (Shi et al. 2017). A recent study has analyzed the relative role of the two alpha helices in the properties of the Myomerger insertions and myoblast fusion and the dependence of Myomerger insertion on the lipid composition of the membranes (Gamage et al. 2022). Removal of the membrane-proximal amphipathic helix-1 or mutation of arginine residues to serine was found to abolish the peptides' cell fusion activity, suggesting that helix-1 is indispensable for the function of the Myomerger. Similarly, both helix-1 removal and mutation significantly decrease the ability of the peptide to affect FRET between fluorescent dyes in the lipid, indicating the importance of helix-1 for positive curvature generation. Removal or mutation of helix-2 also significantly diminishes the fusion activity of Myomerger, but some residual activity remains, indicating that helix-1 has some activity on its own (Gamage et al. 2022). The activity of Myomerger ectodomain depends on anionic lipid PS, which, as discussed above, is normally confined to the inner monolayer of PM but transiently released to the outer monolayer at the time of myoblast fusion (reviewed in Whitlock and Chernomordik 2021)). PS promotes the folding of helix-1 of the Myomerger ectodomain, binding to the membrane, generation of positive, spontaneous curvature, and promotion of complete fusion in the myoblasts arrested at the hemifusion stage. The current model proposes that hydrophobic helix-2 is inserted into the membrane from the start. At the same time, initially, disordered helix-1 becomes ordered and shallowly inserts into the membrane in response to PS externalization. In this scenario, PS externalization is a trigger, and helix-1 insertion into the lipid bilayer is the primary driver of generating the positive monolayer curvature and transition from hemifusion to pore formation.

It remains to be seen whether the mechanism in which Myomerger promotes the hemifusion-to-fusion pore transition through generating positive contributions to the spontaneous curvature of the outer monolayers of PMs is utilized by proteins that mediate other biological fusion processes. Published mechanisms of protein-mediated fusion primarily focus on explaining how fusion proteins initiate the formation of hemifusion intermediates by bringing proximal monolayers of two strongly curved fusing membranes into close contact. The transition to the formation and expansion of a fusion pore in the HD is most often explained by the specific properties of the transmembrane domains of the fusion proteins. However, several studies reported fusion pore opening even when transmembrane domains of certain fusion proteins (SNAREs and influenza hemagglutinin) were replaced with a lipid anchor (Markosyan et al. 2000; Xu et al. 2011). On the other hand, many protein fusogens feature functionally important membrane-interacting amphiphilic regions that may, in analogy to Myomerger, promote fusion completion by raising spontaneous curvature of proximal monolayers and tension in HD. Future work will show whether this mechanism of fusion pore formation is specific for myoblasts or is also utilized in other membrane fusion processes.

Interestingly, the role of the generation of positive, spontaneous curvature in the induction of the fusion pore formation has also been suggested for the membrane-proximal amphipathic helix domain of FAST proteins (Duncan 2019). Membrane-proximal amphipathic helices of FAST proteins are curvature sensing and prefer positive membrane curvature. Note that in contrast to Myomerger helices, FAST protein helices are in the cytosolic domains of the protein and can be expected to promote fusion pore formation directly since they are inserted into the inner monolayers of PMs, which are coming together in the HD. Therefore, despite the shared dependence on the positive curvature generations, FAST proteins, and Myomerger appear to promote fusion pore opening by distinct mechanisms.

Post-Fusion Stages

Opening the fusion pore signifies the completion of a bona fide fusion reaction. However, the biological functions of multinucleated myotubes, osteoclasts, and syncytiotrophoblast require complete unification of the volumes of the fusing cells. At later stages of cell-cell fusion, nascent fusion pores of a few nanometers in diameter expand to yield an open lumen of cell-size diameter either by displacement of membrane material toward the periphery of the contact zone (Chen et al. 2008) or by vesiculation removing the excess of the membrane material into cells (Doberstein et al. 1997). The existence of transient tunneling nanotubes that dynamically connect different types of cells but do not expand to yield syncytia (Kolba et al. 2019; Pepe et al. 2022) indicates that the opening of a fusion pore does not necessarily lead to its expansion. In order to identify the cells that formed early fusion intermediates but did not proceed through the late stage of cell reshaping that completes the formation of a multinucleated cell, it is essential to complement syncytium formation assay—the conventional way of quantifying cell-cell fusion by scoring numbers and sizes of the syncytia—with an assay in which early stages of fusion are detected as either lipid mixing or redistribution of small aqueous probes between the cells.

Several studies on cell-cell fusion mediated by viral fusogens (influenza hemagglutinin and baculovirus gp64), as well as on myoblast fusion and osteoclast fusion, suggest that downstream of formation of early fusion intermediates, the final expansion of the fusion pores leading to the formation of a multinucleated cell depends on phosphatidylinositol(4,5)bisphosphate content, and cell metabolism (Richard et al. 2011; Leikina et al. 2013; Verma et al. 2014). These findings indicate that while fusion initiation is catalyzed by protein machinery located at the surface of the cells, late stages of cell-cell fusion are controlled by ATP-dependent intracellular machinery. The dependence of fusion pore expansion on metabolic energy can be explained by the fact that as long as the pore expands within an area of tight intercellular contact, the lipid bilayer at the rim of the pore has a strong positive curvature (Figure 2C). Expansion of the fusion pores does not spontaneously proceed because the elastic energy accumulated in the rim increases with the growth of the pore perimeter. Pore formation can be promoted by positive curvature-generating intracellular protein. Indeed, in the experiments on cell-cell fusion initiated by viral fusogens, overexpression or introduction of the ENTH domain of epsin, GRAF1 BAR domain, or FCHo2 F-BAR domain does not affect early fusion stages detected as lipid mixing but promotes the formation of multinucleated cells (Richard et al. 2011). Although these results do not indicate the role of these specific proteins in syncytia formation, they establish that positive curvature-generating proteins can promote syncytia formation. Since the curvature of the fusion pore is comparable to that of the highly curved intracellular tubular and vesicular compartments, one can suggest that these positive curvature-inducing domains bind to the fusion pore edge, lower line tension of the pore and thus promote its expansion. Late stages of cell-cell fusion also depend on one of the best characterized positive curvature-generating proteins, dynamin. Blocking the activity of dynamin GTPase, suppression of dynamin 2 expression, and expression of dominant negative mutants of dynamin does not inhibit hemifusion and opening of a fusion pore but inhibits syncytium formation initiated by viral and developmental cell-cell fusogens (Leikina et al. 2013; Richard et al. 2011; Verma et al. 2014).

Concluding Remarks

While mechanisms underlying cell-to-cell fusion in normal physiology and in diseases such as viral infections and cancer remain to be clarified, it is already clear that dynamic changes in the shapes of the cells play an important role throughout the entire multistep fusion process. Bending deformations in the pre-fusion, fusion, and post-fusion stages of the formation of multinucleated cells have different time and length scales. The dynamic changes in the membrane shapes during the formation of hemifusion and nascent fusion pore intermediates, as well as at the rim of the expanding fusion pore, are likely too fast and too localized (nanometers scale) to be resolved by light microscopy but have been a subject of many theoretical studies carried out by modeling the membranes as macroscopic continuous films and by computer simulations. In contrast, the pre-fusion deformations of cell membranes, including the formation of actin-rich protrusions bringing membranes into close contact, as well as post-fusion changes in the shapes of the cells and micron-sized aqueous connections between the cells, are readily amenable for experimental analysis. We expect future work on the molecular mechanisms by which proteins shape membranes at different stages of cell-cell fusion processes to bring about novel approaches for understanding and treating musculoskeletal disorders, pregnancy complications related to abnormalities and dysfunction of the placenta, and cancer and other diseases linked to insufficient or excessive fusion.

Acknowledgments

The authors wish to thank Dr. Michael Kozlov, Tel Aviv, for numerous enlightening discussions over the years. The research in the L.V.C. laboratory was supported by the Intramural Research Program of the Eunice Kennedy Shriver National Institute of Child Health and Human Development, National Institutes of Health.

References

Bi, P., A. Ramirez-Martinez, H. Li et al. 2017. Control of muscle formation by the fusogenic micropeptide myomixer. Science 356(6335): 323–327.

Brukman, N.G., B. Uygur, B. Podbilewicz et al. 2019. How cells fuse. J. Cell Biol. 218(5): 1436–1451.

Chan, K.M.C., A.L. Arthur, J. Morstein et al. 2021. Evolutionarily related small viral fusogens hijack distinct but modular actin nucleation pathways to drive cell-cell fusion. Proc. Natl. Acad. Sci. U.S.A. 118(1).

Chan, K.M.C., S. Son, E.M. Schmid et al. 2020. A viral fusogen hijacks the actin cytoskeleton to drive cell-cell fusion. Elife 9.

Chen, A., E. Leikina, K. Melikov et al. 2008. Fusion-pore expansion during syncytium formation is restricted by an actin network. J. Cell Sci. 121 (Pt 21): 3619–28.

Chernomordik, L.V. and M.M. Kozlov. 2003. Protein-lipid interplay in fusion and fission of biological membranes. Annu. Rev. Biochem. 72: 175–207.

Doberstein, S.K., R.D. Fetter, A.Y. Mehta et al. 1997. Genetic analysis of myoblast fusion: blown fuse is required for progression beyond the prefusion complex. J. Cell Biol. 136 (6): 1249–61.

Duncan, R. 2019. Fusogenic Reoviruses and Their Fusion-Associated Small Transmembrane (FAST) Proteins. Annu. Rev. Virol. 6(1): 341–363.

Gamage, D.G., E. Leikina, M.E. Quinn et al. 2017. Insights into the localization and function of myomaker during myoblast fusion. J. Biol. Chem. 292(42): 17272–17289.

Gamage, D.G., K. Melikov, P. Munoz-Tello et al. 2022. Phosphatidylserine orchestrates Myomerger membrane insertions to drive myoblast fusion. Proc. Natl. Acad. Sci. U.S.A. 119(38): e2202490119.

Golani, G., E. Leikina, K. Melikov et al. 2021. Myomerger promotes fusion pore by elastic coupling between proximal membrane leaflets and hemifusion diaphragm. Nat. Commun. 12(1): 495.

Kim, J.H. and E.H. Chen. 2019. The fusogenic synapse at a glance. J. Cell Sci. 132(18).

Kolba, M.D., W. Dudka, M. Zareba-Koziol et al. 2019. Tunneling nanotube-mediated intercellular vesicle and protein transfer in the stroma-provided imatinib resistance in chronic myeloid leukemia cells. Cell Death Dis. 10(11): 817.

Kozlov, M.M. and L.V. Chernomordik. 2002. The protein coat in membrane fusion: lessons from fission. Traffic. 3(4): 256–67.

Kozlov, M.M. and J.W. Taraska. 2023. Generation of nanoscopic membrane curvature for membrane trafficking. Nat. Rev. Mol. Cell Biol. 24(1): 63–78.

Kweon, D.H., B. Kong and Y.K. Shin. 2017. Hemifusion in synaptic vesicle cycle. Front Mol. Neurosci. 10: 65.

Lee, D.M. and E.H. Chen. 2019. Drosophila myoblast fusion: invasion and resistance for the ultimate union. Annu. Rev. Genet. 53: 67–91.

Leikina, E., D.G. Gamage, V. Prasad et al. 2018. Myomaker and myomerger work independently to control distinct steps of membrane remodeling during myoblast fusion. Dev. Cell 46(6): 767–780 e7.

Leikina, E., K. Melikov, S. Sanyal et al. 2013. Extracellular annexins and dynamin are important for sequential steps in myoblast fusion. J. Cell Biol. 200(1): 109–23.

Leikina, E., A. Mittal, M.S. Cho et al. 2004. Influenza hemagglutinins outside of the contact zone are necessary for fusion pore expansion. J. Biol. Chem. 279(25): 26526–32.

Markosyan, R.M., F.S. Cohen and G.B. Melikyan. 2000. The lipid-anchored ectodomain of influenza virus hemagglutinin (GPI-HA) is capable of inducing nonenlarging fusion pores. Mol. Biol. Cell 11(4): 1143–52.

Millay, D.P., J.R. O'Rourke, L.B. Sutherland et al. 2013. Myomaker is a membrane activator of myoblast fusion and muscle formation. Nature 499(7458): 301–5.

Pepe, A., S. Pietropaoli, M. Vos et al. 2022. Tunneling nanotubes provide a route for SARS-CoV-2 spreading. Sci. Adv. 8(29): eabo0171.
Richard, J.P., E. Leikina, R. Langen et al. 2011. Intracellular curvature-generating proteins in cell-to-cell fusion. Biochem. J. 440(2): 185–93.
Shi, J., P. Bi, J. Pei et al. 2017. Requirement of the fusogenic micropeptide myomixer for muscle formation in zebrafish. Proc. Natl. Acad. Sci. U.S.A. 114(45): 11950–11955.
Shilagardi, K., S. Li, F. Luo et al. 2013. Actin-propelled invasive membrane protrusions promote fusogenic protein engagement during cell-cell fusion. Science 340(6130): 359–63.
Vanslembrouck, B., J.H. Chen, C. Larabell et al. 2022. Microscopic Visualization of Cell-Cell Adhesion Complexes at Micro and Nanoscale. Front Cell Dev. Biol. 10: 819534.
Verma, S.K., E. Leikina, K. Melikov et al. 2014. Late stages of the synchronized macrophage fusion in osteoclast formation depend on dynamin. Biochem. J. 464(3): 293–300.
Whitlock, J.M. and L.V. Chernomordik. 2021. Flagging fusion: Phosphatidylserine signaling in cell-cell fusion. J. Biol. Chem. 296: 100411.
Whitlock, J.M., E. Leikina, K. Melikov et al. 2023. Cell surface-bound La protein regulates the cell fusion stage of osteoclastogenesis. Nat. Commun. 14(1): 616.
Xu, H., M. Zick, W.T. Wickner et al. 2011. A lipid-anchored SNARE supports membrane fusion. Proc. Natl. Acad. Sci. U.S.A. 108(42): 17325–30.
Zhang, Q., A.A. Vashisht, J. O'Rourke et al. 2017. The microprotein Minion controls cell fusion and muscle formation. Nat. Commun. 8: 15664.

13

Lipid Bilayers as Platforms for Understanding Biological Memory and the Development of Neuromorphic Computing

C. P. Collier,[1] *D. Bolmatov*[2,3] and *J. Katsaras*[2,3,4,*]

Introduction

Evidence for the existence of the plasma membrane (PM) dates back to the 18th century (Nollet 1752). However, it was not until 1899 that Charles Overton (Overton 1899) and later on, Gorter and Grendel, using red blood cells (Gorter and Grendel 1925), determined that the boundary layer surrounding cells (i.e., the PM) contained lipids. Since then, the PM has evolved from the concept of a "lipoid film" with adsorbed proteins forming a protein-lipid-protein "sandwich" structure (Danielli and Davson 1935) to the fluid mosaic model of the PM, where proteins were now embedded in a two-dimensional fluid lipid bilayer (Singer and Nicolson 1972). Since those developments, one of the most important advances in PM structure has been the concept of lipid rafts, dynamic assemblies of proteins and lipids that can form platforms that control various biological functions (e.g., neurotransmission) (Simons and Ikonen 1997).

It is now understood that lipids are the underlying organizing structure of biological membranes, with mammalian membranes containing hundreds of lipid species (Symons et al. 2021). Simply, these amphipathic molecules, which differ in their hydrocarbon chains and headgroups, have central roles in a range of biological functions, such as storing energy, absorbing vitamins and other biomolecules,

[1] Center for Nanophase Materials, Oak Ridge National Laboratory, Oak Ridge, TN, 37831.

[2] Department of Physics and Astronomy, University of Tennessee, Knoxville, TN, 37996.

[3] Shull Wollan Center, Oak Ridge National Laboratory, Oak Ridge, TN, 37831.

[4] Labs and Soft Matter, Neutron Scattering Division, Oak Ridge National Laboratory, Oak Ridge, TN, 37831.

* Corresponding author: katsarasj@ornl.gov

signaling, etc. (Horn and Jaiswal 2019). However, in addition to their biological relevance, lipids have recently been shown to behave as memory capacitors (memcapacitors), i.e., two-terminal memory elements (memelements) capable of co-locating signal processing and short-lived memory at the nanoscale (Najem et al. 2019).

In 1971, Leon Chua introduced a "missing" memelement, namely a new two-terminal circuit element which he termed a memory resistor or memristor, because it "remembered" the amount of current that passed through it (Chua 1971)—this concept was later extended to include memory capacitors and memory inductors (Di Ventra et al. 2009). In the case of ideal memristors, when plotting current vs. voltage, one observes a pinched hysteresis loop described by Chua (1971) and later observed in a range of solid-state devices (Shin et al. 2010; Zidan et al. 2014; Karimi and Rezai 2019). These solid-state memristors emulate some synaptic functions but with nonbiological switching mechanisms that have prevented them from achieving the interconnectivity and energy efficiency of the human brain. However, lipid bilayers have recently exhibited memelement behaviors (Najem et al. 2018; Najem et al. 2019; Scott et al. 2022). Here, we will discuss current lipid bilayer research, which connects biomimetic membranes as platforms for understanding how the brain transmits and stores information at the molecular level and for the development of neuromorphic computing, whose potential promises to achieve the complexity, performance, and energy efficiency of the human brain.

Droplet Interface Bilayer (DIB) and Memristance

Using the droplet interface bilayer (DIB) platform, which involves the spontaneous formation of a phospholipid bilayer when two lipid-coated water droplets in a hydrocarbon solvent come into contact (Bayley et al. 2008), we have shown that phospholipid bilayer membranes doped with the ion channel, alamethicin, exhibit memristive (memory resistance) behavior, emulating key synaptic functions related to short-term potentiation, including paired-pulse facilitation (PPF) and depression (PPD) (Najem et al. 2018). However, unlike state-of-the-art solid-state memristive devices (Wang et al. 2017), two-terminal biomolecular memristors featured similar structures (biomembrane), switching mechanism (ion channels), and ionic transport modalities as those found in biological synapses (Sheng et al. 2012), while operating at considerably lower power levels. In Figure 1A, voltage-dependent thresholds for alamethicin ion channel formation in DIBs function analogously to Ca^{2+}-dependent activation of α-amino-3-hydroxy-5-methyl-4-isoxazolepropionic acid (AMPA) and N-methyl-D-aspartate (NMDA) receptors (Sigworth 1994). These devices have exhibited reversible pinched hysteresis in plots of current versus voltage, a hallmark of memristive systems.

Figure 1B shows the time-dependent characteristics and persistence of this biomolecular memristor in response to voltages above the switching threshold (130–140 mV). This system also exhibited short-term synaptic plasticity (STP) in the form of pair pulse facilitation (PPF) (Figure 1C) and depression (PPD) (Stevens and Wang 1995), which previously required sophisticated materials and fabrication protocols for solid-state devices. Simulations based on the time and

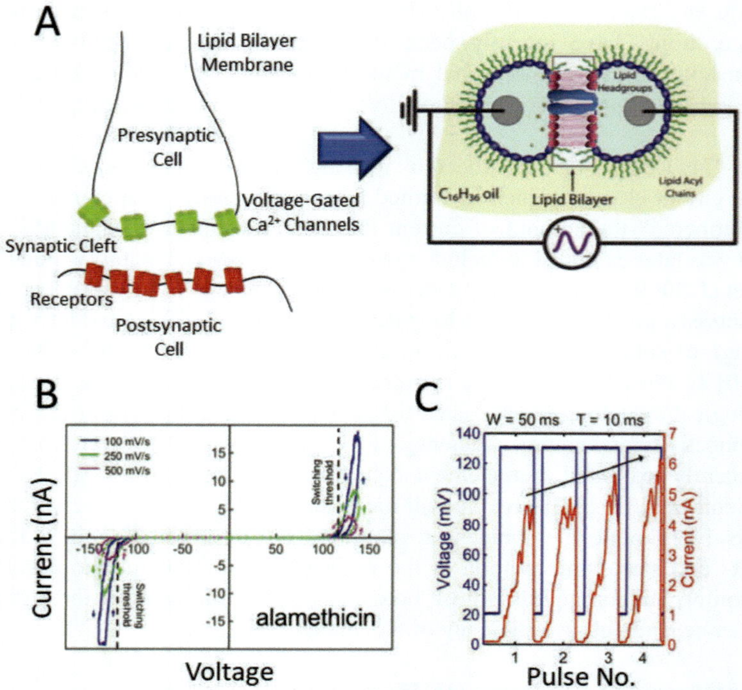

Figure 1. Biomolecular memristors resemble biological synapses and exhibit nonlinear memristive behavior and bipolar threshold switching. **(A)** Biological synapse (left) vs. a biomembrane synaptic mimic in a droplet interface bilayer (DIB) (right). **(B)** Pinched hysteresis in current vs. voltage measurements is a signature of a memristive system. **(C)** Paired-pulse facilitation (PPF) from a memristor to a series of 130 mV, 50 ms pulses separated by 10 ms. (Figure adapted from Najem et al. 2018.)

switching parameters of these memristors indicated the state equation consisted of the coupling of numerous conductive pores per unit area above the voltage threshold for alamethicin, with voltage-dependent changes in bilayer membrane area due to electrowetting, which is analogous to the expression of additional AMPA receptors in biosynapses. Simulations have also shown the generation of spike rate-dependent plasticity (SRDP), which, when connected to a nonvolatile memristor, spike timing dependent plasticity (STDP) is observed, analogous to STDP-based neural signaling in the brain (Sigworth 1994).

Memcapacitance

Recently, we observed memcapacitive (memory capacitance) behavior in DIB membranes with no ion channels (Najem et al. 2019). Nonlinear dynamics and capacitive memory were governed by two implicitly coupled, voltage-dependent state variables—i.e., membrane radius and thickness are shown schematically in Figure 2A. Voltage changed these variables through electrowetting, by increasing the membrane area (as described earlier for the memristor), and through electrocompression that drives out residual alkane oil used in DIBs from the hydrophobic center of the bilayer, leading to a decrease in membrane thickness as a function of voltage. Collectively,

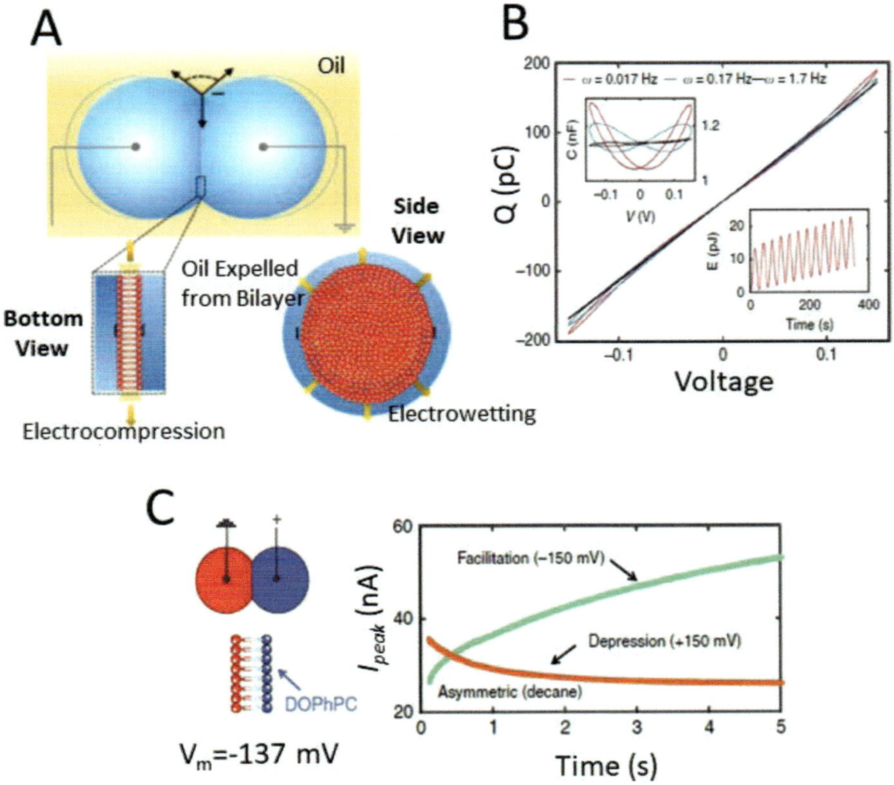

Figure 2. (A) Schematic describing the geometrical changes in the DIBs lipid bilayer caused by voltage. Changes consist of electrowetting-driven formation of bilayers with increased area and a concurrent decrease in thickness due to removing residual oil from the membrane by electrocompression. **(B)** Frequency dependence of charge-voltage hysteresis and corresponding capacitance (top inset) and the energy dissipated during repeated reconfiguration of the bilayer with voltage (bottom inset). **(C)** An asymmetric membrane with an intrinsic bias of −137 mV due to differences in dipole potentials of the two opposing leaflets, leading to PPF at negative polarity and PPD at positive polarity. (Figure adapted from Najem et al. 2019.)

electrowetting (Poulos et al. 2009) and electrocompression (El-Beyrouthy et al. 2019) are two examples of what is commonly known as electrostriction of lipid bilayers (Hamill and Martinac 2001).

This system was capable of tunable signal processing and learning through STP, like the above-described memristor. However, unlike the memristor, which is a purely dissipative element, the memcapacitor can store energy (Di Ventra et al. 2009). Like the memristor, the electrical signature of the memcapacitor involved pinched hysteresis, but in this case, in the charge-voltage plane. This is shown in Figure 2B, where we plotted the frequency dependence of the pinched hysteresis in charge vs. voltage and the capacitance vs. voltage (top inset) in response to a sinusoidal voltage. The lower inset of the figure shows the energy dissipated by the reconfigurable membrane at 17 mHz as a linearly increasing slope of the sinusoidal capacitive energy ($E = CV^2/2$). Moreover, short-term facilitation (STF) and depression (STD) (Jackman and Regehr 2017) could be tuned through judicious choices of, for example, the alkane oil used (decane vs. hexadecane) in the DIB platform and bilayer symmetry (both leaflets contain the same or different chemical composition lipids). Figure 2C shows the response of an asymmetric bilayer in decane consisting of phospholipids, where an ether lipid in one bilayer leaflet (1,2-diphytanoyl-sn-glycero-3-phospho-choline, DPhPC) was replaced with an ester lipid (1,2-di-O-phytanoyl-sn-glycero-3-phospho-choline, DOPhPC). The difference in dipole potential results in an additional intrinsic bias of –137 mV (Najem et al. 2018). In this case, PPF and PPD depended on the polarity of the voltage signal, with STF taking place at negative voltages and STD at positive voltages.

Memcapacitance in lipid bilayers will likely impact two areas of science and technology: (1) modular, low-power materials that could interface with cells and tissues for biosensing and processing of biological signals due to their soft and biocompatible properties (Villar et al. 2013); (2) as a model system to study capacitive excitability in neuronal membranes (Weinberg 2015).

Dynamic Electrochemical Impedance Spectroscopy (dEIS)

Until recently, we considered the memristance and memcapacitance properties of lipid bilayer membranes to be independent due to, for example, the lack of ion channels in purely memcapacitive membranes. This is a common assumption held for both solid-state and soft-matter memelements. However, there is no *a priori* requirement for such a strict delineation between memristors and memcapacitors. Indeed, it is not unreasonable to expect memristance and memcapacitance to be coupled in complex biological membranes comprised of hundreds of different lipids, ion channels, receptors, and small molecules, all contributing to enhanced adaptation and plasticity. Even without ion channels, memristive behavior can occur if the dielectric loss of the bilayer (excluding transmembrane charge transport) is altered because of changes in lipid headgroup reorganization or acyl tail rearrangement. In general, however, the contributions of coupled memristance and memcapacitance to lipid bilayers determined solely from linear electrical biasing measurements (current-voltage curves) cannot be easily deconvolved.

To address this coupling between memristance and memcapacitance, we used dynamic electrochemical impedance spectroscopy (dEIS) to explicitly probe the bilayer's response to individual sinusoidal frequencies and its time-dependent impedance (Sacci et al. 2022). Electrochemical impedance spectroscopy (EIS) measures the frequency responses of electrochemical systems, which include energy storage and dissipation components (Lasia 2002). However, EIS measures the electrical response of the membrane only under steady-state conditions and thus cannot be used for time-dependent dynamics. In contrast, dEIS probes the *temporal development of dynamics* in electrochemical systems, including DIBs, by providing a multi-sine probe to measure the frequency-resolved electrical responses (complex impedance) of membranes as functions of time (Sacci et al. 2014). Importantly, dEIS measurements can deconvolve time-dependent memristance and memcapacitance, allowing the frequency-dependent separation of molecular from geometrical

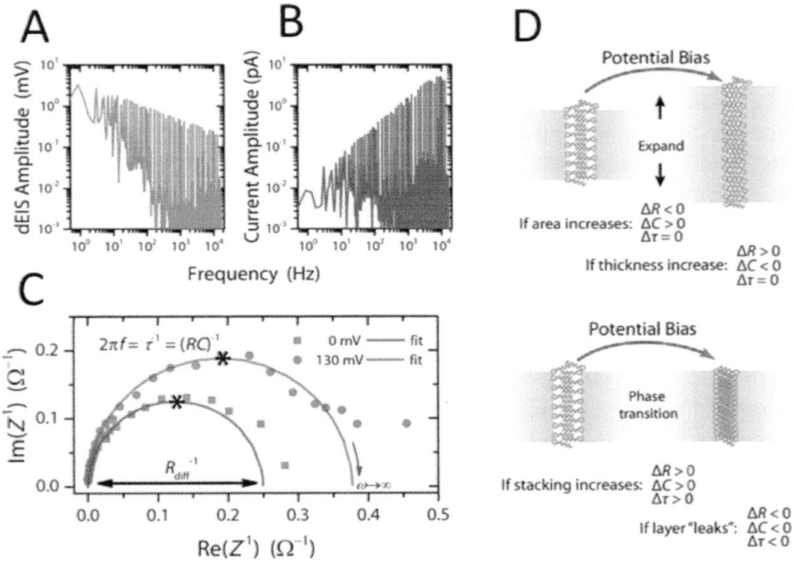

Figure 3. **(A)** Fourier transformations of the multi-sine voltage (A) and current **(B)** waveforms to the frequency domain using dynamic electrochemical impedance spectroscopy (dEIS). **(C)** Representative dEIS spectra were measured at two potentials during a 10 mHz biasing scan and the corresponding *RC* circuit fitting results. The measured data points deviate from the calculated semicircle behaviors at higher frequencies—beyond what is predicted by the RC-circuit model used to fit the data—indicating that the lipid bilayer undergoes a cascade of relaxation processes at the molecular level. (D) Structural evolution of the lipid bilayer membrane with applied voltage, including (1) geometrical responses based on voltage-dependent migration of oil in the membrane and changes in interfacial tension of the bilayer and (2) structural responses corresponding to time- and voltage-dependent "phase transitions" based on changes to the lipid bilayer at the molecular scale. (Figure adapted from Sacci et al. 2022.)

changes in the membrane. Specifically, the geometrical response corresponds to electrowetting (Najem et al. 2018) due to interfacial tension and inter-drop adhesion and to electrocompression (Najem et al. 2019) due to the voltage-dependent migration of oil in the DIB membrane. The molecular response corresponds to a time- and voltage-dependent reorganization of the lipid bilayer at the molecular scale.

Figure 3 shows Fourier transformations of the multi-sine voltage (A) and current (B) waveforms to the frequency domain using dEIS. The dynamics can be fitted to the time constant of an equivalent $\tau = RC$ circuit to obtain changes to the differential resistance ΔR_{diff} and differential capacitance ΔC_{diff}, both of which can be separated using the time-dependent complex impedance of dEIS. C_{diff} is related to the frequency dependence of the impedance, and R_{diff} is related to the shift in the real component of the impedance. Assuming no charge transfer reactions or desolvation, resistors and capacitors can be added to give a linear RC circuit independent of applied voltage and history, resulting in a single, time-independent time constant $\tau = RC$. However, dEIS probes the time dependence of the differential resistance ΔR_{diff} and capacitance ΔC_{diff}, resulting in a time-dependent time constant $\tau = R_{diff} C_{diff}$.

Figure 3C shows that changes in R_{diff} (related to the reciprocal diameter of the semicircle) can be indicative of changes in dielectric loss or charge transport through the DIB. In contrast, a change in C_{diff} (related to the frequency at the semicircles' maxima, indicated by asterisks) is associated with a structural change in the bilayer interface or thickness. The data points beyond the semicircle fit the data indicate that the lipid bilayer experiences changes to its interfaces, consisting of degradation of its dielectric and charge transport properties and altered thickness fluctuations in the high-frequency limit ($w \rightarrow \infty$) (see Figure 3C). Moreover, the R_{diff} data beyond the fit to the 0-mV data (blue semicircle, Figure 3C) can be attributed to collective lipid bilayer motions, namely the propagation of compression (longitudinal phonons) and shear waves (transverse phonons). The lipid bilayer can support these motions for limited periods (a few picoseconds), during which time a reference lipid that is surrounded by its neighboring lipids experiences a continuous pressure induced by compression and shear forces, whereby the lipid undergoes cage-like dynamics. The result is a nanoscopic lipid domain that is more solid-like and is surrounded by transient nanoscopic pores. Shear forces eventually become negligible—due to disordering effects in the bilayer—enabling the reference lipid to self-diffuse out from its "cage" due to being continuously exposed to compression waves (Bolmatov et al. 2020). The existence of solid-like nanoscopic domains surrounded by nanopores, with subsequent relaxation events, is transient and re-emergent, appearing at random places and times in the bilayer. This molecular level mechanism in the bilayer underpins the transmembrane transport of small molecules and solutes (e.g., H_20, CO_2, and ions), as was discovered using inelastic X-ray scattering (Bolmatov 2022).

The existence of transient nanopores enables the passive transport of ions across the lipid bilayer, increasing its porosity (data points beyond the blue semicircle in Figure 3C). In the presence of a cyclic electric field (130 mV), the passive transport of ions across the bilayer is enhanced, facilitating the re-emergent behavior of transient nanopores. A recent study using the Van Hove function has revealed that bilayer thickness fluctuations exist due to collective headgroup dipole motions in zwitterionic lipid membranes (Bolmatov et al. 2023). Headgroup dipoles comprised of phosphate and choline functional groups continuously experience elastic deformations, such as dipole squeezing and stretching. Moreover, headgroup dipoles undergo collective motions that generate lipid bilayer surface undulations, enabling the bilayer dipole potential and surface potential to fluctuate continuously. Such intrinsic bilayer surface undulations define their flexoelectric and piezoelectric properties (i.e., the conversion of mechanical energy into electric energy) in the absence of external stimuli and can be enhanced when a biased voltage is applied.

Figure 3D shows the structural evolution of the lipid bilayer membrane with applied voltage, including its (1) geometrical responses based on voltage-dependent migration of oil in the membrane and changes in interfacial tension of the bilayer and (2) structural responses corresponding to time- and voltage-dependent "phase transitions" based on changes to the lipid bilayer at the molecular scale.

We found that biasing voltages with higher frequencies (1 Hz–10 mHz) and moderate amplitudes (< 150 mV) resulted in a single-function, namely a purely memcapacitive behavior in the lipid bilayer (Sacci et al. 2022), as reported by Najem et al. (2019). However, lower frequencies with increased amplitudes introduced a dual functionality involving an increased memristance that had been decoupled from memcapacitance. Thus, the lipid bilayer can store two types of information, one based on geometrical changes to capacitance (memcapacitance) and one based on changes to the fitted time constant of an equivalent RC_{tot} circuit (memristance). In this manner, dEIS allows for the simultaneous extraction of multiple memory elements, namely, memristance and memcapacitance from the same two-terminal device, and can be used to multiplex circuit functionality and potential computational output in a manner that leads to new memory behaviors and enhanced understanding of design principles leading to biological and artificial intelligence.

Emergent Behavior in the Form of Long-Term Potentiation

Above, we described a system capable of tunable signal processing and learning through STP (Najem et al. 2019), where postsynaptic efficacy is dependent on the history of presynaptic activity (Deng and Klyachko 2011). Importantly, there are two types of STP, PPF and PPD, which have opposite effects on synaptic efficacy (Rossbroich et al. 2021). However, STP memory is volatile and cannot be stored or accessed over extended periods, which precludes learning. On the other hand, long-term (LTP) is a process involving persistent strengthening of synapses that leads to a long-lasting increase in signal transmission between neurons and is an important mechanism for learning and memory (Bliss and Lømo 1973). Recently, it was shown (Scott et al. 2022) that exposing DPhPC DIBs to a novel electrical stimulation training protocol consisting of repetitive sinusoidal voltage cycling resulted in an emergent metastable lipid bilayer with persistent synaptic plasticity in the form of LTP. The emergent LTP behavior was attributed to bilayer structural reorganization from the buildup of an asymmetric charge distribution across the lipid bilayer. Interestingly, the reversal of the voltage stimulus polarity resulted in LTP erasure, with lipid bilayer memory capacitive stored energy values returning to levels like those prior to the lipid bilayer being exposed to the training protocol (repetitive sinusoidal voltage cycling). These results highlight the different types of memory stored in lipid bilayers, as well as informing the next generation of neuromorphic devices (Scott et al. 2022).

Subsequently, we extended the Scott et al. (2022) to include other salts (NaCl, LiCl, and $TmCl_3$) (Scott et al. 2023). We showed that LiCl and $TmCl_3$ had a profound effect on depression LTP, relative to KCl, with NaCl having an intermediate effect on LTP, somewhere between KCl and LiCl and $TmCl_3$ (Figure 5A). Based on those results, Scott and co-workers discussed how LTP is developed. Namely, the membrane is continuously pumped (i.e., training) into a non-equilibrium steady state, where it undergoes dynamic changes, increasing the system's potential energy (PE). The PE is then dissipated as heat, changing the membrane's structure and dielectric properties through changes to the orientational polarization of the adsorbed permanent water dipoles in the Stern layer. At the molecular level, energy dissipation takes place via repetitive shearing of the membrane, flexing, and molecular reorganization (Taylor et al. 2014; Amador et al. 2021). Taken together, the various forces acting on the PC headgroups during the training period induce long-lived piezoelectric and flexoelectric responses (Lozano et al. 2016; Ermolina et al. 2001; Petrov 2006).

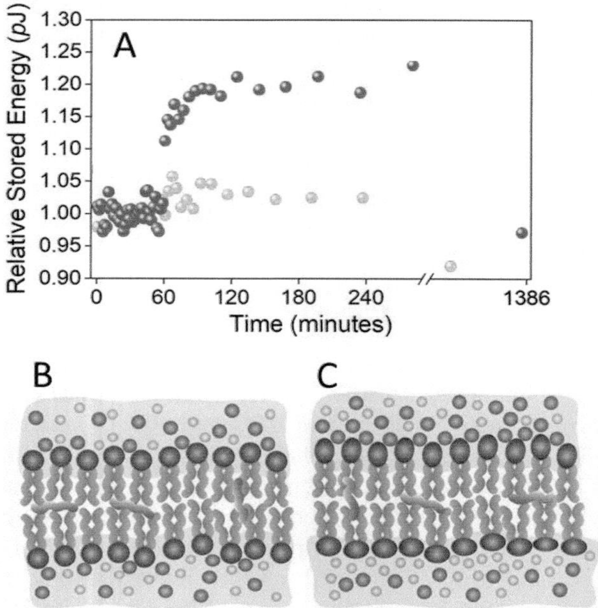

Figure 4. **(A)** The scaled capacitive memory stored energy values obtained from integrating the positive lobe areas of the q-V pinched hysteresis loops (Figure 2B) as functions of time, for both the training (dark dots) and no training (light dots) data. Note that the elevated capacitive memory stored energy plateaus and remains relatively constant over prolonged periods, indicative of LTP. Schematic showing the distribution of K^+ (dark spheres) and Cl^- (light spheres) ions before training (B), and after training (**C**), using a low-frequency sinusoidal voltage waveform. Prior to training, cations and anions are equally distributed on both sides of the bilayer (**B**). However, upon training, K^+ ions preferentially associate with one side of the bilayer, while the Cl^- counterions preferentially associate with the other side. This results in an asymmetrically charged, metastable lipid bilayer with almost twice the capacitive memory stored energy of the initial lipid bilayer. (Figure adapted from (Scott et al. 2022.)).

In contrast to the 2022 paper, the 2023 paper by Scott et al. shows that LTP results from the differential capacitance, dq/dV.

Scott et al. (2023) then went on to show how the different cations (K^+, Na^+, Li^+, and Tm^{3+}) interact very differently with the zwitterionic PC headgroups. Namely, K+ cations interact with the headgroups by disrupting the ensemble-averaged dipole moment densities, altering the dipole potentials at the headgroup region of the bilayer.

Specifically, K+ cations not in the vicinity of the headgroups maintain their levels of hydration and do not couple with the headgroups. However, when in proximity, they shed their hydration waters and interact with the PC headgroups, causing them to order through the production of significant electric fields (10^9–10^{10} V/m) across the membrane, strengthening the net, orientationally averaged dipole moment and producing a well-defined electrochemical double layer (Figure 5B). In contrast, Na^+, Li^+, and Tm^{3+} disrupt the level of hydration at the membrane interface, resulting in poorly aligned PC headgroups and a weakening of the net dipole moment. The result is an ill-defined electrochemical double layer and Nernst potential.

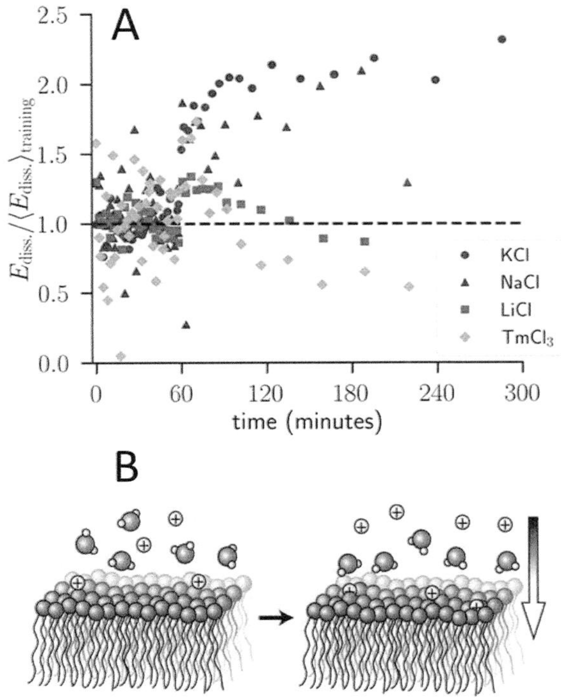

Figure 5. **(A)** Normalized system dissipation energies ($E_{dissipation\_norm}$) of the different salts plotted as a function of time. Compared to KCl, LTP for the other cations was of shorter duration, following the sequence KCl > NaCl > LiCl > $TmCl_3$. The dashed line indicates the times when NaCl, LiCl, and $TmCl_3$ returned to pre-training levels. **(B)** Schematic showing alignment of the permanent water dipoles in the Stern layer due to increasing cationic charge density during training. (Figure adapted from Scott et al. 2023.)

Concluding Remarks

Lipid bilayers with (Katsaras et al. 1992; Killian and Nyholm 2006) and without (Tardieu et al. 1973) peptides/proteins have been studied for over seven decades. However, it has only been in the last few years that they have been used as neuromorphic platforms to understand the formation of short- and long-term memories. The different science discussed here supports the notion that the membrane provides a model for understanding how and where biological memory is stored and is a target for treating brain diseases that cannot be cured or controlled by the current drugs targeting proteins.

Future work will include neutron and X-ray scattering experiments, providing a more detailed picture of bilayer static and dynamic structures under the same conditions used in the current study. These studies will be challenging to carry out with the current sample preparation protocol and may require the development of other sample preparation approaches. However, if successful, the complementary information obtained from neutron and/or X-ray scattering experiments will enable us to understand better the structural reorganization giving rise to metastable states associated with synapse-like plasticity and LTP observed in model lipid membranes. As they relate to neuromorphic computing, lipid bilayers can be used as platforms emulating the brain's efficiency and flexible cognitive capabilities to develop artificial neural networks using crossbar architectures of two-terminal passive circuit elements.

Acknowledgments

C.P.C. and J.K. are supported through the Scientific User Facilities Division of the Department of Energy (DOE) Office of Science, sponsored by the Basic Energy Science (BES) Program, DOE Office of Science, under contract no. DE-AC05-00OR22725. D.B. is supported through the National Science Foundation, Division of Molecular and Cellular Biosciences (MCB), under contract no. 2219289. Part of the work was performed at the Center for Nanophase Materials Sciences (CNMS), a US DOE Office of Science User Facility.

References

Amador, G.J., D. Van Dijk, R. Kieffer et al. 2021. Hydrodynamic shear dissipation and transmission in lipid bilayers. Proc. Natl. Acad. Sci. USA 118(21): e2100156118.

Bayley, H., B. Cronin, A. Heron et al. 2008. Droplet interface bilayers. Mol. Biosyst. 4(12): 1191–1208.

Bliss, Tim V.P. and T. Lømo. 1973. Long-lasting potentiation of synaptic transmission in the dentate area of the anaesthetized rabbit following stimulation of the perforant path. J. Physiol. 232(2): 331–356.

Bolmatov, D. 2022. The phonon theory of liquids and biological fluids: developments and applications. J. Phy. Chem. Lett. 13(31): 7121–7129.

Bolmatov, D., C.P. Collier, D. Zav'yalov et al. 2023. Real space and time imaging of collective headgroup dipole motions in zwitterionic lipid bilayers. Membranes 13(4): 442.

Bolmatov, D., D. Soloviov, M. Zhernenkov et al. 2020. Molecular picture of the transient nature of lipid rafts. Langmuir 36(18): 4887–4896.

Chua, L. 1971. Memristor-the missing circuit element. IEEE Trans. Circuit Theory 18(5): 507–519.

Danielli, J.F. and H. Davson. 1935. A contribution to the theory of permeability of thin films. J. Cell. Comp. Physiol. 5(4): 495–508.

Deng, P.-Y. and V.A. Klyachko. 2011. The diverse functions of short-term plasticity components in synaptic computations. Comm. Integr. Biol. 4(5): 543–548.

Di Ventra, M., Y.V. Pershin and L.O. Chua. 2009. Circuit elements with memory: memristors, memcapacitors, and meminductors. Proc. IEEE 97(10): 1717–1724.

El-Beyrouthy, J., M.M. Makhoul-Mansour, G. Taylor et al. 2019. A new approach for investigating the response of lipid membranes to electrocompression by coupling droplet mechanics and membrane biophysics. J. R. Soc. Interface 16(161): 20190652.

Ermolina, I., A. Strinkovski, A. Lewis et al. 2001. Observation of liquid-crystal-like ferroelectric behavior in a biological membrane. J. Phys. Chem. B. 105(14): 2673–2676.

Gorter, E. and F.J.E.M. Grendel. 1925. On bimolecular layers of lipoids on the chromocytes of the blood. J. Exp. Med. 41(4): 439.

Hamill, O.P. and B. Martinac. 2001. Molecular basis of mechanotransduction in living cells. Physiol. Rev. 81(2): 685–740.

Horn, A. and J.K. Jaiswal. 2019. Structural and signaling role of lipids in plasma membrane repair. Curr. Top. Membr. 84: 67–98.

Jackman, S.L. and W.G. Regehr. 2017. The mechanisms and functions of synaptic facilitation. Neuron 94(3): 447–464.

Karimi, A. and A. Rezai. 2019. Novel design for memristor-based n to 1 multiplexer using new imply logic approach. IET Circuits, Devices Syst. 13(5): 647–655.

Katsaras, J., R.S. Prosser, R.H. Stinson et al. 1992. Constant helical pitch of the gramicidin channel in phospholipid bilayers. Biophys. J. 61(3): 827–830.

Killian, J.A. and T.K.M. Nyholm. 2006. Peptides in lipid bilayers: the power of simple models. Curr. Opin. Struct. Biol. 16(4): 473–479.

Lasia, A. 2002. Electrochemical impedance spectroscopy and its applications: Springer.

Lozano, M.M., J.S. Hovis, F.R. Moss III et al. 2016. Dynamic reorganization and correlation among lipid raft components. J. Am. Chem. Soc. 138(31): 9996–10001.

Najem, J.S., Md S. Hasan, R.S. Williams et al. 2019. Dynamical nonlinear memory capacitance in biomimetic membranes. Nat. Commun. 10(1): 1–11.

Najem, J.S, G.J. Taylor, R.J. Weiss et al. 2018. Memristive ion channel-doped biomembranes as synaptic mimics. ACS Nano 12(5): 4702–4711.

Nollet, J.-A. 1752. Recherches sur les causes du bouillonnement des liquides. Hist. Acad. Roy Sci. 1: 57–104.

Overton, C.E. 1899. On the general osmotic properties of the cell, their probable origin, and their significance for physiology. Vierteljahrsschr Naturforsch Ges Zurich 44: 88–135.

Petrov, A.G. 2006. Electricity and mechanics of biomembrane systems: flexoelectricity in living membranes. Anal. Chim. Acta 568(1-2): 70–83.

Poulos, J.L., W.C. Nelson, T.-J. Jeon et al. 2009. Electrowetting on dielectric-based microfluidics for integrated lipid bilayer formation and measurement. Appl. Phys. Lett. 95(1): 013706.

Rossbroich, J., D. Trotter, J. Beninger et al. 2021. Linear-nonlinear cascades capture synaptic dynamics. PLoS Comput. Biol. 17(3): e1008013.

Sacci, R.L., H.L. Scott, Z. Liu et al. 2022. Disentangling memristive and memcapacitive effects in droplet interface bilayers using dynamic impedance spectroscopy. Adv. Electron. Mater. 8(5): 2200121.

Sacci, R.L., F. Seland and D.A. Harrington. 2014. Dynamic electrochemical impedance spectroscopy, for electrocatalytic reactions. Electrochim. Acta 131: 13–19.

Scott, H.L., D. Bolmatov, P.T. Podar et al. 2022. Evidence for long-term potentiation in phospholipid membranes. Proc. Natl. Acad. Sci. U.S.A 119(50): e2212195119.

Scott, H.L., D. Bolmatov, U.I. Premadasa et al. 2023. Cations control lipid bilayer memcapacitance associated with long-term potentiation. ACS App. Mater. Interfaces.

Sheng, J., L. He, H. Zheng et al. 2012. Calcium-channel number critically influences synaptic strength and plasticity at the active zone. Nat. Neurosci. 15(7): 998–1006.

Shin, S., K. Kim and S.-M. Kang. 2010. Memristor applications for programmable analog ICs. IEEE Trans. Nanotechnol. 10(2): 266–274.

Sigworth, F.J. 1994. Voltage gating of ion channels. Q. Rev. Biophys. 27(1): 1–40.

Simons, K. and E. Ikonen. 1997. Functional rafts in cell membranes. Nature 387(6633): 569–572.

Singer, S.J. and G.L. Nicolson. 1972. The Fluid mosaic model of the structure of cell membranes: Cell membranes are viewed as two-dimensional solutions of oriented globular proteins and lipids. Science 175(4023): 720–731.

Stevens, C.F. and Y. Wang. 1995. Facilitation and depression at single central synapses. Neuron 14(4): 795–802.

Symons, J.L., K.-J. Cho, J.T. Chang et al. 2021. Lipidomic atlas of mammalian cell membranes reveals hierarchical variation induced by culture conditions, subcellular membranes, and cell lineages. Soft Matter 17(2): 288–297.

Tardieu, A., V.L. and F.C. Reman. 1973. Structure and polymorphism of the hydrocarbon chains of lipids: a study of lecithin-water phases. J. Mol. Biol. 75(4): 711–733.

Taylor, R.W., F. Benz, D.O. Sigle et al. 2014. Watching individual molecules flex within lipid membranes using SERS. Sci. Rep. 4(1): 5940.

Villar, G., A.D. Graham and H. Bayley. 2013. A tissue-like printed material. Science 340(6128): 48–52.

Wang, Z., S. Joshi, S.E. Savel'ev et al. 2017. Memristors with diffusive dynamics as synaptic emulators for neuromorphic computing. Nat. Mater. 16(1): 101–108.

Weinberg, S.H. 2015. Membrane capacitive memory alters spiking in neurons described by the fractional-order Hodgkin-Huxley model. PloS One 10(5): e0126629.

Zidan, M.A., H. Omran, A. Sultan et al. 2014. Compensated readout for high-density MOS-gated memristor crossbar array. IEEE Trans. Nanotechnol. 14(1): 3–6.

Index

L

M

N

O

P

R

S

T

V

X

For Product Safety Concerns and Information please contact our
EU representative GPSR@taylorandfrancis.com Taylor & Francis
Verlag GmbH, Kaufingerstraße 24, 80331 München, Germany